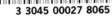

Electric Motors

AUDEL®

Electric Motors

by Edwin P. Anderson and Rex Miller

An Audel® Book

Macmillan Publishing Company
New York

Collier Macmillan Canada
Toronto

Maxwell Macmillan International
New York   Oxford   Singapore   Sydney

Foreword

Electric motors play a very important part in furnishing power for all types of domestic and industrial applications. Their versatility, dependability, and economy of operation cannot be equaled by any other form of motive power. It is estimated that electric motors are utilized in over 90 percent of industrial applications. This figure would be higher except for the absence of power lines in some remote areas. The application of electric motors in the average home also has reached a high degree of utilization, ranging from the smallest units found in electric clocks to the larger units in air conditioners, heating plants, etc. It is a rare individual, indeed, whose daily life is not affected in some way by electric motors.

This book has been prepared to give practical guidance in the selection, installation, operation, and maintenance of electric motors. Electricians, industrial maintenance personnel, and installers should find the clear descriptions and illustrations, along with the simplified explanations, a ready source of information for the many problems they will encounter. Both technical and nontechnical persons who desire to gain knowledge of electric motors will benefit from the theoretical and practical coverage this book offers.

Rex Miller
Edwin P. Anderson

Contents

CHAPTER 1

Motor Principles

In order to obtain a clear concept of the principles on which the electric motor operates, it is necessary first to understand the fundamental laws of magnetism and magnetic induction. It is not necessary to have a great number of expensive laboratory instruments to obtain this knowledge — instead, children's toy magnets, automobile acessories, etc., will suffice. It is from these principles that the necessary knowledge about the behavior of permanent magnets and the magnetic needle can be obtained.

In our early schooldays, we learned that the earth is a huge permanent magnet with its north magnetic pole somewhere in the Hudson Bay region, and that the compass needle points toward the magnetic pole. The compass is thus an instrument that can give an indication of magnetism.

The two spots on the magnet that point one to the north and the other to the south are called the *poles* — one is called the north-seeking pole (N) and the other the south-seeking pole (S).

Magnetic Attraction and Repulsion

If the south-seeking, or S, pole of a magnet is brought near the S pole of a suspended magnet, as in Fig. 1-1, the poles repel each other. If the two N poles are brought together, they also repel each other. But if an N pole is brought near the S pole of the moving magnet, or an S pole toward the N pole, the two unlike poles attract each other. In other words, *like poles repel each other,* and *unlike poles attract each other.* It can also be shown by experiment that these attractive or repulsive forces between magnetic poles vary inversely as the square of the distance between the poles.

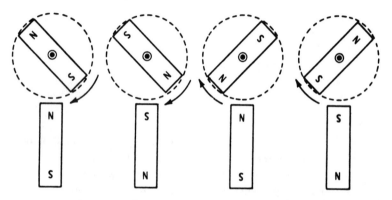

Fig. 1-1. Illustrating that like poles of permanent magnets repel each other and unlike poles attract each other.

Effects of an Electric Current

As a further experiment, connect a coil to a battery as shown in Fig. 1-2. Here it will be noticed that the compass points to one end of the coil, but if the battery connections are reversed, the compass points away from that end. Thus, the direction of the current through the coil affects the compass in a manner similar to the permanent magnet in the previous experiment.

In the early part of the 19th century, Oersted made the discovery showing the relationship existing between magnetism and electricity. He observed that when a wire connecting the poles of a battery was held *over* a compass needle, the north pole of the needle was deflected as shown in Fig. 1-3. A wire placed *under* the

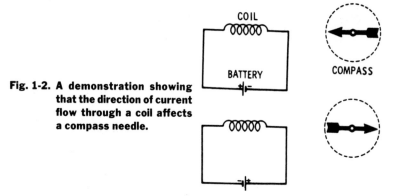

Fig. 1-2. A demonstration showing that the direction of current flow through a coil affects a compass needle.

compass needle caused the north pole of the needle to be deflected in the opposite direction.

Magnetic Field of an Electric Current

Inasmuch as the compass needle indicates the direction of magnetic lines of force, it is evident from Oersted's experiment that an electric current sets up a magnetic field at right angles to the conductor. This can be shown by the experiment illustrated in Fig. 1-4. If a strong current is sent through a vertical wire which passes through a horizontal piece of cardboard on which iron filings are placed, a gentle tap of the board causes the iron filings to arrange

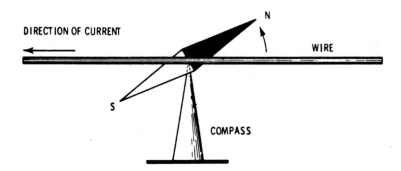

Fig. 1-3. A demonstration showing that current flowing through a wire will cause a compass needle to deflect.

themselves in concentric rings about the wire. A compass placed at various positions on the board will indicate the direction of these lines of force, as shown in Fig. 1-4.

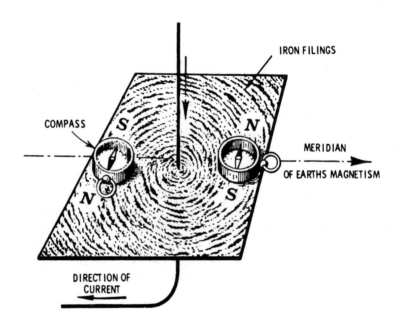

Fig. 1-4. An experiment to show the direction of lines of force that surround a conductor carrying current.

A convenient rule for remembering the direction of the magnetic flux around a straight wire carrying current is the so-called *left-hand rule*. With reference to Fig. 1-5, it can be seen that if the wire is held by the left hand, with the thumb pointing in the direction of the current, the fingers will point in the direction of the magnetic field.

Conversely, if the direction of the magnetic field around a conductor is known, the direction of the current in the conductor can be found by applying this rule.

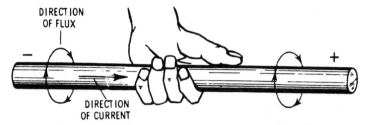

DIRECTION
OF FLUX

DIRECTION
OF CURRENT

Fig. 1-5. Using the left-hand rule to determine the direction of magnetic lines of force (flux) around a conductor through which current is flowing.

Electromagnets

A soft-iron core surrounded by a coil of wire is called an *electromagnet*. The electromagnet owes its great utility not so much to its great strength as to its ability to change its magnetic strength with the strength of the applied current. An electromagnet is a magnet only when the current flows through its coil. When the current is interrupted, the iron core returns almost to its natural state. This loss of magnetism is not absolutely complete, however, since a very small amount, called *residual magnetism,* remains.

An electromagnet is a part of nearly all electrical devices, including electric bells, telephones, motors, and generators.

The polarity of an electromagnet may be determined by means of the left-hand rule used for a straight wire as follows: *Grasp the coil with the left hand so that the fingers point in the direction of the current in the coil, and the thumb will point to the north pole of the coil.* See Fig. 1-6.

The strength of an electromagnet depends on the strength of the current (in amperes) times the number of loops of wire (turns) — that is, the *ampere-turns* of the coil (Fig. 1-7). In practical electromagnets, it is customary to make use of both poles by bending the iron core and the coil in the form of a horseshoe. It is from this form that the name *horseshoe magnet* is derived.

Induced Currents

If the ends of a coil of wire having many turns are connected to a sensitive galvanometer, as shown in Fig. 1-8, and the coil is moved

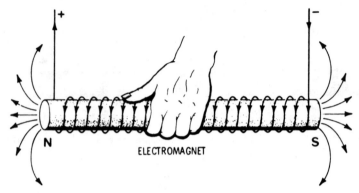

Fig. 1-6. Using the left-hand rule to determine the polarity of an electromagnet.

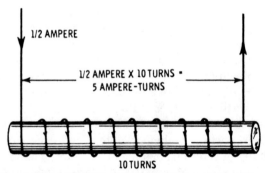

1/2 AMPERE

1/2 AMPERE X 10 TURNS =
5 AMPERE-TURNS

10 TURNS

Fig. 1-7. Ampere turns of a coil are equal to the product of the current (in amperes) flowing through the windings multiplied by the number of turns in the coil.

up and down over one pole of a horseshoe magnet, a deflection of the galvanometer pointer will be observed. It will also be noted that in lowering the coil, the deflection of the galvanometer pointer will be in a direction opposite that of the needle when the coil is raised. When the coil is lowered and held down, the galvanometer pointer returns to zero. This experiment shows that it is possible to produce a momentary electric current without an apparent electrical source.

The electrical current produced by moving the coil in a magnetic field is called an *induced current*. It is evident from the experiment that the current is induced only when the wire is moving, and that

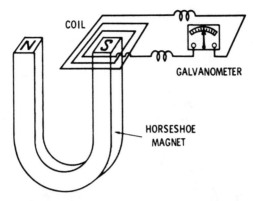

Fig. 1-8. Demonstrating how the movement of a coil in a magnetic field generates an electric current.

the direction of the current is reversed when the motion changes direction. Since an electric current is always made to flow by an electromotive force (*emf*), the motion of a coil in a magnetic field must generate and produce an induced electromotive force.

The direction of an induced current may be stated as follows: *An induced current has such a direction that its magnetic action tends to resist the motion by which it is produced.* This is known as *Lenz's law.*

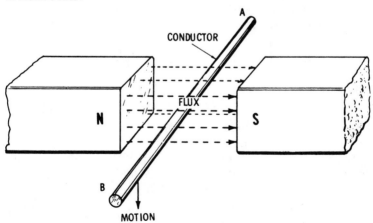

Fig. 1-9. A voltage is induced when a wire cuts through magnetic lines of force.

The most useful application of induced currents is in the construction of electrical machinery of all sorts, the most common of which are the generator and motor.

A simple way to obtain a fundamental understanding of the generator is to think of the induced electromotive force *(emf)* produced in a single wire when it is moved across a magnetic field. Suppose wire *AB* in Fig. 1-9 is pushed down through the magnetic field. An induced emf is set up in *AB*, making point *B* at a higher potential than point *A*. This can be shown by connecting a voltmeter from *A* to *B*.

As long as the wire remains stationary, no current flows. In fact, even if the wire moves parallel to the lines of force, no current flows. Briefly, a wire must move so as to cut lines of magnetic force in order to have an emf induced in it.

Direct-Current (DC) Generators

Theory of Operation

A *generator* is a machine that converts mechanical energy into electrical energy. This is done by rotating an *armature*, which contains *conductors*, through a *magnetic field*. The movement of the conductors through a magnetic field produces an induced emf in the moving conductors. In any generator, a relative motion between the conductors and the magnetic field will always exist when the shaft is rotated.

Parts of a DC Generator

The principal parts of a DC generator are an armature, commutator, field poles, brushes and brush rigging, yoke or frame, and end bells or end frames (Fig. 1-10A).

Armature — The armature is the structure upon which the coils are mounted. These coils cut the magnetic lines of force. The armature is attached to a shaft. The shaft is suspended at each end of the machine by bearings set in the end bells, as shown in Fig. 1-11. The armature core, which is circular in cross section, consists of many sheets of soft iron. The edge of the laminated core is

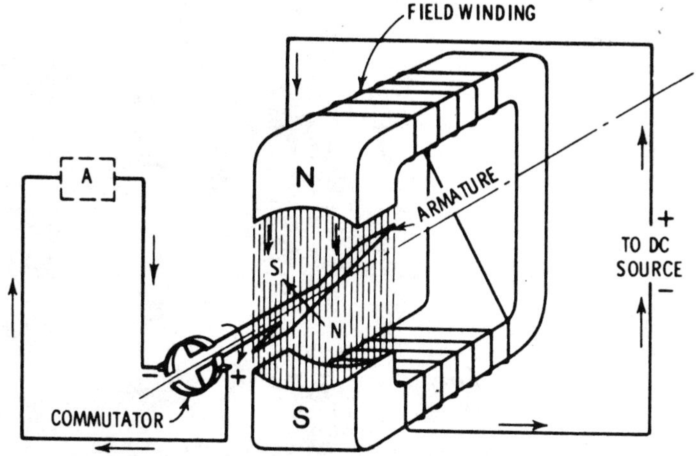

Fig. 1-10A. An elementary direct-current (DC) generator. A commutator keeps the current flowing in the same direction in the load circuit (A).

slotted (Fig. 1-12). Coil windings fit into these slots. The windings are held in place and in their slots by wooden or fiber wedges. Sometimes steel bands are also wrapped around the completed armature to provide extra support. On small generators, the laminations of the armature core are usually pressed onto the armature shaft.

Commutator — The commutator is that part of the generator that rectifies the generated alternating current to provide direct current output (Fig. 1-11). It also connects the stationary output terminals to the rotating armature. A typical commutator consists of *commutator bars*. The bars are wedge-shaped segments of hard-drawn copper. These segments are insulated from each other by thin strips of mica. Commutator bars are held in place by steel V-rings or clamping flanges, as shown in Fig. 1-13. These are bolted to the commutator sleeve by hexagonal cap screws. The *commutator sleeve* is keyed to the shaft that rotates the armature. A mica collar or ring insulates the commutator bars from the commutator sleeve. The commutator bars usually have risers or flanges to which the leads from the associated armature coils are soldered. These risers serve as a shield for the soldered connections when the commutator bars become worn. When risers are not provided, it is necessary to solder the leads from the armature

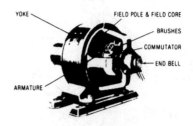

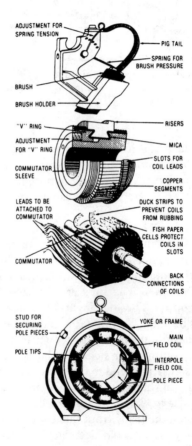

Fig. 1-10B. Parts of a DC generator, stationary type.

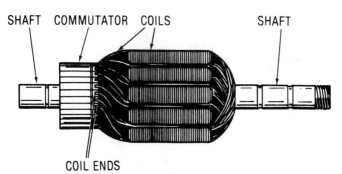

SHAFT COMMUTATOR COILS SHAFT

COIL ENDS

Fig. 1-11. Armature of a DC generator.

SLOTS

Fig. 1-12. Unwound armature core on a shaft.

coils to short slits in the ends of the commutator bars. The brushes contact the commutator bars. The brushes collect the current generated by the armature coils. The brush holders transfer the current to the main terminals. The commutator bars are insulated from each other. Thus, each set of brushes, as it contacts the commutator bars, collects current of the same polarity. This results in a continuous flow of direct current. The finer the division of the commutator bars, the less the ripple that appears in the current, and therefore, the smoother the flow of the DC output.

Field Pole and Frame — The frame or yoke of a generator serves two purposes. It provides mechanical support for the machine. It is also a path for the completion of the magnetic circuit. The lines of force that pass from the north to the south pole through the armature are returned to the north pole through the

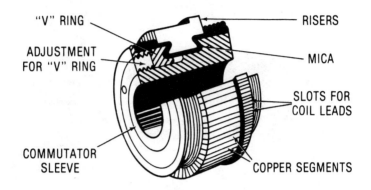

Fig. 1-13. Commutator construction.

frame. Frames are made of electrical-grade steel. The method of construction of field poles and frames varies with the manufacturer. Fig. 1-14 shows the magnetic circuit of a two-pole generator.

Field Windings — The field windings are connected so that they produce alternate north and south poles, as shown in Fig. 1-15. Connection is done that way to obtain the correct direction of emf in the armature conductors. The field windings form an electromagnet that establishes the generator field flux. These field windings may receive current from an external DC source; or they may be connected directly across the armature, which then becomes the source of voltage. When the windings are energized, they establish magnetic flux in the field yoke, pole pieces, air gap, and armature core (Fig. 1-14).

Brushes and Brush Holders — The brushes carry the current from the commutator to an external circuit. Usually they are a mixture of graphite and metallic powder. Brushes are designed to slide freely in their holders because the commutator surface is usually uneven, and the brushes and commutator wear. The freedom thus allows the brushes to have good contact with the commutator despite wear or uneven surfaces.

Proper pressure of the brushes against the commutator is maintained by means of springs. This pressure is usually about 1½ to 2 pounds per square inch of brush contact area. A low resistance connection — usually braided copper wire — is provided between the brushes and brush holders.

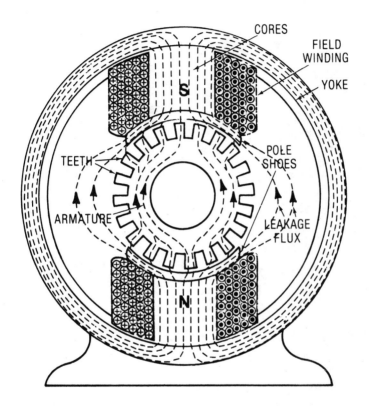

Fig. 1-14. Magnetic circuit of a two-pole generator.

Armature Windings

The simplest generator armature winding is a *loop* or *single coil*. Rotating this loop in a magnetic field induces an emf. The strength of the magnetic field and the speed of rotation of the conductor determine the emf produced.

A *single-coil generator* is shown in Fig. 1-16. Each coil terminal is connected to a bar of a two-segment metal ring. The two segments of the split rings are insulated from each other and the shaft. This forms a simple commutator. The commutator mechanically reverses the armature coil connections to the external circuit at the

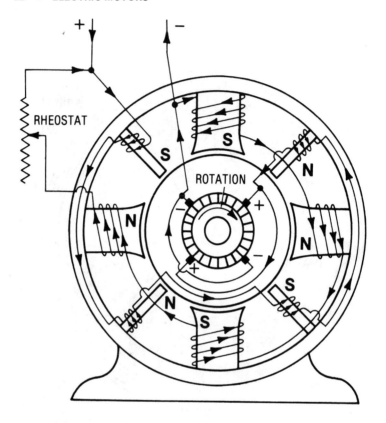

Fig. 1-15. Schematic wiring diagram of a shunt generator.

same instant that the direction of the generated voltage reverses the armature coil. This process is known as commutation.

Fig. 1-17 is a graph of a *pulsating current* (DC) for one rotation of a single-loop, two-pole armature. A pulsating current or *direct voltage* has ripple. In most cases, this current is not usable. More coils have to be added.

Effect of More Coils — The heavy black line in Fig. 1-18 shows the DC output of a two-loop (coil) armature. A great reduction in voltage ripple is obtained by using two coils instead of one. Since there are now four commutator segments in the commutator and only two brushes, the voltage cannot fall lower than point A. Therefore, the ripple is limited to the rise and fall between points

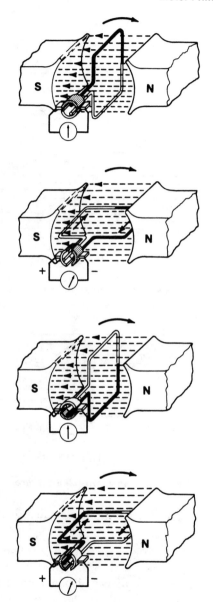

Fig. 1-16. Single-coil generator with commutator.

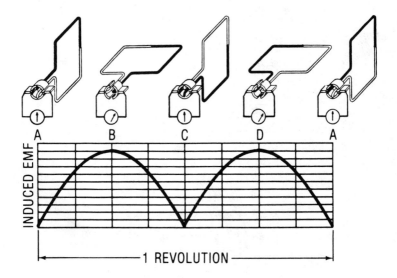

Fig. 1-17. Output of a single-coil DC generator.

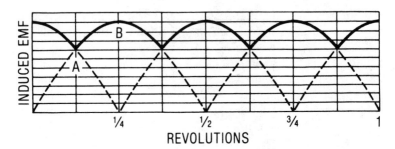

Fig. 1-18. Output voltage from a two-coil armature.

A and B. Adding more armature coils will reduce the ripple even more.

Armature Losses

There are three losses in every DC generator armature. One is the copper loss in the winding. The second is the eddy current loss in the core. The third is the hysteresis loss caused by the friction of the revolving magnetic particles in the core.

Copper Losses — Copper loss is the power lost in heat in the windings due to the flow of current through the copper coils. This loss varies directly with the armature resistance and the square of the armature current. The armature resistance varies inversely with the cross-sectional area.

Armature copper loss varies mainly because of the variation of electrical load on the generator and not because of any loss occurring in the machine. This is because most generators are constant-potential machines supplying a current output that varies with the electrical load across the brushes. The limiting factor in load on a generator is the allowable current rating of the generator armature.

The armature circuit resistance includes the resistance of the windings between brushes of opposite polarity, the brush contact resistance, and the brush resistance.

Eddy Current Losses — If a DC generator armature core were made of solid iron and rotated rapidly in the field, as shown in Fig. 1-19A, excessive heating would develop even with no-load current in the armature windings. This heat would be the result of a generated voltage in the core itself. As the core rotates, it cuts the lines of magnetic field flux at the same time the copper conductors of the armature cut them. Thus, induced currents alternate through the core, first in one direction and then in the other. These currents cause heat.

Such induced currents are called *eddy currents*. They can be minimized by sectionalizing *(laminating)* the armature core. For instance, a core is split into two equal parts, as shown in Fig. 1-19B. These parts are insulated from each other. The voltage induced in each section of iron is thus one-half of what it would have been if it remained solid. The resistance of the eddy current paths is doubled. That is because resistance varies inversely with the cross-sectional area of the lamination.

If the armature core is subdivided into many sections or laminations, as in Fig. 1-19C, the eddy current loss can be reduced to a negligible value. Reducing the thickness of the laminations reduces the magnitude of the induced emf in each section. It also increases the resistance of the eddy current paths. Laminations in small generator armatures are usually $1/64$ inch thick. Often the laminations are insulated from each other by a thin coat of lacquer.

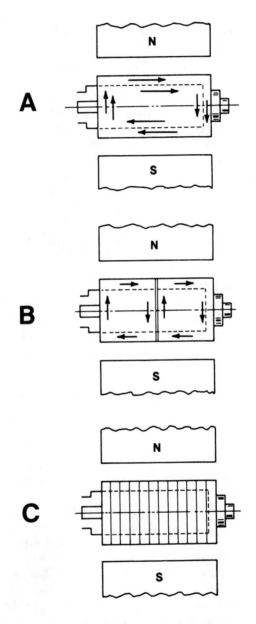

Fig. 1-19. Eddy currents in a DC generator armature core.

Sometimes they are insulated simply by the oxidation of the surfaces caused by contact with the air while the laminations are being annealed. The voltages induced in laminations are small; thus the insulation need not be great.

All electrical rotating machines and transformers are laminated to reduce eddy current loses.

Eddy current loss is also influenced by speed and flux density. The induced voltage, which causes the eddy currents to flow, varies with the speed and flux density. Therefore, the power loss,

$$P = \frac{E^2}{R}$$

varies as the square of the speed and the square of the flux density.

Hysteresis Losses — When an armature revolves in a stationary magnetic field, the number of magnetic particles of the armature that remain in alignment with the field depends on the strength of the field. If the field is that of a two-pole generator, these magnetic particles will rotate, with respect to the particles not held in alignment, one complete turn for each revolution of the armature. The rotating of the magnetic particles in the mass of iron produces friction and heat.

Heat produced this way is called *magnetic hysteresis loss.* The hysteresis loss varies with the speed of the armature and the volume of iron. The *flux density* varies from approximately 50,000 lines per square inch in the armature core to 130,000 lines per square inch in the iron between the bottom of adjacent armature slots (the *tooth root*). Heat-treated silicon steel having a low hysteresis loss is used in most DC generator armatures. The steel is formed to the proper shape. Then the laminations are heated to a dull red heat and allowed to cool. This annealing process reduces the hysteresis loss to a low value.

Armature Reaction

Armature reaction in a generator is the effect on the main field of the armature acting as an electromagnet. With no armature current, the field is undistorted (Fig. 20A). This flux is produced entirely by the ampere-turns of the main field windings. The neutral plane *AB* is perpendicular to the direction of the main field

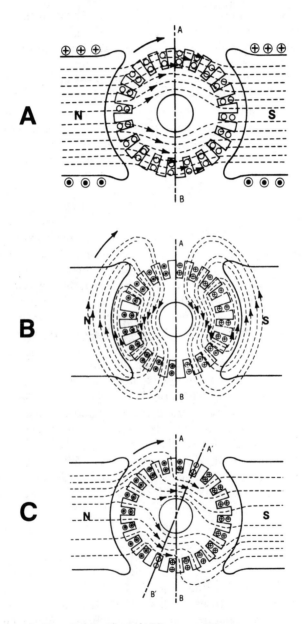

Fig. 1-20. Flux distribution in a DC generator.

flux. When an armature conductor moves through this plane, its path is parallel to the undistorted lines of force. Thus, the conductor does not cut through any flux, and no voltage is induced in the conductor. The brushes are placed on the commutator so that they short-circuit coils passing through the neutral plane. With no voltage generated in the coils, no current will flow through the local path formed momentarily between the coils and segments spanned by the brush. Therefore, no sparking at the brushes will result.

When a load is connected across the brushes, armature current flows through the armature conductors. The armature itself becomes a source of magnetomotive force. The effect of the armature acting as an electromagnet is shown in Fig. 1-20B. The main field coils are de-energized, and full-load current is applied to the armature circuit from an external source. The conductors on the left of the neutral plane all carry current toward the observer. Those on the right carry current away from the observer. These directions are the same as those that the current would flow if it were under the influence of the normal emf generated in the armature with normal field excitation.

These armature-current-carrying conductors establish magnetomotive force that is perpendicular to the axis of the main field. In Fig. 1-20B the force acts downward. This magnetizing action of the armature current is called *cross magnetization*. It is present only when current flows through the armature circuit. The amount of cross magnetization produced is proportional to the armature current.

When current flows in both the field and armature circuits, the two resulting magnetomotive forces distort each other (Fig. 1-20C). They twist in the direction of rotation of the armature. The mechanical (no-load) neutral plane, AB, is now advanced to the electrical (load) neutral plane A'B'. When the armature conductors move through plane A'B', their paths are parallel to the distorted field. The conductors cut no flux. Thus, no voltage is induced in them. The brushes must, therefore, be moved on the commutator to the new neutral plane. They are moved in the direction of armature rotation. The absence of sparking at the commutator indicates the correct placement of the brushes. The amount that the neutral plane shifts is proportional to the load on the generator. That is because the amount of cross-magnetizing magnetomotive force is directly proportional to the armature current.

When the brushes are shifted into the electrical neutral plane $A'B'$, the direction of the armature magnetomotive force is downward and to the left, instead of vertically downward (Fig. 1-21A). The armature magnetomotive force may now be resolved into two components (Fig. 1-21B).

The conductors at the top and bottom of the armature within sectors BB produce a magnetomotive force that is directly in opposition to the main field and weakens it. This component is called the *armature-demagnetizing mmf*. The conductors on the right and left sides of the armature within sector AA produce a *cross-magnetizing mmf* at right angles to the main field axis. This cross-magnetizing force tends to distort the field in the direction of rotation. As mentioned, the distortion of the main field of the generator is the result of armature reaction. Armature reaction occurs in the same manner in multipolar generators.

Compensating for Armature Reaction — The effects of armature reaction are reduced in DC machines by the use of *high-flux density pole tips*, a *compensating winding*, and *commutation poles*.

The cross-sectional area of the pole tips is reduced by building field poles with laminations having only one tip. These laminations are alternately reversed when the pole core is stacked so that a space exists between alternate laminations at the pole tips. The reduced cross section of iron at the pole tips increases the flux density. Thus, they become saturated. Cross-magnetizing and demagnetizing forces of the armature do not affect the flux distribution in the pole face as much as they would reduced flux densities.

The compensating winding consists of conductors embedded in the pole faces parallel to the armature conductors. The winding is connected in series with the armature. It is arranged so that the ampere-turns are equal in magnitude and opposite in direction to those of the armature. The magnetomotive force of the compensating winding, therefore, neutralizes the armature magnetomotive force, and armature reaction is almost eliminated. Compen-

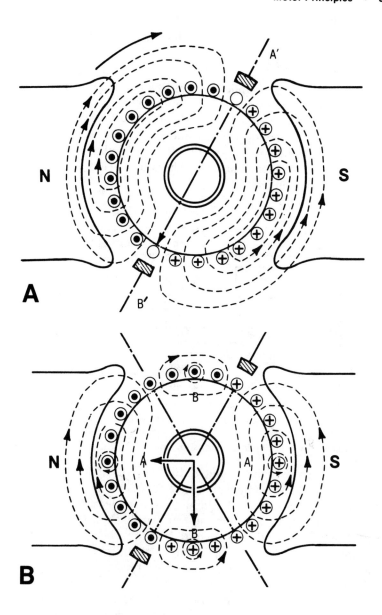

Fig. 1-21. Effect of brush shift on armature reaction.

sating windings are costly, so they are generally used only on high-speed and high-voltage large-capacity generators.

Motor Reaction in a Generator

When a generator supplies current to a load, the load current creates a force that opposes the rotation of the generator armature. An *armature conductor* is represented in Fig. 1-22. When a conductor is moved downward and the circuit is completed through an external load, current flows through the conductor in the direction indicated. This causes lines of force around it in a clockwise direction.

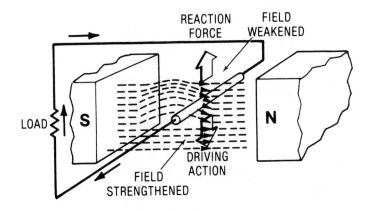

Fig. 1-22. Motor reaction in a generator.

The interaction of the conductor field and the main field of the generator weakens the field above the conductor and strengthens it below the conductor. The field consists of lines that act like stretched rubber bands. Thus an upward reaction force is produced that opposes the downward driving force applied to the generator armature. If the current in the conductor increases, the reaction force increases. More force must then be applied to the conductor to keep it from slowing down.

With no armature current, no magnetic reaction exists. Therefore, the generator input power is low. As armature current increases, the reaction of each armature conductor against rotation increases. The driving power to maintain the generator armature speed must be increased. If the prime mover driving the generator is a gasoline engine, this effect is accomplished by opening the throttle of the carburetor. If the prime mover is a steam turbine, the main steam-admission valve is opened wide so that more steam can flow through the turbine.

Types of DC Generators

DC generators are classified by how excitation current is supplied to the field coils. There are two major classifications:

1. Separately excited.
2. Self-excited.

Self-exciting generators are further classified by the method of connecting the field coils. These include series-connected, shunt-connected, and compound-connected generators.

Separately Excited Generators

A separately excited generator is one for which the field current is supplied by another generator, by batteries, or by some other outside source. Fig. 1-23 shows a typical circuit.

Fig. 1-24 shows the voltage characteristics of a separately excited generator. When operated at constant speed with constant field

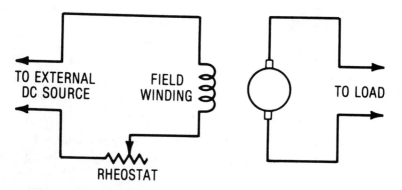

Fig. 1-23. Connection of a separately excited DC generator.

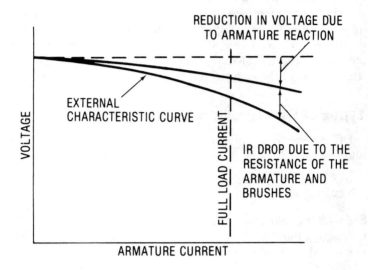

Fig. 1-24. Voltage characteristics of a separately excited DC generator.

excitation but not supplying current, the terminal voltage of this type of generator equals the generated voltage. When the unit is delivering current, the terminal voltage is less than the generated voltage. The total amount of voltage drop equals the drop due to armature reaction plus the voltage drop due to the resistance of the armature and the brushes. Separately excited generators, however, are seldom used.

Self-Excited Generators

There are three types of self-excited generators:
1. Series.
2. Shunt.
3. Compound.

There are some variations of the compound type.

Series Generators — When all the windings are connected in series with the armature, a generator is series-connected. See Fig. 1-25 for the typical series-connected circuit. Fig. 1-26 shows the voltage characteristics of a series generator. With no load, the only voltage present is due to the cutting of the flux established by

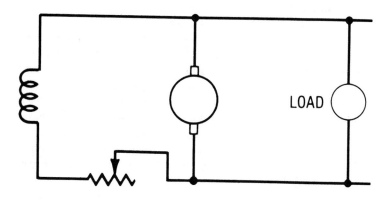

Fig. 1-25. A typical series DC generator circuit.

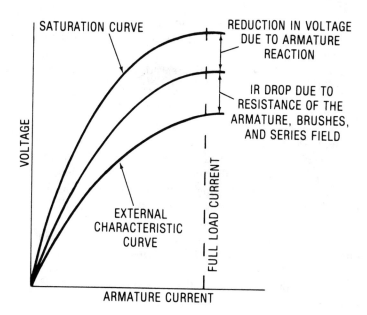

Fig. 1-26. Voltage characteristics of a series DC generator.

residual magnetism. (Residual magnetism is magnetism retained by the poles of a generator when it is not in operation.) However, when a load is applied or increased, the current through the field coil increases the flux. Therefore, the generated voltage increases. The voltage generated tends to increase directly as the current increases, but three factors lessen the voltage increase. One factor is saturation of the field core. If field excitation is increased beyond the point at which the flux produced no longer increases directly as the exciting current, the core is said to be saturated. The second factor is armature reaction. The effect of this reaction increases as the current load increases. The third factor is loss in terminal voltage. This loss is caused by ohmic resistance of the armature winding, brushes, and series field. This loss increases as the unit is loaded. Since the terminal voltage of a series generator varies under changing load conditions, it is generally connected in a circuit that demands constant current. When used that way, it is sometimes referred to as a constant-current generator, even though it does not tend to maintain a constant current itself. Constant current is achieved by connecting a variable resistance in parallel with the series field. The variable resistance can be manually or automatically controlled. Thus, as the load is increased, the resistance of the shunt path is decreased. This permits more of the current to pass through it and maintains a relatively constant field.

Shunt Generators — When the field windings are connected in parallel with the armature, the generator is shunt-connected. Fig. 1-27 is a typical circuit of a shunt generator; Fig. 1-28 shows the voltage characteristics of a shunt generator. A comparison of the voltage characteristics of a shunt generator shows they are similar to those of a separately excited generator. In both instances, the terminal voltage drops from the no-load value as the load is increased. But note that the terminal voltage of the shunt generator remains fairly constant until it approaches full load. This is true even though the graph of the shunt generator has an extra factor that causes the terminal voltage to decrease: the weakening of the field as the current approaches full load. It is, therefore, better to use a shunt generator, and not a separately excited or a series generator, when a constant voltage with a varying load is required. Shunt generators are readily adaptable to applications where the speed of the

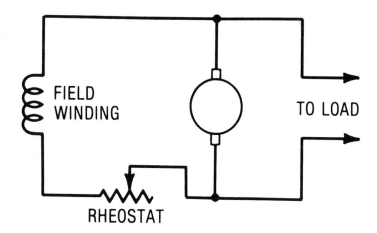

Fig. 1-27. Connection of a shunt DC generator.

prime mover cannot be held constant. Aircraft and automobile engines are typical examples of variable-speed prime movers that require a constant voltage. Constant voltage is obtained by controlling generator field current, which is accomplished by varying the shunt field resistance to compensate for changes in speed of the prime mover.

Compound Generators — If both a series and a shunt field are included in the same unit, it is possible to obtain a generator with a voltage-load characteristic somewhere between those of a series and a shunt generator. Fig. 1-29 shows typical circuits of compound-wound (series-shunt) DC generators. Fig. 1-29A is a cumulative compounded generator. Its series and shunt fields are wound to aid each other. Fig. 1-30 shows the voltage characteristics of a compound-wound DC generator. By changing the number of turns in the series field it is possible to obtain three distinct types of compound generators.

Overcompounded — An overcompounded generator is one in which there are more turns in the series field than necessary to give about the same voltage at all loads. Thus the terminal voltage at

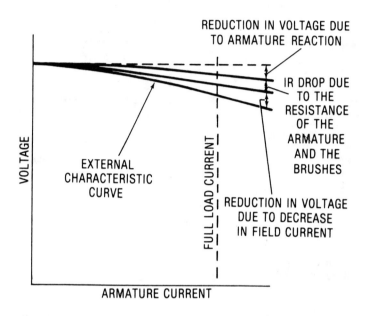

Fig. 1-28. Voltage characteristics of a shunt DC generator.

full load will be higher than the no-load voltage. This is desirable when power must be transmitted a long distance. The higher generated voltage compensates for the voltage loss in the transmission line.

Flat Compounded — A flat-compounded generator is one in which the relationship of the turns in the series and shunt fields is such that their terminal voltage is about the same over the entire load range.

Undercompounded — An undercompounded generator is one in which the series field does not have enough turns to compensate for the voltage drop of the shunt field. The voltage at full load is less than the no-load voltage. In an undercompounded generator, the series and shunt fields are connected so as to oppose rather than to aid one another. It is referred to as being differentially compounded. The terminal voltage of this type of generator decreases rapidly as the load increases. Undercompounded gener-

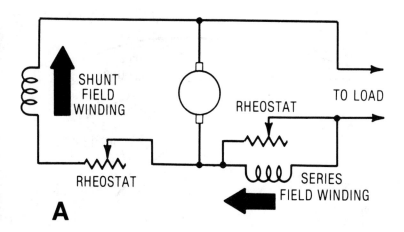

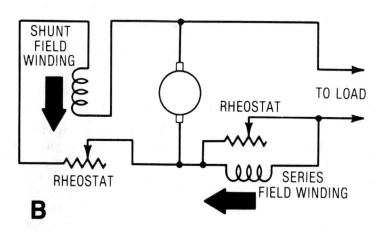

Fig. 1-29. Typical compound-wound DC generator circuits.

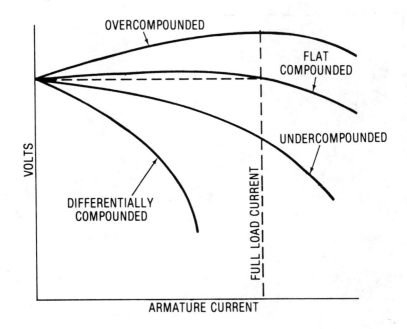

Fig. 1-30. Voltage characteristics of a compound-wound DC generator.

ators are used in applications where a short circuit might occur, as in an arc welder.

Control of DC Generators

Generally, the DC generator is controlled by a resistor that produces variable resistance, called a rheostat. After the generator is brought up to speed by the prime mover, the rheostat is adjusted. The rheostat may be manually or automatically operated. The adjustment of the rheostat controls the amount of excited current fed to the field coils. Metering requires the use of a DC voltmeter and ammeter of appropriate ranges in the generator output circuit. Matched sets of shunt-wound or compound-wound generators with series-field equalizer connections are used for parallel operation. Precautions must be observed when connecting the machines to generator buses.

Voltage Regulation

Series Generator — The series generator is classified as a constant-speed generator. It can be used to supply series motors, series arc-lighting systems, and voltage boosting on long DC feeder lines. The series generator is excited entirely by low-resistance field coils connected in series with the armature terminals and the load. The circuit of a DC series generator is shown in Fig. 1-25. The voltage increases with load because the load current provides the necessary additional field excitation. Low-resistance shunts may be used across series field coils to obtain desired voltage characteristics. The series field of the generator is adjusted so the output voltage may be maintained at a constant value. Because series generators have poor voltage regulation, only a few are in use.

Shunt Generator — The shunt DC generator can be called a constant-potential generator. It is seldom used for lighting and power because of its poor voltage regulation. The field coils in this type of generator have a comparatively high resistance; they are connected across the armature terminals in series with the rheostat. A DC shunt generator circuit is shown in Fig. 1-27. Shunt generators sometimes have separate excitation. This prevents reversal of the generator polarity and allows better voltage regulation. Shunt generators are frequently used with automatic voltage regulators as exciters for AC generators.

Compound Generator — The compound generator is the most widely used DC generator. The speed of a compound generator affects its generating characteristics. Therefore, the compounding can be varied. The engine governor can be adjusted for the proper no-load voltage. The range of the shunt-field rheostat and the engine characteristic limit the amount of speed variation that can be obtained. Compound generators can be connected either cumulatively or differentially.

Direct-Current Motors

A machine that converts electrical energy into mechanical energy is called a *motor*. The functions of a DC generator and a DC motor are interchangeable in that a generator may be operated as a motor, and vice versa. Structurally, the two machines are identical. The motor, like the generator, consists of an electromagnet, an armature, and a commutator with its brushes.

Fig. 1-10A will serve to illustrate the operation of a direct-current motor as well as a generator. The magnetic field, as indicated, will be the same for a motor because of current flowing in the field windings. Now, let the outside current at A have a voltage applied that causes a current to flow in the armature loop, as indicated by the arrows.

It must be remembered that any current flowing in a loop or coil of wire produces a magnetic field. This is exactly what happens in the armature of this motor. In addition, a second magnetic field is produced, with poles N and S perpendicular to the armature loop. The north pole of the main magnetic field attracts the south pole of the armature, and since the loop is free, it will revolve. At the instant the north and south poles become exactly opposite, however, the commutator reverses the current in the armature, making the poles of the field and the armature opposite, and the loop is then repelled and forced to revolve further. Again the armature current is reversed when unlike poles approach, and the armature is free to revolve. This continues as long as there is current in the armature and field windings.

It should be observed that, in an actual motor, there is more than one loop (called an *armature coil*), each with its terminals connected to adjacent commutator segments (Fig. 1-10B). Hence, the attracting and repelling action is correspondingly more powerful and also more uniform than that of the weak and unstable action obtained with the single-loop armature described here.

The various types of direct-current motors as well as their operating characteristics and control methods are fully treated in a later chapter.

Alternating-Current Motors

When a coil of wire is rotated in a magnetic field, the current changes its direction every half turn. Thus, there are two alternations of current for each revolution of a bipolar machine. As previously noted, this alternating current is rectified by the use of a commutator in a direct-current generator. In an alternating-current generator, also termed an *alternator,* the current induced in the armature is led out through *slip rings* or *collector rings,* as shown in Fig. 1-31.

A magnetic field is established between the north pole and the south pole by means of an exciting current flowing in winding W. A loop of wire, L, in this field is arranged so that it can be rotated on axis X, and the ends of this loop are brought out to slip rings SS, on which brushes BB can slide.

This circuit, of which the rotating loop is a part, is completed through the slip rings at A. When the loop is rotating, voltage is produced in conductors F and G, which will cause a current to flow out to A, where the circuit is completed.

The simplified machine represented in Fig. 1-31 is a two-pole, single-phase, revolving-armature, alternating-current generator. The magnetic field — the coils of wire and iron core — are called simply the field of the generator. The rotating loop in which the voltage is induced is called the armature.

The rotating-armature type of generator is generally used only on small machines, whereas large machines almost without exception are built with rotating fields.

If the voltage completes 60 cycles in one second, the generator is termed a *60-hertz* machine. The current that this voltage will cause to flow will be a 60-hertz current. The term *hertz* indicates cycles per second.

Polyphase Machines

A two-phase generator is actually a combination of two single-phase generators, as shown in Fig. 1-32. The armatures of these two machines are mounted on one shaft and must revolve together, always at right angles to each other. If the voltage waves or curves are plotted as in Fig. 1-33, it will be found that when phase *1* is in such a position that the voltage is at a maximum, phase *2* will be in such a position that the voltage in it is zero. A quarter of a cycle later, phase *1* will be zero and phase *2* will have advanced to a position previously occupied by *1*, and its voltage will be at a maximum. Thus, phase *2* follows phase *1* and the voltage is always

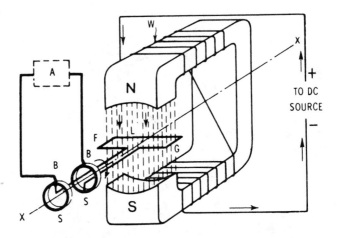

Fig. 1-31. An elementary alternating-current (AC) generator of the rotating-armature type. The slip rings and brushes are used to collect the current from the armature.

exactly a quarter of a cycle behind because of the relatively mechanical positions of the armatures.

It has been found economical to have more than one coil for each pole of the field. Because of this, present-day AC generators are built as three-phase units in which there are three sets of coils on the armature. These three sets of armature coils may each be used separately to supply electricity to three separate lighting circuits.

In a three-phase generator, three single-phase coils (or windings) are combined on a single shaft and rotate in the same magnetic field, as shown in Fig. 1-34. Each end of each coil is brought out through a slip ring to an external circuit. The voltage in each phase alternates exactly one-third of a cycle after the one ahead of it because of the mechanical arrangement of the windings on the armature. Thus, when the voltage in phase *1* is approaching a maximum positive, as shown in Fig. 1-35, the voltage in phase *2* is at a maximum negative, and the voltage in phase *3* is declining. The succeeding variations of these voltages are as indicated.

In practice, the ends of each phase winding are not brought out to separate slip rings, but are connected as shown in Fig. 1-36. This

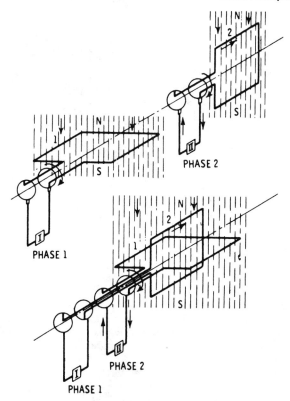

Fig. 1-32. An elementary two-phase AC machine constructed by combining the two single-phase machines shown at the top of the illustration.

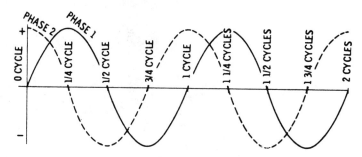

Fig. 1-33. Curves representing the voltages in two separate loops of wire that are positioned at right angles to each other and rotating together.

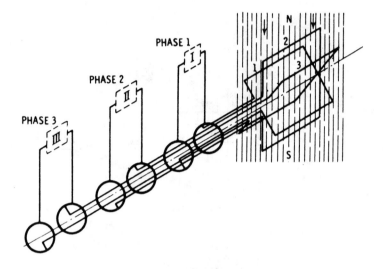

Fig. 1-34. An elementary three-phase AC machine is one in which three separate loops of wire are displaced from one another at equal angles, the loops made to rotate in the same magnetic field, and each loop brought out to a separate pair of slip rings.

arrangement makes only three leads necessary for a three-phase winding, each lead serving two phases. This allows each pair of wires to act like a single-phase circuit that is substantially independent of the other phases.

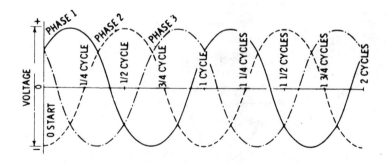

Fig. 1-35. Curves illustrating the voltage variation in a three-phase machine. One cycle of rotation produces one hertz of alternating current.

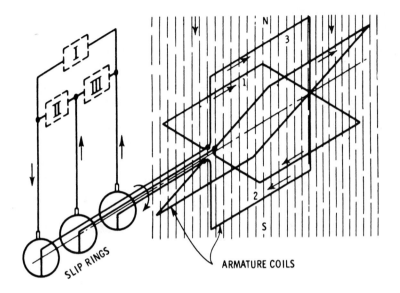

SLIP RINGS

ARMATURE COILS

Fig. 1-36. Commercial three-phase machines usually have the separate loops of wire connected as shown. This requires only three slip rings.

Revolving Magnetic Field

In the diagrams studied thus far, the poles producing the magnetic field have been stationary on the frame of the machine, and the armature in which the voltages are produced rotates. This arrangement is universally employed in direct-current machines, but alternating-current motors and generators generally have revolving fields because they need only two slip rings.

When the revolving-field construction is employed, the two slip rings need only carry the low-voltage exciting current to the field. For a three-phase machine with a rotating armature, at least three slip rings would be required for the armature current, which is often at a high voltage and therefore would require a large amount of insulation, adding to the cost of construction. A schematic of a single-phase AC generator with revolving field is shown in Fig. 1-37. The operation of practically all polyphase alternating-current motors depends on a revolving magnetic field that pulls the rotating part of the motor around with it.

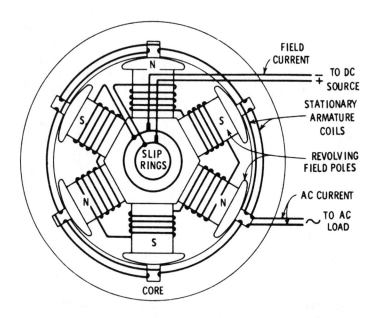

Fig. 1-37. Construction details of an AC generator having six poles and a revolving field. If this generator is to deliver 60 hertz current, it must be driven at a speed of exactly 1200 rpm.

To produce a rotating field, assume that two alternating currents of the same frequency and potential, but differing in phase by 90°, are available. Connect them to two sets of coils wound on the inwardly projecting poles of a circular iron ring, as illustrated in Fig. 1-38. It will be noted that when the current in phase *1* is at a maximum, the current in phase *2* is zero. Poles A and A_1 are magnetized, while poles B and B_1 are demagnetized. The magnetic flux is in a direction from N to S, as indicated by the arrow in the center of diagram I.

Referring to the voltage curves, it will be found that one-eighth of a cycle (45°) later, the current in phase *1* has decreased to the same value to which the current in phase *2* has increased. The four poles are now equally magnetized, and the magnetic flux takes the direction of the arrow shown in the center of diagram II.

One-eighth of a cycle later, the current in phase *1* has dropped to zero, while the current in phase *2* is at its maximum. With reference

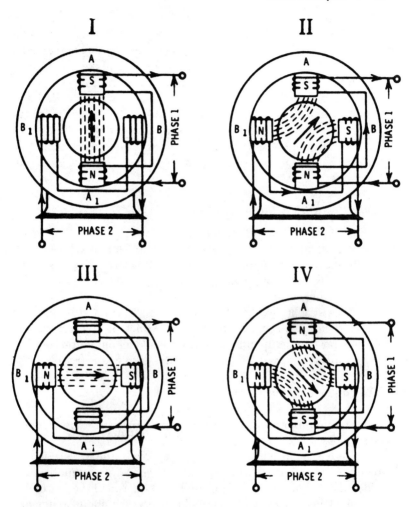

Fig. 1-38. Illustrating how a rotating magnetic field is produced by two currents 90 degrees apart.

to diagram III, in Fig. 1-38, this condition indicates that poles A and A_1 are demagnetized, but that poles B and B_1 are magnetized, with the flux from N to S, as shown by the center arrow.

Continuing the analysis, notice that after an additional one-eighth of a cycle, the current in both phases 1 and 2 has decreased and that the four poles are again equally magnetized, with the magnetic flux in the direction as indicated by the center arrow in diagram IV. If this process is continued at successive intervals during a complete period or cycle of change in the alternating current, the magnetic flux represented by the arrow will make a complete revolution for each cycle of the current.

The action of the current is inducing a rotating magnetic field, which would cause a magnet to revolve on its axis according to the periodicity of the impressed alternating current. This analysis explains the action in a two-phase motor. The rotating magnetic field in a three-phase AC motor having any number of poles can be similarly obtained.

Synchronous Motors

Any AC generator can be employed as a motor, provided that it is first brought up to the exact speed of a similar generator supplying the current to it, and provided that it is then put in step with the alternations of the supplied current. Such a machine is called a *synchronous motor*. However, because of complications in starting, most synchronous motors of late construction are equipped with a *damper* or *amortisseur* winding, which produces a starting torque, permitting them to be started as induction motors.

The speed of a synchronous motor depends on the frequency of the current supplied to it and the number of poles in the motor. The equation for the speed is:

$$\text{revolutions per minute} = \frac{\text{frequency} \times 60}{\text{number of pairs of poles}}$$

Since a synchronous motor runs at exactly this speed, it is a relatively simple matter to calculate the speed of any motor provided that the number of poles and the frequency of the source are known. Thus, for example, an eight-pole synchronous motor operating from a 60-hertz source has a speed of

$$\text{rpm} = \frac{60 \times 60}{4} = 900 \text{ rpm}$$

Induction Motors

Although the synchronous motor is used commercially in certain applications, the induction motor is used more extensively because of its simplicity. There are two principal classes of polyphase motors, namely:

1. Squirrel cage.
2. Wound rotor.

By definition, an induction motor is one in which the magnetic field in the rotor is induced by currents flowing in the stator. The rotor has no connections whatever to the supply line.

Squirrel-Cage Motor — This type of induction motor consists of a *stator*, which is identical to the armature of a synchronous motor, with a "squirrel-cage" rotor with bearings to support it. Because the stator receives the power from the line, it is often called the *primary* and the rotor the *secondary*.

In an induction motor of this type, the squirrel-cage winding takes the place of the field in the synchronous motor. The squirrel cage consists of a number of metal bars connected at each end to supporting metal rings. As in the synchronous motor, a rotating field is set up by the currents in the armature.

As this field revolves, it cuts the squirrel-cage conductors, and voltages are set up in them exactly as though the conductors were cutting the field in any other motor. These voltages cause current to flow in the squirrel-cage circuit, through the bars under the north poles into the ring, back to the bars under the adjacent south poles, into the other ring, and back to the original bars under the north pole, completing the circuit.

The current flowing in the squirrel cage, down one group of bars and back in the adjacent group, makes a loop that establishes magnetic fields with north and south poles in the rotor core. This loop consists of one turn, but there are several conductors in parallel and the currents may be large. These poles in the rotor, attracted by the poles of the revolving field, set up the currents in the armature winding, and follow them around in a manner similar to that in which the field poles follow the armature poles in a synchronous motor.

There is, however, one interesting and important difference between the synchronous motor and the induction motor — the rotor of the latter does not rotate as fast as the rotating field in the armature. If the squirrel cage were to go as fast as the rotating field, the conductors in it would be standing still with respect to the rotating field, rather than cutting across it. Thus, there could be no voltage induced in the squirrel cage, no currents in it, no magnetic poles set up in the rotor, and no attraction between it and the rotating field in the stator. The rotor revolves just enough slower than the rotating field in the stator to allow the rotor conductors to cut the rotating magnetic field as it slips by, and thus induces the necessary currents in the rotor windings.

This means that the motor can never rotate quite as fast as the revolving field, but is always slipping back. This difference in speed is called the *slip*. The greater the load, the greater the slip will be — that is, the slower the motor will run — but even at full load, the slip is not too great. In fact, this motor is commonly considered to be a constant-speed device. The various classes of squirrel-cage motors, and their operation and control, are given in a later chapter.

Wound-Rotor Motor — This type of induction motor differs from the squirrel-cage type in that it has wire-coil windings in it

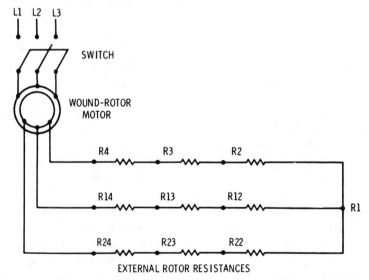

Fig. 1-39. Wiring diagram showing the connections between the slip rings and external resistances for a wound-rotor motor. The resistances are connected to a drum controller, the drum rotation of which determines the amount of resistance in the circuit and thus the speed of the motor.

instead of a series of conducting bars in the rotor. These insulated coils are grouped to form definite polar areas having the same number of poles as the stator. The rotor windings are brought out to slip rings whose brushes are connected to variable external resistances (Fig. 1-39).

By inserting external resistance in the rotor circuit when starting, a high torque can be developed with a comparatively low starting current. As the motor accelerates up to speed, the resistance is gradually reduced until, at full speed, the rotor is short-circuited. By varying the resistance at the rotor circuit, the motor speed can be regulated within practical limits.

This method of speed control is well-suited for the wound-rotor motor because it is already equipped with a starting resistance in each phase of the rotor circuit. By making these resistances of large enough current-carrying capacity to prevent dangerous heating in continuous service, the same resistances can also serve to regulate the speed. Although effective speed control is best secured by the use of direct-current motors, the wound-rotor motor, because of its adjustable rotor resistance, possesses one of the few methods of speed control available for alternating-current motors.

Slip

The speed of a synchronous motor is constant for any given frequency and number of poles in the motor. In an induction motor, however, this exact relationship does not exist, because the rotor slows down when the load is applied. The ratio of the speed of the field (relative to the rotor) to synchronous speed is termed the *slip*. It is usually written:

$$s = \frac{N_s - N}{N_s}$$

where

s is the slip (usually expressed as a percentage of synchronous speed),

N_s is the synchronous speed,

N is the actual rotor speed.

For example, a six-pole, 60-hertz motor would have a synchronous speed of 1200 rpm. If its rotor speed were 1164 rpm, the slip would be:

$$s = \frac{1200 - 1164}{1200} = 0.03, \text{ or } 3\%$$

Single-Phase Motors

Single-phase induction motors may be divided into two principal classes, namely:

1. Split-phase.
2. Commutator.

Split-phase motors are further subdivided into resistance-start, reactor-start, split-capacitor, and capacitor-start motors. Commutator-type motors are subdivided into two groups, series and repulsion, and each of these is further subdivided into several types and combinations of types.

These two classes and subdivisions represent various electrical modifications of single-phase induction motors, where one modification must be used to produce the necessary starting torque.

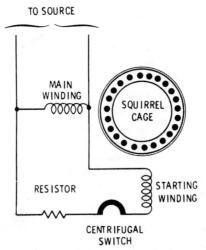

Fig. 1-40. Schematic diagram showing the winding arrangement of a resistance-start, split-phase, induction motor.

All methods serve to increase the phase angle between the main winding and the starting winding so as to produce a rotating magnetic field similar to that in a two-phase motor.

Resistance-Start Motors

A resistance-start motor is a form of split-phase motor having a resistance connected in series with the starting (sometimes called *auxiliary*) winding. A schematic diagram of this type of motor may be represented as in Fig. 1-40. This shows a resistance connected in series with the starting winding to provide a two-phase rotating-field effect for starting. When the motor reaches approximately 75 percent of its rated speed, a centrifugal switch opens to disconnect the starting winding from the line. This motor is known as resistance-start, split-phase type and is commonly used on washing machines and similar appliances. It is not practical to build such motors for the heavier types of starting duty.

Split-Capacitor Motors

In a split-capacitor motor (Fig. 1-41), two stationary windings are connected to a single-phase line. The capacitor has the peculiar characteristic of shifting the phase of the current in coil 2 with respect to the current in coil 1. This provides the same action as in the two-phase motor discussed previously, producing the rotating

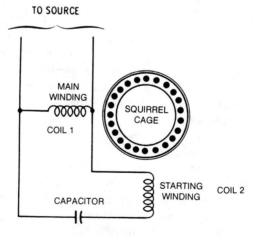

TO SOURCE

MAIN WINDING

COIL 1

SQUIRREL CAGE

STARTING WINDING

COIL 2

CAPACITOR

Fig. 1-41. Schematic diagram of a split-capacitor motor.

TO SOURCE

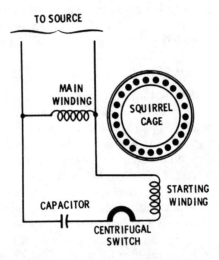

Fig. 1-42. Winding connections in a capacitor motor.

field effect to rotate the squirrel-cage rotor. The capacitor is mounted permanently in the circuit. Because capacitors for continuous duty are expensive and somewhat bulky, it is not practical to make this motor for heavy-duty starting.

Capacitor-Start Motors

In applications where a high starting torque is required, a motor such as that shown in Fig. 1-42 is employed. This is only another form of split-phase motor having a capacitor (or condenser, as it was once called) connected in series with the starting winding. The construction is similar to the split-capacitor motor, but differs mainly in that the starting winding is disconnected at approximately 75 percent of rated speed by a centrifugal switch, as in the case of the resistance-start motor.

The centrifugal switch is mounted on the motor shaft and, as the name implies, works on the centrifugal principle, disconnecting the starting winding when the speed at which the switch is set is reached. The capacitor-start motor has a greater starting ability than the resistance-start motor. Because the capacitor is in use only during the starting period, a high capacity can be obtained economically for this short-term duty.

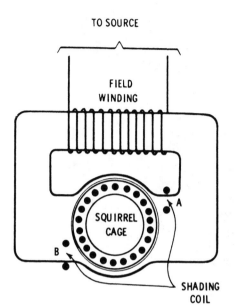

TO SOURCE

FIELD
WINDING

SQUIRREL
CAGE

A

B

SHADING
COIL

Fig. 1-43A. Arrangement of the windings in a skeleton-type shaded-pole motor.

Shaded-Pole Motors

Another type of single-phase induction motor is schematically represented in Fig. 1-43A. This motor consists principally of a squirrel-cage rotor and two or more coils with an iron core to increase the magnetic effect. Part of one end of this core is surrounded by a heavy copper loop known as the *shading ring*. This ring has the characteristic of delaying the flow of magnetism through it. With alternating current applied to the coil, the magnetism is strong first at *A*, and then slightly later at *B*. This gives a rotating-field effect that causes rotation of the squirrel cage in the direction in which the shading ring points. A motor thus constructed is known as a shaded-pole motor.

Because of the limitations of force and current possible in shading poles, it is not feasible to build efficient motors of this type larger than approximately 1/20 hp (or 37.3 W). These motors are used principally on small fans, agitators, and timing devices (Fig. 1-43B).

Fig. 1-43B. Small shaded-pole fan motor.

Repulsion-Start Induction Motors

Repulsion-start induction motors are operated in various ways. In the running position, the brushes may or may not be raised. If the same rotor winding is used for both starting and running, the commutator is short-circuited at about 75 percent of rated speed to obtain a rotor winding approximating the squirrel cage in its functioning. Other designs have two rotor windings — that is, a squirrel cage and a wound winding for running and starting, respectively. In this type, no rotor mechanism is required because the magnetic conditions automatically transfer the burden from one winding to the other as the motor comes up to speed.

Repulsion starting may best be explained by the action of a wire connected to a battery and moved across the face of a magnet. Here, there is a force on the wire that, for example, tends to move it upward or downward, depending on the direction in which the current is flowing. It can thus be demonstrated that a current-carrying wire in a magnetic field has a force acting on it tending to move it in a certain direction. Also, if the direction of the current flowing through the wire is reversed, the force and motion are also reversed.

Repulsion starting operates on this principle. Current is caused to flow in the wires of the rotor winding, and these wires are affected by a magnetic field.

Fig. 1-44 shows a stationary C-shaped iron core on which is mounted a coil connected to a single-phase supply line. In the opening of the C is a ring of iron on which is wound a continuous and uniform coil. The path of the magnetism produced by the coil wound on the C-shaped core is around through the C-shaped core and, dividing equally, half of the magnetism passes through each half of the iron ring.

The winding and the magnetism are identical in both halves of the ring. Thus, any effect that the magnetism may have on the winding between E and G will be the same as that produced in windings E and H. This can be proven by connecting an ammeter between points G and H. It will be found that no current is flowing between these two points. By further tests it can be shown that maximum current will flow when a wire is connected between E and F. Thus, the first requirement of our principle has been satisfied — with a wire connecting E and F, there is a current flowing in the rotor winding.

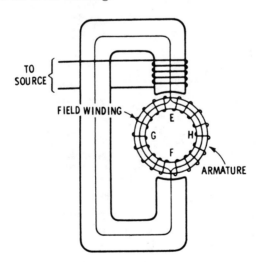

Fig. 1-44. Illustration to show the operating principles of a repulsion-start induction motor.

Assume that the current flows upward in this wire from F to E. At point E, it divides equally, half going to the winding to the left of E, and the other half to the right. Referring to the wires on the outer surface of the ring, those on the right have the current flowing toward the observer, while those on the left have the current flowing away. Thus the magnetic field from the C-shaped core tends to force the wires on the right in one direction and those on the left in the other direction. The forces are equal and opposite so that they neutralize each other and no motion takes place.

In order for the rotor to rotate, it is necessary to add a magnetic field that can effectively react with the current in the rotor winding. This is done readily by adding another C-shaped core with its own coil, as shown in Fig. 1-45. The rotor-winding current under each tip of this C-shaped core is all in the same direction, and rotation is obtained. The wire from E to F in Fig. 1-44 has been replaced with stationary brushes so that a connection is maintained as the rotor turns.

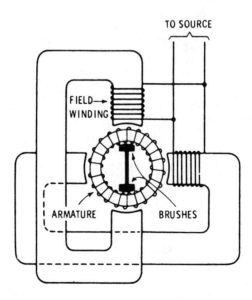

Fig. 1-45. Schematic diagram of a repulsion-start induction motor.

Motor Control

Although the function of motor control is fully covered later in this book, a brief outline of its essentials will be of aid in further study of the various types of motors and their associated control circuits.

The elementary functions of control are starting, stopping, and reversing of the motor. These, however, are only a few of the many contributions that the control renders to efficient operation of industrial motors.

The most common control functions of industrial motors are:

1. To limit torque on the motor and machine.
2. To limit motor starting current.
3. To protect the motor from overheating.
4. To stop the motor quickly.
5. To regulate speed.
6. Miscellaneous functions.

Limiting Torque

One example of the need for limiting torque is that of a belt-driven motor-operated machine throwing the belt when the motor is started. The pulleys may be correctly lined up and the belt tension may be correct; yet the belt is thrown off in starting. This is the result of applying torque too quickly at standstill, and can be avoided by limiting the torque on the motor in starting. As another example, the blades on centrifugal fans can be sheared off if too much torque is applied to the fan in starting.

Limiting Starting Current

It is a common sight to see DC motors flash over at the commutator when too much current is applied to the motor in bringing it up to speed. Also, it is common to see lights blink when a motor on the same power circuit is started. True, this blinking of lights can be reduced by selecting a motor with the right characteristics, but usually the real solution is the selection of a control that limits the starting current, either by inserting resistance in the circuit or by using a reduced-voltage source of power.

Protection from Overheating

Motors are designed to produce full-load torque for a definite period without overheating. While the motor is capable of exceeding its normal output for limited periods, there is nothing inherent in the motor to keep its temperature within safe limits. It is therefore the function of the control to prevent the motor from overheating excessively without shutting it down unnecessarily.

Quick Stopping

Where a driven machine has high inertia, it will continue to run for a considerable time after the power has been disconnected. There are several types of controls, such as electric brakes, to stop a motor quickly. The one most generally used on AC motors is the plugging switch. To plug a motor, it is necessary only to disconnect it from the line, and then reconnect it so that the power applied to the motor tends to drive it in the opposite direction. This brakes the motor rapidly to a standstill, at which time the plugging switch cuts off the reverse power.

Speed Regulation

Fans are sometimes run at various speeds, depending on the ventilation requirements. For some applications, it is advisable to use a two-speed motor, but where a greater variety of speeds is required, a motor with variable speed control may be the best solution to the problem.

Miscellaneous Control Functions

Adequate control equipment covers various other protective functions, which are not as common as those enumerated previously. Among these are *reverse-phase protection,* which prevents a motor from running in the wrong direction if a phase is inadvertently reversed; *open-phase protection,* which prevents the motor from running on single phase in case a fuse blows, and *undervoltage protection,* which prevents a motor from starting after a power failure unless started by the operator.

Summary

If the south-seeking (S) pole of a magnet is brought near the S pole of a suspended magnet, the poles repel each other. Likewise, if the two north-seeking (N) poles are brought together, they repel each other. However, if an N pole is brought near the S pole or if an S pole is brought near the N pole, the two unlike poles attract each other. In other words, *like poles repel each other, and unlike poles attract each other.* Experiments have shown these attracting or repelling forces between magnetic poles to vary inversely as the square of the distance between the poles.

Oersted discovered the relation between magnetism and electricity in the early 19th century. He observed that when a wire connecting the poles of a battery was held *over* a compass needle, the N pole of the needle was deflected in one direction when the current flowed, and a wire placed *under* the compass needle caused the N pole of the needle to be deflected in the opposite direction. The compass needle indicates the direction of the magnetic lines of force, and an electric current sets up a magnetic field at right angles to the conductor. The so-called *left-hand rule* is a convenient method for determining the direction of the magnetic flux around a straight wire carrying a current — *if the wire is held in the left hand, with the thumb pointed in the direction of the current, the fingers will point in the direction of the magnetic field.* Conversely, if the direction of the magnetic field around a conductor is known, the direction of the current in the conductor can be determined by applying the rule.

An *electromagnet* is a soft-iron core surrounded by a coil of wire. The magnetic strength of an electromagnet can be changed by changing the strength of its applied current. When the current is interrupted, the iron core returns to its natural state. This loss of magnetism is not complete, however, because a small amount of magnetism, or *residual magnetism*, remains. The electromagnet is used in many electrical devices, including electric bells, telephones, motors, and generators. The polarity of an electromagnet can be determined by means of the left-hand rule, as follows: *Grasp the coil with the left hand; with the fingers pointing in the direction of the current in the coil, the thumb will point to the north pole of the coil.*

If a coil of wire having many turns is moved up and down over one pole of a horseshoe magnet, a momentary electric current without an apparent electrical source is produced. This current produced by moving the coil of wire in a magnetic field is called an *induced current.* Lenz's law states that an induced current has such a direction that its magnetic action tends to resist the motion by which it is produced. The generator and the motor are examples of useful applications of induced currents.

A *generator* converts mechanical energy into electrical energy. Its essential parts are a magnetic field, usually produced by permanent magents, and a moving coil or coils called the *armature.*

The DC generator is classified either as a separately excited or a self-excited type. The separately excited has very little practical use; the self-excited is the one most often used. The self-excited DC generator can be broken down into a number of classifications: the series, shunt, compound, overcompound, flat compound, and undercompound. Each type has particular advantages and disadvantages according to its load and speed of rotation. Some of these generator types cannot be regulated in terms of a constant voltage output; they are used for other purposes where voltage regulation is not so important.

A *motor* converts electrical energy into mechanical energy. The motor, like the generator, consists of an electromagnet, an armature, and a commutator with its brushes.

The two principal classes of polyphase induction motors are the *squirrel-cage* motor and the *wound-rotor* motor. By definition, an induction motor is one in which the magnetic field in the rotor is induced by currents flowing in the stator. The rotor has no connection whatever to the supply line.

Single-phase motors can be divided into two principal classes as follows:
1. Split-phase.
    a. Resistance-start.
    b. Split-capacitor.
    c. Capacitor-start.
    d. Repulsion-start.
2. Commutator.
    a. Series.
    b. Repulsion.

Review Questions

1. How can the direction of the magnetic field around a straight wire carrying a current be determined?
2. Describe the basic construction of an electromagnet.
3. How can the polarity of an electromagnet be determined?
4. How is an "induced current" produced?
5. What is the basic difference between a generator and a motor?
6. What is the function of a DC generator?
7. How are the commutator segments of a DC generator insulated?
8. What materials are used to make brushes for a DC generator and/or motor?
9. How is the neutral plane of a generator shifted?
10. How are DC generators classified?
11. What is the only type of compound generator commonly used?
12. What is the name of the mechanical power source used to drive generators?
13. How are the effects of armature reaction overcome or reduced permanently in a generator?
14. Why will a shunt generator build up to full terminal voltage with no external load connected?
15. What is the name given to power lost in heat in the windings of a generator due to the flow of current through the copper?
16. As armature current of a generator increases, what happens to the motor reaction force?
17. How can compound generators be connected?
18. What is the name given to the part of a DC generator into which the working voltage is induced?
19. What are the two principal types of induction motors?
20. What are the two principal types of single-phase motors?

CHAPTER 2

AC Generators (Alternators)

Principles of Operation

The fundamental principle for the generation of an emf in an AC generator (also called an *alternator*) is the same as in a DC generator. The generation of an emf in an armature conductor depends solely on a relative motion between the conductor and the magnetic field. Two constructions are possible. The magnetic field may be stationary and the armature may rotate. In this case, the magnetic field is called the stator. The armature is called the rotor. Or, the magnetic field may rotate and the armature may be stationary. In this case, the magnetic field is called the rotor and the armature is called the stator.

In almost all DC generators, the field is stationary and the armature is rotated. But in almost all AC generators, the armature is stationary and the field is rotated. This latter type of construction provides several advantages. A rotating armature requires slip rings to carry current to the external load. Such rings are difficult to insulate. They are frequent sources of trouble, often causing open and short circuits. A stationary armature needs no slip rings. Thus, armature leads can be continuously insulated conductors from the armature coils to the bus bars. It is more difficult to insulate conductors in a rotating armature than in a stationary armature because of the centrifugal force that results from rotation. Also, the stationary armature allows alternators to operate with higher voltages than DC generators.

Inasmuch as the AC motor will not operate without a source of alternating current, it is important that we take a closer look at the device that provides the power for the operation of all AC motors. There are some similarities, which will be pointed out as we progress in the chapter with the AC motor. DC motors and generators are compared or contrasted from time to time and will be referenced as facts to keep in mind as the operation of the AC motor is studied.

Fig. 2-1 shows a simplified drawing of an AC generator. Certain features of this generator are basic to the design of all AC generators:

1. A magnetic field is necessary. In this case, the north and south poles of a permanent magnet are used.

2. A coil must be rotated through the magnetic field.

3. A coil consists of two coil sides in which emf's are induced.

4. A coil of more than one turn presents a coil side of more than one conductor. The total emf induced in a coil is equal to the emf induced in one conductor multiplied by the number of conductors.

5. At any instant, the emf induced in one coil side is equal and opposite in direction to that induced in the other coil side. The two emf's appear in series between the collector rings because of the back connection.

6. Each coil side is connected to a metallic collector ring.

7. A brush is continually in contact with each collector ring. The brushes conduct the current in the coil to the load.

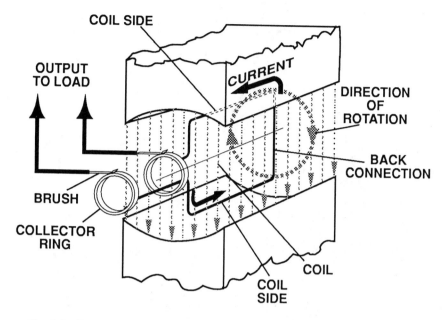

Fig. 2-1. Simple alternator with one coil in a magnetic field.

The voltage produced by this generator is an alternating voltage. One complete revolution of the coil will produce one hertz of voltage. That is, the voltage builds up from zero to a maximum. Then it drops to zero. Then it builds up again in the opposite direction to a maximum. Finally, the voltage drops to zero to complete the hertz. Such a hertz of alternating current or voltage is represented by a *sine wave*. A sine wave is shown in Fig. 2-2.

AC Generator Components

Rotor

Recall that the rotating part of the generator is called the rotor. The field of the AC generator, which is placed on the rotor, is either the salient-pole type or the turbo type. Fig. 2-3 shows a salient-pole rotor and Fig. 2-4 shows a turbo-type rotor.

When the AC generator is to be driven by a slow-speed diesel engine or by a water turbine (up to about 720 rpm), the salient-pole or projecting-pole rotor is used. The field poles are formed by

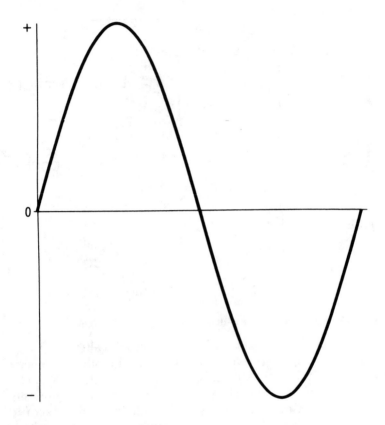

Fig. 2-2. Sine wave output of an alternator.

fastening a number of steel laminations to a spoked frame or *spider*. The heavy pole pieces produce a flywheel effect on the slow-speed rotor. This helps to keep the angular speed constant. It also reduces variation in the voltage and frequency of the generator output. In high-speed alternators (up to 3,600 rpm), the smooth-surface turbo-type rotor is used for two major reasons. One, it has less air-friction (heating) loss. Two, the windings can be placed so that they can withstand the centrifugal forces developed at high speeds. The turbo-type rotors are a solid-steel forging, a number of steel discs fastened together with the field coils locked

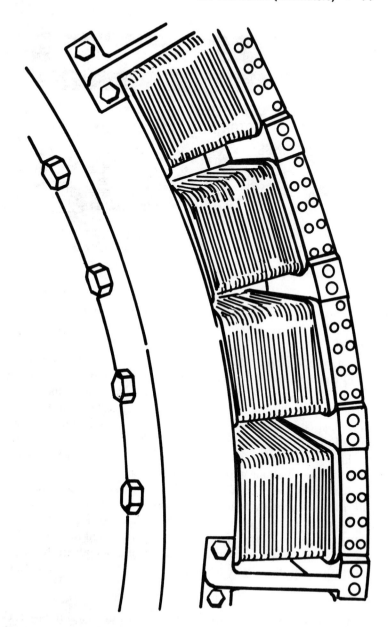

Fig. 2-3. Salient-pole rotor for an alternator.

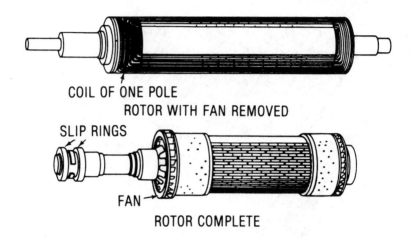

COIL OF ONE POLE
ROTOR WITH FAN REMOVED
SLIP RINGS
FAN
ROTOR COMPLETE

Fig. 2-4. Turbo-type rotor for an alternator.

in slots. These field coils are usually placed so they distribute the field flux evenly around the rotor, as shown in Fig. 2-4.

Stator

In a rotating-field AC generator, the armature windings are stationary and are therefore the stator. The armature iron, being in a moving magnetic field, is laminated in order to reduce eddy current losses. A typical AC generator stator is shown in Fig. 2-5. In high-speed turbo-type generators, the stator laminations are ribbed to provide sufficient ventilation. This is necessary because the high temperature developed in the windings cannot be dissipated in the small air gap between the rotor and the stator. Fig. 2-6 illustrates the close tolerance. In some large installations, the alternators are totally enclosed and cooled by hydrogen gas under pressure, which has greater heat-dissipating properties than air. Stator coils in high-speed alternators must be well braced. Bracing prevents coils from being pulled out of place when the alternator is operating with heavy loads.

Exciter

Like many DC generators, AC generators need a separate DC source for their fields. This DC field current must be obtained

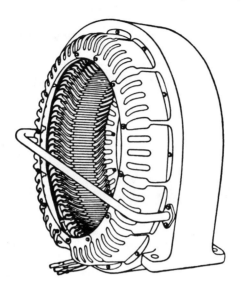

Fig. 2-5. Typical alternator stator.

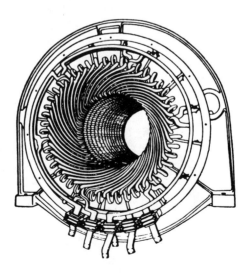

Fig. 2-6. Note how close together the windings are.

from an external source called an exciter. The exciter used to supply this current is usually a flat, compound-wound DC generator designed to furnish from 125 to 250 volts. The exciter armature may be mounted directly on the rotor shaft of the AC generator or it may be belt driven. Fig. 2-7 shows an exciter armature and generator field mounted on the same shaft.

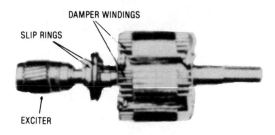

Fig. 2-7. Exciter armature and generator field mounted on the same shaft.

Brushless exciters are also used to provide the DC fields. The brushless exciter is an AC generator that converts the AC power to DC. It does so by means of a diode rectifier assembly, which is attached to, but insulated from, the generator shaft. The brushless exciter has no friction-producing parts such as brushes, brush holders, commutators, or slip rings. It needs very little maintenance. Fig. 2-8 shows a brushless rotor.

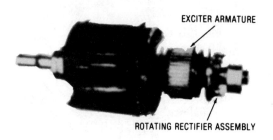

Fig. 2-8. A brushless rotor.

Static Exciter

Another method of field excitation commonly used is the static exciter. It is called a static exciter because it contains no moving parts. A portion of the AC current from each phase of the generator output is fed back through a system of transformers, rectifiers, and reactors to the field windings as DC excitation current. With this system an external source of DC current is necessary for initial excitation of the field windings. On engine-driven generators, the initial "field flash" may be obtained from storage batteries, which are also used to start the engine.

Frame and Shaft

The frame and shaft of an AC generator serve the same purpose as in the DC generator. The frame completes the magnetic circuit of the field. It also supports the parts and windings. The shaft upon which the rotor turns is supported by the end bells or end frames.

Types of Alternators or AC Generators

Single-Phase Alternator

Single-phase alternators are seldom used except for special purposes. They are used as emergency generators and for construction crews. As a rule, this type of alternator is low powered and self-excited. Fig. 2-9 shows a typical alternator. Its construction is similar to a DC generator with an auxiliary AC winding on the DC armature. The DC winding on the rotating member is of usual lap- or wave-type construction. The winding is connected to the commutator bars in the usual way. The DC winding output provides the current for DC field excitation and other DC power applications. A

Fig. 2-9A. Top view of alternator driven by an 8-hp gasoline engine to produce 3 kW of AC for emergency use.

Fig. 2-9B. Side view of the alternator location on the emergency power unit.

second open-wave winding is laid in the slots of the rotating member on top of the DC winding. This second winding, which is connected to slip rings, supplies the AC output.

Two-Phase Alternator

Multiphase or polyphase AC generators have two or more single-phase windings symmetrically spaced around the stator. In a two-phase alternator, there are two single-phase windings physically spaced so that the AC voltage induced in one is 90° out-of-phase with the voltage induced in the other. When one winding is being cut by maximum flux, the other is being cut by no flux.

Fig. 2-10A is a schematic diagram of a two-phase, four-pole alternator. This stator consists of two single-phase windings (phases) separated from each other. Each phase consists of four windings. These windings are connected in series so that their voltages add. The rotor is identical with the rotor used in the single-phase alternator. Note in Fig. 2-10B the waveforms produced by this type of generator.

The two phases of a two-phase alternator can be connected to each other as shown in Fig. 2-11A. Only three leads are brought out from the generator for connection to a load. This type of power is seldom used. However, in some parts of Europe this type of power is available for home and commercial use. It has an advantage when starting motors: AC motors that use two-phase power do not need a start winding or a switch to remove the winding from the circuit once the motor has reached operating speed, as is the case with single-phase motors.

Three-Phase Alternator

The three-phase alternator, as the name indicates, has three single-phase windings. These windings are spaced so that the voltage induced in each winding is 120° out of phase with the voltage in the other two windings. A schematic diagram of a three-phase generator showing all three coils is complex.

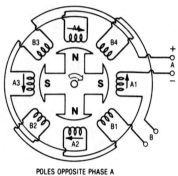

POLES OPPOSITE PHASE A

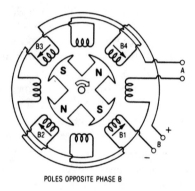

POLES OPPOSITE PHASE B

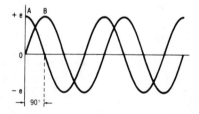

Fig. 2-10A. A two-phase, four-pole alternator and the results of its generated waveform.

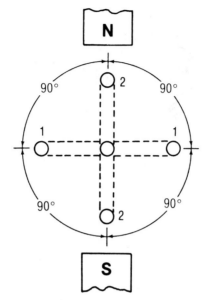

GENERATION OF TWO-PHASE VOLTAGE

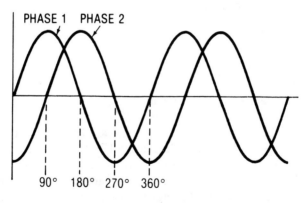

TWO-PHASE WAVEFORM

Fig. 2-10B. Generation of two-phase voltage.

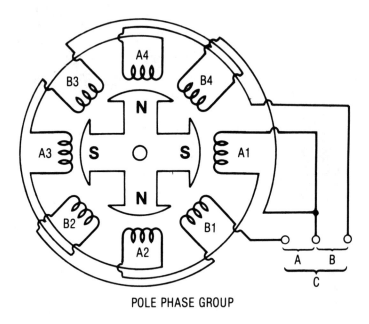

POLE PHASE GROUP

Fig. 2-11A. The two phases of a two-phase alternator can be connected together to produce single-phase power.

Fig. 2-12A shows the connections for various load options on the three-phase generator. Fig. 2-12B shows the output waveforms of the alternator. Note that there are 120 degrees of separation between each phase of the output. Electrical power generated by power companies for use in homes and business is all produced as three-phase. The three phases are then divided by three separate transformers into single-phase power for three different subdivisions or three different customers. Some three-phase power is used by businesses to drive large motors. Three-phase motors do not require as much maintenance as single-phase motors.

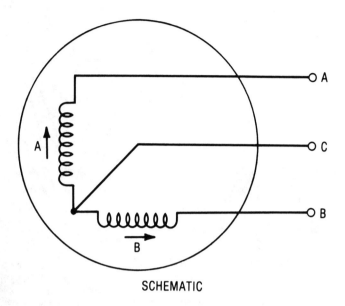

SCHEMATIC

Fig. 2-11B. Two-phase power brought out with all three connections available for connection to consuming devices.

Wye Connection

Instead of six leads coming out of the three-phase alternator, one of the leads from each phase may be connected to form a common junction. The stator is then wye- or star-connected. Fig. 2-13 shows a wye connection. The common lead may or may not be brought out of the machine. If it is brought out, it is called the *neutral*. One advantage of the neutral is the balancing of the load between or among all coils. The neutral serves as a common return circuit from all three phases. It maintains a voltage balance across the loads. No current flows in the neutral when the loads are balanced. The three-phase four-wire system is widely used in industry and for aircraft AC power systems.

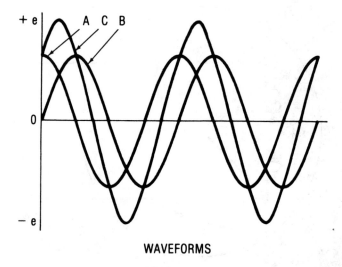

WAVEFORMS

Fig. 2-11C. Waveforms produced by a two-phase alternator with all three connections brought out separately.

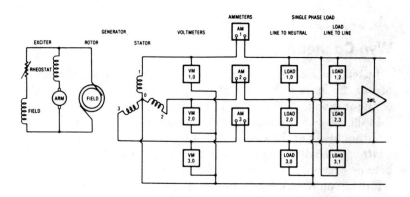

Fig. 2-12A. Various load options for the three-phase alternator.

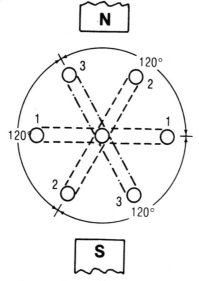

GENERATION OF THREE-PHASE VOLTAGE

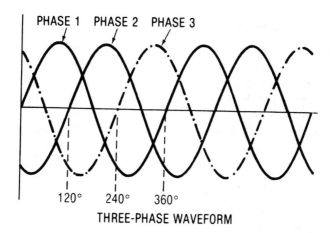

THREE-PHASE WAVEFORM

Fig. 2-12B. Generation of three-phase voltage and the resultant waveforms.

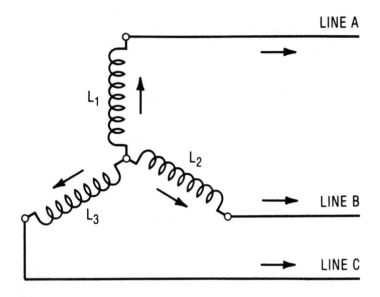

Fig. 2-13. Current flow in three-phase windings, wye-connected.

Delta Connection

A three-phase stator may also be connected in a delta configuration. In a delta-connected alternator, one phase winding and the start end of another are connected to the finish end of the third. The start end of the third is connected to the finish end of the first. The three junction points are connected to the line wires leading to the load. Fig. 2-14 shows a delta connection. When the generator phases are properly connected in delta, no appreciable current flows within the delta loop when there is no external load connected to the alternator. If any one of the phases is reversed with respect to its correct connection, a short-circuit current will flow within the windings on no load. This will caus damage to the windings.

To avoid connecting a phase in reverse, it is necessary to test the circuit before closing the delta. This may be done by connecting a

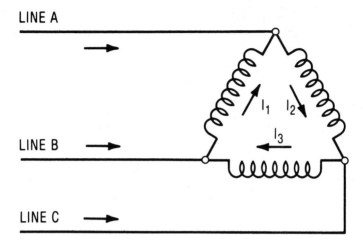

Fig. 2-14. Current flow in delta-connected three-phase alternator windings.

voltmeter or fuse wire between the two ends of the delta loop before closing the delta. The two ends of the delta loop should never be connected if there is an indication of any appreciable current or voltage between them when no load is connected to the alternator.

Power in a Balanced Wye Connection

The power delivered by a balanced three-phase wye-connected system is equal to three times the power delivered by each phase. The total true power is

$$P_t = 3 \times E_{phase} \times I_{phase} \times \cos \angle\theta$$

Since

$$E_{phase} = \frac{E_{line}}{\sqrt{3}} \text{ and } I_{phase} = I_{line}$$

the total true power is

$$P_t = \frac{E_{line}}{\sqrt{3}} \times I_{line} \times \cos \angle\theta$$

Power in a Balanced Delta Connection

The power delivered by a balanced three-phase delta-connected system is also three times the power delivered by each phase (Fig. 2-15).

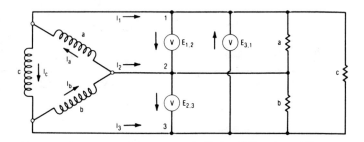

Fig. 2-15A. Power delivered by a balanced three-phase delta-connected system is three times the power delivered by each phase.

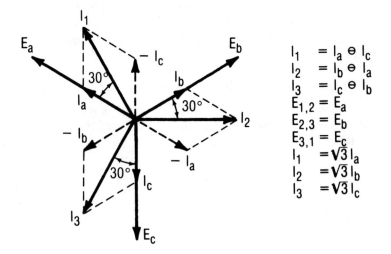

$$I_1 = I_a \ominus I_c$$
$$I_2 = I_b \ominus I_a$$
$$I_3 = I_c \ominus I_b$$
$$E_{1,2} = E_a$$
$$E_{2,3} = E_b$$
$$E_{3,1} = E_c$$
$$I_1 = \sqrt{3}\, I_a$$
$$I_2 = \sqrt{3}\, I_b$$
$$I_3 = \sqrt{3}\, I_c$$

Fig. 2-15B. Current and voltage relationships in a three-phase alternator.

$$E_{phase} = E_{line} \text{ and } I_{phase} = \frac{I_{line}}{\sqrt{3}}$$

The total true power is

$$P_t = 3E_{line} \times \frac{I_{line}}{\sqrt{3}} \times \cos \angle\theta$$

Thus, the expression for three-phase power delivered by a balanced delta-connected system is the same as the expression for three-phase power delivered by a balanced wye-connected system. Two examples are given to illustrate the phase relationships between current, voltage, and power in:

1. A three-phase wye-connected system.
2. A three-phase delta-connected system.

Example 1 — A three-phase wye-connected alternator has a terminal voltage of 450 volts. It delivers a full load current of 300 amperes per terminal at a power factor of 80 percent. Find the phase voltage, the full-load current per phase, the kilovolt-ampere (kVA) or apparent power rating, and true power output.

1. $E_{phase} = \dfrac{E_{line}}{\sqrt{3}} = \dfrac{450}{1.73} = 259.8$ volts

2. $I_{phase} = I_{line} = 300$ amperes

3. Apparent power = $\sqrt{3} \; E_{line} I_{line}$

   AP = $\sqrt{3} \times 450 \times 300 = 233{,}826.8591$ VA or 233.8268591 kVA

4. True power = $\sqrt{3} \; E_{line} I_{line} \cos \angle\theta$

   TP = $1.732050808 \times 450 \times 300 \times 0.8$

   TP = $187{,}061.4873$ watts or 187.0614873 kW

Note: $\cos \angle\theta$ = Power factor

Example 2 — A three-phase delta-connected alternator has a terminal voltage of 450 volts. The current in each phase is 200 amperes. The power factor of the load is 75 percent. Find the line voltage, the line current, the apparent power, and the true power.

1. E_{phase} = E_{line} = 450 volts

2. I_{line} = $\sqrt{3}\ I_{phase}$ = 1.732050808 $\times$ 200 = 346.4101616 amperes

Apparent power = $\sqrt{3}\ \times\ E_{line}\ \times\ I_{line}$

AP = 1.732050808 $\times$ 450 $\times$ 346.4101616 = 270,000 VA or 270 kVA

True power = $\sqrt{3} \times E_{line} \times I_{line} \times \cos \angle\theta$

TP = 1.732050808 $\times$ 450 $\times$ 346.4101616 $\times$ 0.75 = 202,500 watts
or 202.5 kw

Frequency

The frequency of the alternator voltage depends on the speed of rotation of the rotor and the number of poles. The higher the frequency needed, the faster the alternator must turn. The lower the speed, the lower the frequency. The more poles on the rotor, the higher the frequency for a given speed. When a rotor has rotated through an angle such that two adjacent poles (a north and a south pole) have passed one winding, the voltage induced in that winding will have varied through one complete cycle or hertz. For a given frequency, the more pairs of poles there are, the lower will be the speed of rotation. A two-pole alternator rotates at twice the speed of a four-pole generator for the same frequency of generated voltage. The frequency of the generator in hertz is related to the number of poles and the speed. This is expressed:

$$f \;=\; \frac{P}{2} \times \frac{N}{60} = \frac{PN}{120}$$

where

P is the number of poles
N is the speed in rpm
f is the frequency

Example 1 — What is the frequency of a two-pole, 3,600 rpm alternator? Simply substitute in the formula just developed to get:

$$\% \text{ regulation } = \frac{E_{NL} - E_{fL}}{E_{fL}} \times 100$$

Example 2 — What is the frequency of a four-pole, 1,800 rpm alternator? Repeat the steps used in the previous example:

$$f = \frac{4 \times 1800}{120} \times 60 \text{ hertz}$$

As you can see, the two alternators produce the same frequency. It is possible to operate the alternator at a lower speed if the number of poles is increased.

Load Changes

When the load on an alternator is changed, the terminal voltage carries the load. The amount of variation depends on the design of the generator and the power factor of the load. With a load having a lagging power factor (one with inductance dominating), the drop in terminal voltage with increased load is greater than for unity (1.00) power factor (that is, a totally resistive load). With a load a power factor that is leading the terminal voltage tends to rise. The causes of terminal voltage changes with load changes are the armature resistance, the armature reactance, and the armature reaction.

Armature Resistance

When current flows through a generator armature winding, there is an IR drop (voltage drop) due to the resistance of the winding. This drop increases with load. Thus, the terminal voltage is reduced. The armature resistance drop is small because the resistance is low.

Armature Reactance

The armature current in an alternator varies approximately as a sine wave. The continuously varying current in the generator armature is accompanied by an IX_L voltage drop in addition to the IR drop. Armature reactance in an alternator may be from 30 to 50 times the value of armature resistance, because of the relatively large inductance of the coils in comparison with its resistance.

Armature Reaction

When an alternator supplies no load, the DC field flux is distributed uniformly across the air gap. When an alternator supplies a reactive load, however, the current flowing through the armature conductors produces an armature magnetomotive force (*mmf*). That force influences the terminal voltage by changing the magnitude of the field flux across the air gap. When the load is inductive, the armature mmf opposes the DC field and weakens it. Thus the terminal voltage decreases. When a leading current flows in the armature, the DC field is aided by the armature mmf. The flux across the air gap is increased; thus, the terminal voltage increases.

Voltage Regulation

Voltage regulation of an alternator is the change of voltage from full-load to no-load. This is expressed in percentage of full-load volts with a constant speed and DC field current.

Example — The no-load voltage of an alternator is 250 volts. Its full-load voltage is 220 volts. What is the percent of regulation?

$$\frac{250 - 220}{220} = 13.6\%$$

Summary

The fundamental principle for the generation of an emf in an AC generator (alternator) is the same as in a DC generator.

The output of an alternator is called a sine wave voltage or current. The current and voltage alternate. Salient-pole or projecting-pole rotors are used in alternators that are slow in speed. The higher speed turbo-driven alternators have a smaller diameter rotor to lessen friction and better withstand centrifugal force.

In a rotating-field AC generator, the armature windings are stationary and are the stator. Bracing prevents coils from being pulled out of place when the alternator is operating with heavy loads.

AC generators (alternators) need a separate source of DC to excite the field to produce the magnetic field needed for producing an emf.

There are single-phase (1ϕ), two-phase (2ϕ), and three-phase (3ϕ) alternators. The phase has to do with the relationship between the outputs of the coils next to one another on the stator. The number of physical degrees displaced makes the difference in the output of the alternator.

Three-phase alternators may have their windings connected in either delta or wye. The output of any 3ϕ alternator winding connected in delta has an advantage of current inasmuch as the available current is 1.732 (that is, $\sqrt{3}$) times the single-winding output.

The number of poles and the speed of rotation determine the frequency of the output voltage and current of an alternator.

Armature reactance, armature reaction, and armature resistance all affect the output voltage of the alternator.

Review Questions

1. What is the armature on an alternator?
2. How much separation (in degrees) is there between the windings of a two-phase alternator?

3. What is the closure voltage necessary when connecting a 3φ delta operation?
4. What is the AC field in an alternator?
5. When is the induced voltage in an alternator at its maximum?
6. What controls the frequency of the output of an alternator?
7. What is the purpose of a DC generator mounted on the same shaft of the alternator as the armature?
8. Does a 3φ wye-connected alternator produce an advantage in voltage or current?
9. Where is the output of a rotating-field alternator taken?
10. Why is the revolving-field AC generator most widely used?
11. How much current flows in a 3φ four-wire system neutral wire when the load is balanced?
12. How are alternators classified?

CHAPTER 3

Synchronous Motors

By definition, a synchronous motor is one that is in unison or in step with the phase of the alternating current that operates it. This condition is only approximated in practice because there is always a slight phase difference. Any single-phase or polyphase alternator will operate as a synchronous motor when supplied with current at the same potential, frequency, and waveshape that it produces as an alternator, the essential condition in the case of an alternator being that it must be speeded up to synchronism before being put in the circuit.

Construction

In construction, synchronous motors are almost identical with the corresponding alternators, and consist essentially of two elements:

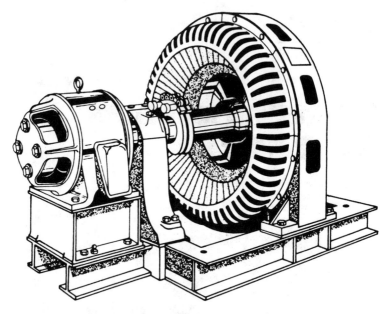

Fig. 3-1. Synchronous motor with a directly-connected exciter.

1. Stator (armature).
2. Rotor (field).

A synchronous motor may have either a revolving armature or a revolving field, although most synchronous motors are of the revolving-field type. The stationary armature is attached to the stator frame, while the field magnets are attached to a frame that revolves with the shaft.

The field coils are excited by direct currents, either from a small DC generator (usually mounted on the same shaft as the motor and called an *exciter*), or from some other source. Fig. 3-1 shows a directly connected exciter.

Principles of Operation

When a balanced polyphase alternating current (Fig. 3-2) is supplied to the armature of a synchronous motor, it produces a rotating magnetic field that rotates in synchronization with that of the

THREE-PHASE CURRENTS

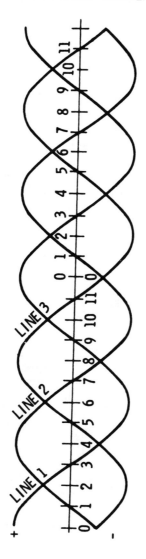

Fig. 3-2. Sinusoidal diagram showing two complete cycles of the current form in a three-phase machine.

supply current. It is this rotating magnetic field acting on the damper or *amortisseur* squirrel-cage winding that produces a starting torque, causing the rotor to rotate. Because the motor starts as a squirrel-cage motor, its speed will be slightly less than synchronous.

A direct current is supplied to the rotor winding, producing alternate north and south poles that lock into position with the rotating field in the armature, and the rotor rotates in step with the field of the supply current. It may easily be visualized, however, that while this locking action is powerful if the field system is already up to synchronous speed when the armature current is flowing, there can be no torque when the field poles (assumed to constitute the rotor) are originally stationary. The reason is that, under this condition, as the armature poles sweep across the field poles they will tend to pull the latter first in one direction and then in another; as a result, the starting torque is zero. Thus, the synchronous motor is not inherently a self-starting machine, but must of necessity be equipped with some form of auxiliary starting device, such as a squirrel-cage winding (which will be described in a later part of this chapter).

A better understanding of the operating principles of the synchronous motor may be had if the diagrams in Fig. 3-3 are examined. These diagrams represent twelve conditions occurring at regular intervals throughout a cycle.

The magnetic field is set up by the alternating current in the windings. Thus, for example, when the current rises in the first phase, a magnetic field is produced only by the first winding. As the current decreases in this winding and increases in the second, the magnetic field shifts along a little until it is all produced by the second winding. When the third winding has maximum current, the field has shifted a little more. The windings are so distributed that this shifting is uniform and continuous throughout.

The thickness of the lines in the diagrams of Fig. 3-3 is proportional to the current flowing in the winding — that is, the thicker the line, the larger the current. It should be noted that an absence of lines indicates no current in the conductor. The arrows in the diagrams indicate the direction of the currents. When one cycle is completed, another is started. In one cycle, the magnetic field rotates from one pole to the next one of similar polarity.

Excitation

In the case of a synchronous motor pulling a constant load, a variation in the field current is followed by a variation in the stator current, giving the V-curve pattern shown in Fig. 3-4. For a given load (with a given motor), there is a single value of field current that will give unity power factor at the motor terminals. Increasing or decreasing the field current from this value will give a power factor less than unity — increasing the field current will give a leading power factor, and decreasing the field current will give a lagging power factor. Stated another way, for a given load with constant voltage, if the field current is changed either way from the unity power-factor value, reactive current will be produced, causing the line current to increase, as shown by the V curves. This reactive current will be leading if the field excitation is increased, or lagging if decreased.

Field excitation for a synchronous motor is obtained from a separate exciter set driven by an induction motor, from a direct-connected or belted exciter, or from a constant DC voltage supply such as a station bus. Standard excitation voltage is either 125 or 250 volts, but the motor field winding is designed for an excitation voltage approximately 10 percent below this, to allow for voltage drop in the line.

In operation, the motor field current is usually set at the value of DC excitation amperes stamped on the motor nameplate, and is kept at this value at all loads. Maintaining full-load nameplate excitation on the motor field maintains the full pull-out torque of the motor, and at the same time gives the maximum amount of reactive kVA for power factor correction. If the motor is to be run for long periods of time at a considerable reduced load, it may be desirable (in the interest of higher efficiency) to reduce the field current to a value below the nameplate marking that will give the nameplate power factor for the actual reduced-load point of operation.

For a unity power factor with the motor operating at part load, the field current is adjusted until the stator current (as read on an AC ammeter) is minimum, as illustrated by the V curves of Fig. 3-4. At minimum stator current, the motor is operating at unity power factor. If this field-current value is then held constant, reference to

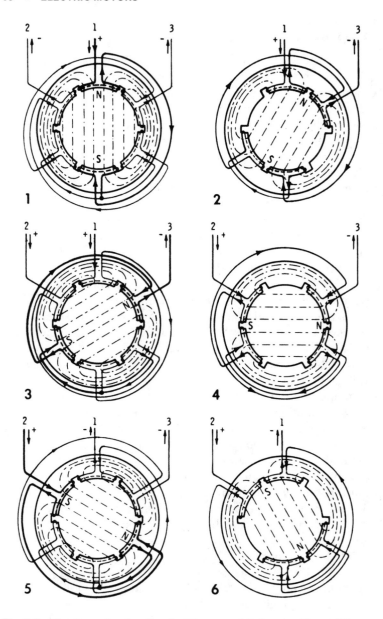

Fig. 3-3. Diagrams showing the electric current and magnetic conditions number of each diagram indicates the corresponding instant

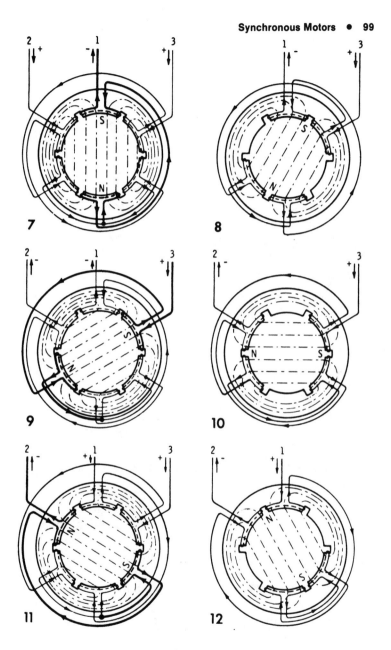

in a two-pole, three-phase motor for each 30° of a complete cycle. The
shown in Fig. 3-2.

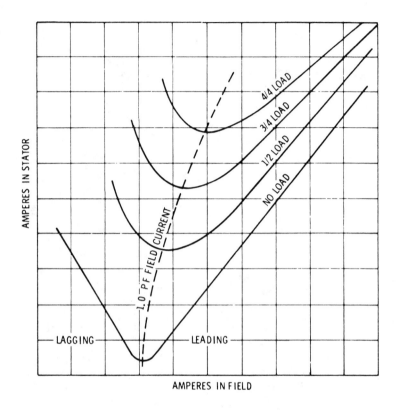

Fig. 3-4. Typical V-pattern characteristics of a synchronous motor.

Fig. 3-5 will show the amount of leading kVA supplied to the line. Note that at full load, the field excitation is sufficient to operate the motor at unity power factor, supplying no reactive kVA. However, at half load, it supplies about 22 percent of the numerical value of the motor's horsepower rating expressed in leading reactive kVA.

For a leading power factor with the motor operating at part load, the field current is first adjusted for minimum stator current (or unity power factor), and the minimum line currents noted. The field current is then increased so that the stator current rises along the V curve to the value required for the desired leading power factor, as determined by the following formula:

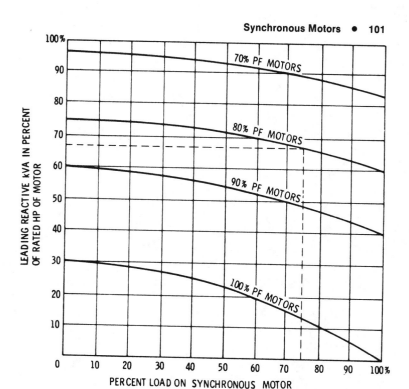

Fig. 3-5. Approximate leading kVA drawn by high-speed synchronous motors operating at partial loads and with full-load excitation maintained.

$$\text{stator current for desired PF} = \frac{\text{minimum stator current}}{\text{desired PF}}$$

Thus, if the minimum stator-current reading is 100 amperes and the desired leading power factor is 0.8, the field current should be increased until the stator current is 100 ÷ 0.8 = 125 amperes.

The necessary field-current adjustment is made by the exciter-field rheostat (Fig. 3-6), where a separate exciter set (direct-connected or belted) is used or by a motor-field or series rheostat where bus excitation is used. The field current should not exceed the rated DC excitation amperes. It should be remembered, however, that when field current is reduced, the motor pull-out torque will be reduced in proportion, and the motor will therefore pull out of step more easily.

Speed

The speed of a synchronous motor is determined by the frequency of the supply current and the number of poles of the motor. Thus, the operating speed is constant for a given frequency and number of poles. The equation for the determination of motor speed is:

$$\text{rpm} = \frac{\text{frequency} \times 120}{P}$$

where
  P is the number of poles of the motor.

All motors are built with an even number of poles, so the available speeds on 60 hertz range from 3,600 rpm for a two-pole machine down to 80 rpm from a machine containing 90 poles. This allows the motor to be directly connected to its load, even at lower speeds where induction motors cannot be used advantageously because of low operating efficiency and power factor.

If the preceding equation is solved with respect to the rpm at standard frequencies and with various numbers of even poles, the speeds in Table 3-1 will be obtained.

Some motors are required to operate at more than one speed, but are constant-speed machines at a particular operating speed. For example, when a speed ratio of 2:1 is required, a single-frame, two-speed synchronous motor may be suitable. Four-speed motors are used when two speeds that are not in the ratio of 2:1 are desired.

The single-frame, two-speed motor is usually of the salient-pole type of construction, with the number of poles corresponding to the low speed. High speed is obtained by regrouping the poles so as to obtain two adjacent poles of the same polarity, follwed by two poles of opposite polarity. This gives the effect of reducing the number of poles on the rotor by one-half for high-speed operation. Corresponding changes in the stator connections are also made. This switching is usually accomplished automatically by means of magnetic starters, by manually operated pole-changing equipment.

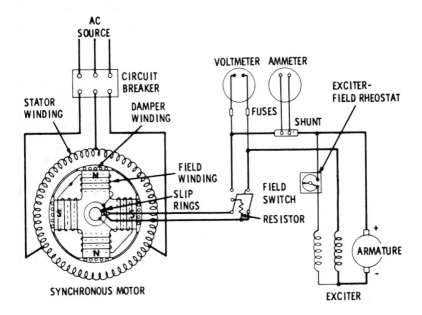

Fig. 3-6. Connections of a synchronous motor and exciter with the exciter-field rheostat, field switch, and exciter-field meters.

Standard voltages of 220, 440, 550 and 2,200 are usually used for synchronous motors up to a rating of 200 hp (149.2 kW). Larger motors have standard voltage ratings of 2,300, 4,000, 6,600, 11,000 and 13,200 volts.

Torque

The torque required to operate the driven machine at every moment between the initial breakaway and the final shutdown is important in determining the motor characteristics. The various torques associated with synchronous motors are termed starting torque, running torque, pull-in torque, and pull-out torque.

Table 3-1. Synchronous Motor Speeds (rpm)

POLES	60 HERTZ	50 HERTZ	40 HERTZ	25 HERTZ
2	3600	3000	2400	1500
4	1800	1500	1200	750
6	1200	1000	800	500
8	900	750	600	375
10	720	600	480	300
12	600	500	400	250
14	514.2	428.6	343	214.3
16	450	375	300	187.5
18	400	333.3	266.6	166.6
20	360	300	240	150
22	327.2	272.7	218.1	136.3
24	300	250	200	125
26	277	230.8	184.5	115.4
28	257.1	214.3	171.5	107.1
30	240	200	160	100
32	225	187.5	150	93.7
34	212	176.5	141.1	88.2
36	200	166.6	133.3	83.3
38	189.5	157.9	126.3	78.9
40	180	150	120	75
42	171.5	142.8	114.2	71.4
44	163.5	136.3	109	
46	156.6	130.5	104.3	
48	150	125	100	
50	144	120	96	
52	138.5	115.4	92.3	
54	133.3	111.1	88.9	
56	128.6	107.2	85.7	
58	124.1	103.5	82.8	
60	120	100	80	
62	116.1	96.8	77.4	
64	112.5	93.7	75	
66	109	90.8	72.7	
68	105.9	88.2	70.6	
70	102.8	85.7		
72	100	83.3		
74	97.3	81		
76	94.7	78.9		
78	92.3	76.9		
80	90	75		
82	87.8	73.2		
84	85.7	71.4		
86	83.7			
88	81.8			
90	80			

Starting Torque

The starting, or *breakaway*, torque required by the driven machine may be as low as 10 percent, as in the case of centrifugal pumps, and as high as 225 or 250 percent of full-load torque, as in the case of loaded reciprocating two-cylinder compressors.

The *starting torque* of a synchronous motor is the torque, or turning effort (usually measured in pound-feet or gram centimeters), that the motor develops when full voltage is applied to the armature winding. The synchronous motor, in itself, has very little starting torque. With modern motors, however, almost any reasonable torque can be obtained by proper design of the damper windings (also called squirrel-cage and amortisseur windings). The different values of starting torque are usually obtained by changes in the resistance and size of the damper windings. As a general rule, the starting inrush of current increases as the starting torque is increased.

Running Torque

The *running torque* is determined by the horsepower and speed of the driven machine, and at any given point, the torque in pound-feet is:

$$T = \frac{5,250 \times \text{hp}}{\text{speed in rpm}}$$

The peak horsepower determines the maximum torque required by the driven machine. The motor must have a breakdown or a maximum running torque in excess of this figure in order to avoid stalling. Certain driven machines, like reciprocating compressors, have load torques that pulsate periodically.

To prevent excessive pulsation in the line current of the motor, proper flywheel effect (WR^2) must be provided, either in the motor or in the driven machine. In the case of synchronous motors, the flywheel effect must be made such that the natural frequency of the motor does not approximate the frequency of any impulses in the load torque. Natural frequency in hertz is:

$$fr = \frac{35.200\, Pr \times f}{\text{rpm}\, WR^2}$$

where

WR^2 is the weight of rotated object in lbs. $\times$ (radius of gyration in ft.)^2(motor and load),

rpm is the speed of motor in revolutions per minute

f is the line frequency,

Pr is the synchronizing power of motor in kilowatts per electrical radian,

fr is natural frequency in hertz.

Pull-In Torque

When a synchronous motor has been started as an induction motor, it will run at from 2 to 5 percent below synchronous speed until the excitation is applied, at which time the rotor will pull into step. The amount of torque or load at which the motor will pull into step is called the *pull-in torque*. Synchronous motors are usually designed for a definite application. The designer should know the nature of the load the motor is required to start so that he can design the motor for the necessary starting and pull-in torque.

Pull-Out Torque

When a synchronous motor is running in synchronism with no load, the individual field poles of the rotor have a fixed position with respect to the rotating magnetic field of the armature. When a mechanical load or resisting torque is applied to the shaft, the motor develops a torque to balance the requirements of the load torque. The increased torque requirement is produced by a backward shift or lag in the position of the field poles with respect to the rotating magnetic field. However, the rotor will maintain its synchronous speed. When, because of the increased torque requirement, the field poles have shifted backward approximately half the distance between adjacent poles, the motor is developing its maximum torque. Any further increase will cause the motor to pull out of step and stop. The maximum torque that the motor will develop without pulling out of step is called the *pull-out torque*. The speed-torque and speed-current characteristics of a typical synchronous motor are shown in Fig. 3-7.

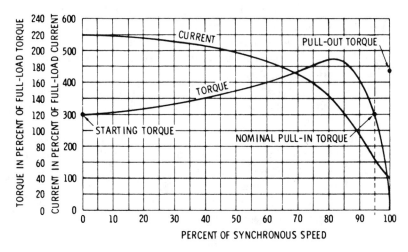

Fig. 3-7. Speed-torque and speed-current characteristics of a typical high-speed synchronous motor.

Power Factor

The power in watts delivered through a single-phase AC circuit is the product of the current, voltage, and power factor, and is written as:

$$W = EI \cos \angle\theta$$

or,

$$\cos \angle\theta = \frac{W}{EI}$$

Since W represents the true power and the product EI the apparent power, the power factor ($\cos \angle\theta$) may be defined as the ratio of the true and apparent power delivered through the circuit.

In a three-phase AC circuit, the power in watts is written as:

$$W = EI \cos \angle\theta\sqrt{3}$$

or,

$$\cos \angle\theta = \frac{W}{EI\sqrt{3}}$$

The fact that a synchronous motor always operates at a leading power factor or unity (100 percent), whereas an induction motor operates at a lagging power factor and receives its excitation from the line, has been made use of for power-factor improvements. Employment of synchronous motors, therefore, helps to improve the plant operating power factor. In many cases, synchronous motors also cost less and have higher efficiencies than corresponding induction motors. This is particularly true for low-speed synchronous motors.

Many rate schedules incorporate a power-factor clause, adjusting the rate according to power factor; others involve a penalty or bonus if the power factor is, for example, below 90-percent or above 90-percent lagging.

A synchronous motor operating at full load and its rated excitation delivers to the power system a leading kVA equal to:

$$kVA = \frac{0.746 \times hp\ rating}{Eff \times \cos \angle\theta} \sqrt{1 - (\cos \angle\theta)^2}$$

It should be noted that, at partial loads and rated excitation, more leading kVA is supplied. The curves in Fig. 3-5 approximate the leading reactive kVA for synchronous motors at four different power-factor ratings and at varying load conditions, based on maintaining full-load rating power excitation at all loads. Thus, at three-quarter load, for example, a 100 hp (74.6 kW) 80-percent power factor synchronous motor driving an air compressor supplies a leading reactive kVA equal to approximately 66 percent of its horsepower rating, or 66 reactive kVA (Fig. 3-5).

The unity power factor synchronous motor, whose curves are shown in Fig. 3-5, supplies a leading reactive kVA only at part load. At full load, however, although providing no leading reactive kVA, it still improves the power factor by adding to the kilowatt load without increasing the reactive-kVA load. A unity power factor synchronous motor costs less than a leading power factor synchronous motor of the same horsepower and rating because less material is used in its windings and magnetic parts. Highest efficiency is also obtained with a unity power factor synchronous motor, but when considerable power factor improvement is required, leading power factor synchronous motors are generally preferable.

A simple approximate rule for estimating the leading kVA requirement for power factor correction is as follows: Each horsepower of an 80-percent power factor synchronous motor, operating at full load and rated excitation, will supply the lagging reactive kVA for a (1 kW) high-speed induction motor. Thus, a 75 hp (55.95 kW) 80-percent power factor synchronous motor will correct to unity power factor appoximately 100 hp (74.6 kW) in high-speed induction motors.

Control

To make a synchronous motor self-starting, a squirrel-cage winding is usually placed on the rotor. After the motor comes up to a speed that is slightly less than synchronous, the rotor is energized. When synchronous motors are started, their DC fields are not excited until the rotor has practically reached full synchronous speed. The starting torque required to bring the rotor up to this speed is produced by induction.

In addition to a DC winding on the field, synchronous motors are generally provided with a damper or amortisseur winding. It consists of short-circuited bars of brass or copper embedded in slots in the pole faces and joined together at either end by means of end rings. This winding, usually termed a squirrel-cage winding, enables the motor to obtain sufficient starting torque for the motor to start under load.

The starting torque necessary to bring the motor up to synchronous speed is termed the pull-in torque. The maximum torque that the motor will develop without pulling out of step is termed the pull-out torque.

When the stator winding in the synchronous motor is being excited by the AC line connection, it immediately sets up a rotating magnetic field. The rotating flux of this field cuts across the damper winding of the rotor and induces secondary currents in the bars of this winding. The reaction between the flux of these secondary currents and that of the rotating stator field produces the torque necessary to start the rotor and to bring it up to speed.

When the motor has been brought up to nearly synchronous speed (as an induction motor because of the damper winding), the

DC field poles are excited and the strong flux of these poles causes them to be drawn into step or full synchronous speed with the poles of the rotating magnetic field of the stator. During normal operation, the rotor continues to revolve at synchronous speed as if the DC poles were locked to the poles of the rotating magnetic field of the stator. Because a synchronous motor has no slip after the rotor is brought up to full speed, no secondary currents are induced in the bars of the damper windings during normal operation.

In starting a synchronous motor as an induction motor, the voltage impressed on the motor should be reduced in starting and while coming up to speed. This reduced starting voltage is usually obtained from a starting compensator (autotransformer) similar to that used in the starting of an induction motor. With this method of starting, the usual practice is to close the starting contactor first. This connects the stator to the reduced voltage. When a speed at near synchronization has been reached, the starting contactor is opened and the running contactor is closed, thus connecting the motor to the full-line voltage. After synchronous speed has been reached, the field switch is closed through a moderate amount of resistance. The field current may now be adjusted in order to make the motor operate at the desired power factor.

It is customary, where the motor must operate at a high starting torque, to use a full-line starting voltage in connection with a time-delay overcurrent relay that will become operative before the surge of starting current can damage the motor windings.

Starting Methods

Methods most commonly used in starting synchronous motors are as follows:

1. Across-the-line.
2. Reduced voltage.
3. Reactance.
4. Resistance.
5. Korndorfer.
6. Part-winding.

Across-the-Line Starting — Across-the-line starting consists simply of closing the main line switch, either manually or by means of a pushbutton arrangement in conjunction with a magnetic type of starter or circuit breaker. After synchronous speed has been attained, the field current is applied as previously described. This method of starting should always be adhered to, unless the driven machine requires two or more increments of starting torque or unless the power-system limitation requires a starting current below that of the full voltage surge. An automatic across-the-line synchronous-motor starting wiring arrangement is shown in Fig. 3-8, with its elementary diagram in Fig. 3-9. The operation and function of the devices shown in Fig. 3-9 are as follows:

Pushing the start button energizes main contactor coil LE, closing the main contacts and connecting L_1, L_2, and L_3 to the motor. An interlock (1) closes at the same time as the main contactor and forms a holding circuit so that the pushbutton may be released.

While the motor is accelerating, the field circuit is closed through a normally closed pole, F_3, of the field contactor, through the discharge resistor and retarding coil FR of the synchronizing relay. This holds the contacts of the synchronizing relay open until the motor approaches synchronization. At the time the main contactor closes, the contactor coil of the synchronizing relay is energized and tends to close the contacts.

As the motor accelerates, the induced field current through the retarding coil of the synchronizing relay decreases, which releases the interlock when the motor nears synchronization. After a few seconds of delay, the interlock closes, energizing the field-contactor coil, causing F_1 and F_2 to close and F_3 and F_4 to open. The closing of F_1 and F_2 applies excitation to the motor field, completing the starting operation. The opening of F_3 opens the discharge circuit of the motor field. Operation of the stop button or the overload relay at any time opens the control circuit and the contactors open, disconnecting the motor from the line.

Some exciters tend to build up voltage slowly. Because of this, the field contactor is equipped with a normally closed interlock that may be used to short-circuit the exciter-field rheostat, thus causing the exciter voltage to build up more quickly.

Power company rules usually limit the starting current that may be drawn by a motor. There are two types of such rules, one type specifying the allowable number of amperes for each horsepower rating. The more recent "increment" type of rule is based on the

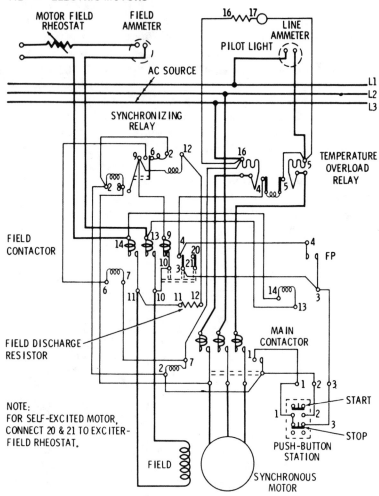

Fig. 3-8. Panel wiring diagram of a typical automatic across-the-line synchronous-motor starter.

capacity to serve a given installation. Usually the total current is not limited, but must be taken in incremental steps, the magnitude of each step being determined by the total connected load.

Reduced-Voltage Starting — Reduced-voltage starting differs from the across-the-line method in that reduced voltage is first applied to the motor by means of a transformer (usually an auto-transformer).

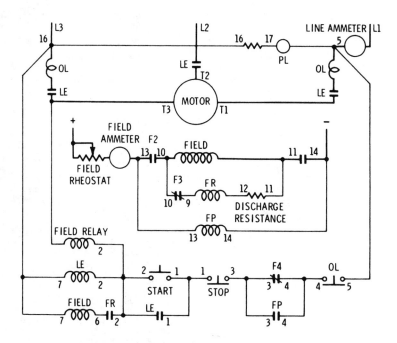

Fig. 3-9. A simplified diagram of the across-the-line motor starter shown in Fig. 3-8.

The starting method shown in Fig. 3-10 utilizes three switches or contactors. Here, the two starting switches, or contactors, are connected to each side of the autotransformer or compensator and are both closed at the same time. After the motor has attained nearly synchronous speed, the starting switches are opened and the running switch is closed. Most starting arrangements of this type are equipped with electrical or mechanical interlocks in order to prevent any manipulating of the switches, except in the correct sequence.

The starting method shown in Fig. 3-11 consists of two switches or contactors. Here, the running switch is closed after the motor speed becomes constant, connecting the motor to the line. The starting switch then opens, being interlocked with the running switch. As soon as the speed again becomes constant, the field switch (now shown) is closed. The controller remains in the running position until the running switch is tripped.

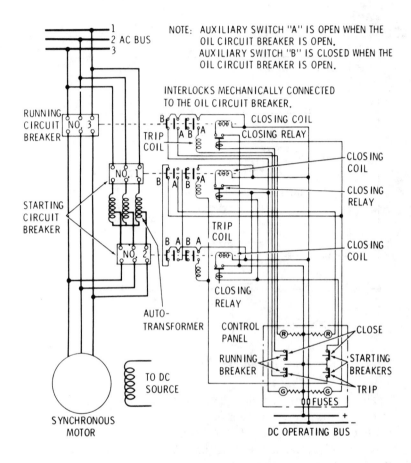

Fig. 3-10. Diagram of a reduced-voltage starter for a synchronous motor.

Another synchronous-motor starting method is illustrated in Fig. 3-12. Switches *1* and *2* are first closed, simultaneously connecting the motor to the line through the star-connected auto-transformers. Power is then supplied to the motor at a reduced voltage. Switch *1* is opened as soon as the speed becomes constant, and switch *3* closes instantly, being electrically interlocked with switch *1*. This connects the motor directly across the line, the terminal voltage having increased to normal without dropping to zero.

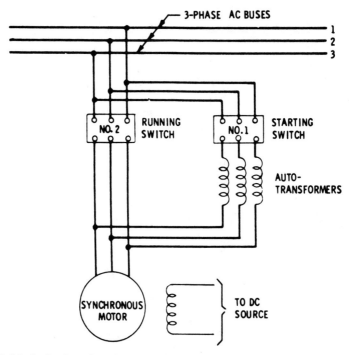

Fig. 3-11. A simple reduced-voltage starter.

Switch 2 is interlocked with switch 3 so that it opens immediately after switch 3 closes. The field switch is closed as soon as the speed again becomes constant. The motor is stopped by switch 3, after which the field switch is opened.

A manual synchronous-motor starting method utilizing two circuit breakers enclosed in one tank is shown in Fig. 3-13. The two autotransformer coils are short-circuited at starting, and the reduced voltage is supplied from the transformer taps as indicated. The autotransformers are usually supplied with several taps in order to provide a suitable range of starting torque.

Reactance Starting — Reactance starting is similar to the previously discussed reduced-voltage starting methods, except that the first step is obtained by reactance in series with the motor armature instead of autotransformers. In the reactance method of starting, more current is required from the line for the same torque

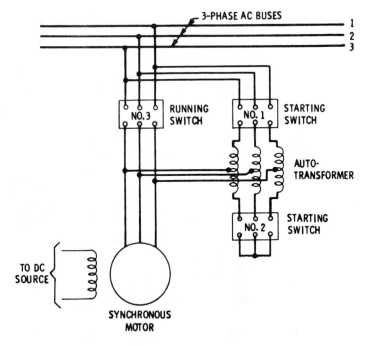

Fig. 3-12. An autotransformer starter for a synchronous motor.

on the first step than when compensators are used. It has an advantage, however, in that no circuit opening is required when the motor is transferred to running voltage, the transfer being accomplished by short-circuiting the reactance.

Resistance Starting — A typical circuit diagram using the resistance method of reduced-voltage starting is shown in Fig. 3-14. Here, switch *1* is closed first, connecting the motor to the line through the entire resistance. Switches *2, 3,* and *4* are then closed, with a time interval between each closing, each switch in turn short-circuiting a part of the resistance. This method of starting is sometimes used when power company rulings require several progressive steps of starting current.

Korndorfer Starting — In common with the reactance and resistance methods of starting, the Korndorfer method permits the motor to be started without opening the motor circuit. The motor is first connected through suitable taps of a compensator, and then

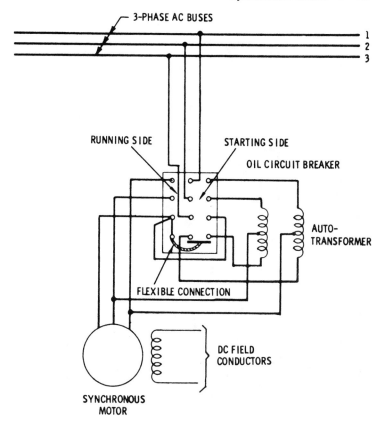

Fig. 3-13. A reduced-voltage starter for a synchronous motor.

started by connecting the compensator to the line. Full voltage is connected by first opening the neutral of the starting compensator, allowing the motor to run with part of the compensator winding in series with the motor, and then short-circuiting the entire compensator winding.

With reference to Fig. 3-15 (showing a typical Korndorfer starting method), switch or contactor *1* is first closed, thus connecting one of the compensator windings to the line. Then switch *2* is closed, completing the motor circuit at reduced voltage. As the motor increases its speed, a timing relay operated by switch *2* opens the circuit of *3*, which in turn opens the transformer neutral. Switch *4* is

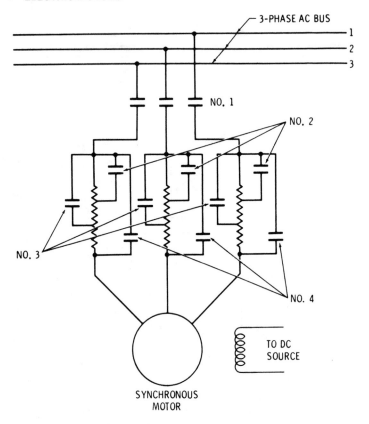

Fig. 3-14. Schematic diagram of a resistance-type synchronous-motor starter.

next closed, connecting the motor to full-line voltage by short-circuiting the compensator sections. By opening switch 2, the reduced-voltage taps of the compensator are disconnected and the permanent running connection to the motor is completed.

Part-Winding Starting — Part-winding starting is sometimes used on synchronous motors. The motor is equipped with various taps on the armature connected to suitable switches. One arrangement of this sort consists in making a two-circuit armature winding such that one circuit can be used to start the motor while the other is left disconnected. This second winding is connected after the motor has been accelerated a required amount.

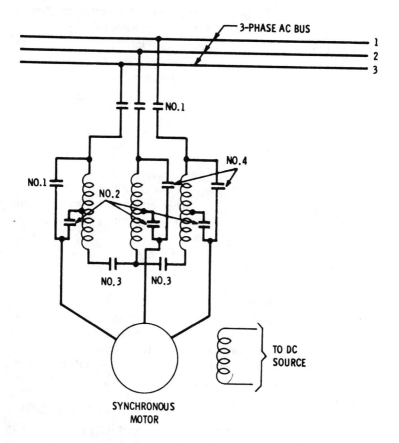

Fig. 3-15. Schematic diagram of a Korndorfer-type synchronous-motor starter.

Another method of synchronous-motor starting is to use an auxiliary prime mover, usually an induction motor. This method of starting is applied to motors having no squirrel-cage winding, or to alternators converted to motor use. The motor can carry no mechanical load while starting.

Application

The correct selection and application of a motor involve a great many factors affecting the installation, operation, and subsequent servicing of the motor. In most cases, the answers to a number of the problems are obvious, while in other instances, the answers are supplied in whole or part by certain intermediaries, such as the manufacturers of motor-equipped devices.

The various classes of service for which synchronous motors are used may be classified as follows:

1. Power-factor correction.
2. Constant-speed, constant-load drives.
3. Voltage regulation.

On power systems employing a large number of induction motors and in other devices having a lagging power factor (such as welders, fluorescent lights, etc.), the power factor may be improved by installation of devices operating with a leading power factor. Such devices are:

1. Synchronous motors.
2. Synchronous capacitors.

When synchronous motors are employed exclusively as power-factor correction devices, they are termed *synchronous capacitors*, because the effect on the power system is the same as that of a static capacitor that also produces a leading current. The power factor of a synchronous motor may be varied by adjustment of the field current. Thus, the power factor may be varied in small steps from a low lagging to a low leading power factor. This makes it possible to vary the power factor over a considerable range and places the characteristics of the motor under the immediate control of the operator at all times.

Because of the higher efficiency possible with synchronous motors, they can be used advantageously on most loads where constant speed is required. Typical applications of high-speed synchronous motors (above 500 rpm) are such drives as centrifugal pumps, DC generators, belt-driven reciprocating compressors, fans, blowers, line shafts, centrifugal compressors, rubber and paper mills, etc.

The field of applications of low-speed synchronous motors (below 500 rpm) are drives such as reciprocating compressors (largest field of use), Jordan engines, centrifugal and screw-type pumps, ball and tube mills, vacuum pumps, electroplating generators, line shafts, rubber and band mills, chippers, metal rolling mills, etc. Synchronous motors are rarely used in sizes below 20 hp (14.92kW). The principal field of application is in sizes of 100 hp (74.6 kW) and larger.

Another field where synchronous motors are suitable is voltage regulation. At the ends of long transmission lines, the voltage tends to vary greatly, especially if large inductive loads are present. If an inductive load is disconnected suddenly, the voltage may rise considerably above normal because of the capacity effect or capacitor action of the line. By installing a synchronous motor with a voltage regulator to control its field, the voltage change on the line may be controlled. The action of the voltage regulator is such that when the voltage drops because of an inductive load, the field strength of the synchronous motor is increased. This raises the power factor and maintains the voltage at the same value. If the voltage tends to rise because of the capacity effect on the line, the regulator weakens the field and causes the motor to have a lagging or inductive load so as to hold the voltage to normal.

Stepper Motors

Stepper motors are used in educational robots that demonstrate the basic operation of robots. They are not presently used in industrial applications, as they have the disadvantage of slipping if overloaded. This means the error created by the slippage can go undetected and ruin whatever is being machined or processed.

The stepper motor is not designed to run continuously. It is designed to rotate in steps in response to electrical pulses. Pulses are fed into the input from a control unit; the motor indexes in precise angular increments. The average shaft speed of the motor, $n_{average}$, is given by the formula:

$$n_{average} = \frac{60 \text{ (pulses per second)}}{\text{number of phases in the winding}} \quad \text{rpm}$$

There are two types of stepper motors. They are the variable-reluctance (VR) and the permanent-magnet (PM) steppers.

The stepper motor is a *synchronous motor* in principle. It usually has three-phase windings or four-phase windings on the stator. The number of poles on the rotor depends on the step size or angular displacement for each input pulse. The rotor is either the *reluctance* type or it may be made of a *permanent magnet*.

Reluctance Type

In the reluctance type, when a pulse is fed to one of the phases on the stator, the rotor tends to align with the magnetic axis of the stator coil. These coils are then sequentially switched. The rotor follows the stator magnetic field in sequence. The variable reluctance stepper motor is shown in Fig. 3-16.

Note how the variable reluctance type of motor has a rotor with eight poles and there are *three* independent eight-pole stators. The stator poles are arranged coaxially with the rotor. When phase *1* is energized, the rotor poles align with the stator poles of phase *1*. Notice from Figure 3-16A that the phase *2* stator is displaced from the phase *1* stator by 15° in the counterclockwise direction. Also note that the phase *3* stator is further displaced from the phase *2* stator by another 15° in the counterclockwise direction. When the current in phase *1* is turned off and phase *2* is energized, the rotor will rotate counterclockwise through 15°. Next, when phase *2* current is turned off and phase *3* is energized, the rotor will turn another 15° in the counterclockwise direction. Then turning off the current in phase *3* and exciting phase *1* will complete one step of 45° in the counterclockwise direction. Additional current pulses in the sequence *1-2-3* will produce further steps in the counterclockwise direction. Reversal of rotation is obtained by reversing the phrase sequence to *1-3-2*.

The stepping motion of the rotor of a stepper motor is characteristic of an undamped system. That means at first the displacement of the rotor overshoots its final position. However, it then gradually settles down to its final position. The stepper motor needs some type of damping for these oscillations, because as the frequency of pulsing increases, the period decreases. When the period is close to the oscillator period, the motor is reaching its operating limit.

The step angle of the stepper motor is determined by the number of poles. Some typical step angles are 15°, 5°, 2°, and 0.72°.

The resolution required for the installation determines the choice of step angle. Operating speed is limited by the degree of system damping. It is possible to obtain speeds up to 200 steps per second. It is also possible to obtain a steady and continuous speed of rotation (slewing) greater than this value. However, the motor is then unable to stop the system in a single step.

As is seen from this description, the stepping motor is used primarily to change electrical pulses into rotary motion that can be used to produce mechanical movement. This makes the stepper motor well-mated to computers — the computer generates pulses needed to operate the stepper motor. Operation of the bipolar stepper motor is accomplished in a four-step switching sequence. Any of the four combinations of switches 1 or 2 will produce an appropriate rotor position location (Fig. 3-17B). After the four combinations of switches have been achieved, the switching cycle repeats itself. Each switching combination causes the motor to move one step.

Permanent Magnet Type

The permanent magnet type of stepper motor is shown in Fig. 3-18A. The stator has a two-phase winding and the rotor has five pole pairs. For the position shown, phase 1 is energized and the rotor turns by (90°–72°) = 18° to align with the pole of phase 2 winding. The switching of the phases for a clockwise rotation is shown in the box in Fig. 3-18B. The rotor achieves an equilibrium position after each step. If disturbed, the rotor will tend to return to its equilibrium position.

Some stepper motors use eight switching combinations to achieve "half stepping." During this type of operation, the motor shaft moves half of its normal step angle for each input pulse applied to the stator. This allows for a very precise and controlled movement. Fig. 3-19 shows a motor with eight stator windings.

Comparison of VR and PM Stepper Motors

VR Motor

No torque with no excitation.
Higher inductance and slower electrical response.
Low rotor inertia and faster mechanical response.

PM Motor
Residual torque with no excitation.
Lower inductance and faster electrical response.
Higher inertia and slower mechanical response.

Each type has its advantages and disadvantages, and motors must be carefully selected depending on the particular application. They are used in process control, machine tools, as computer peripherals, and in some medical equipment.

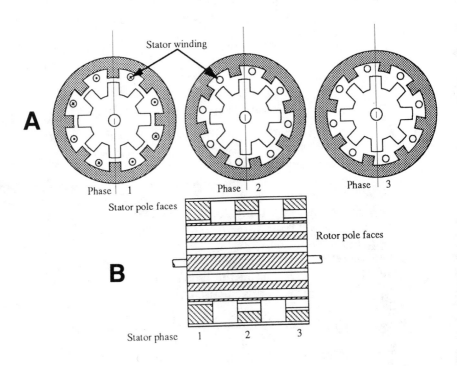

Fig. 3-16. Variable Reluctance Stepper Motor. (A) Note displacement of stator phases 1, 2, and 3. (B) Cross-sectional view of the motor. Note the location of stator phases 1, 2, and 3.

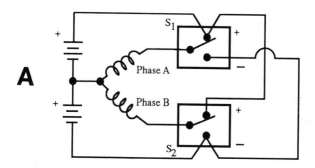

	CW Rotation	
Step	Switch A	Switch B
1	off	—
2	+	off
3	off	+
4	—	off
1	off	—

Fig. 3-17. Switching sequence for clockwise rotation of the motor.

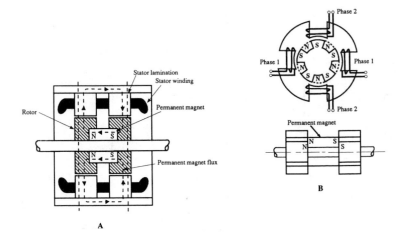

Fig. 3-18. Permanent magnet stepper motor. (A) Cross-sectional view. (B) Schematic — wiring.

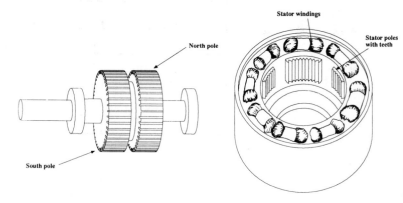

Fig. 3-19. PM stepper with 8-stator poles.

Synchronous Motor Troubleshooting Chart

Symptom and Possible Cause   *Possible Remedy*

Motor Will Not Start

(a) Faulty connection

(b) Open circuit, one phase

(c) Short circuit, one phase

(d) Voltage falls too low

(e) Friction high

(f) Field excited

(g) Load too great

(h) Automatic field relay not working

(i) Wrong direction of rotation

(a) Inspect for open or poor connection.

(b) Test, locate, and repair.

(c) Open and repair.

(d) Reduce the impedance of the external circuit.

(e) Make sure bearings are properly lubricated. Check bearing tightness. Check belt tension. Check load friction. Check alignment.

(f) Be sure field-applying contactor is open and field discharge contactor is closed through discharge resistance.

(g) Remove part of load.

(h) Check power supply to solenoid.

(i) Reverse any two leads.

Motor Will Not Come Up to Speed

(a) Excessive load

(b) Low voltage

(c) Field excited

(a) Decrease the load. Check operation of unloading device (if any) on driven machine.

(b) Increase voltage.

(c) Be sure field-applying contactor is open, and field-discharge contactor is closed through discharge resistance.

Symptom and Possible Cause   *Possible Remedy*

Fails to Pull into Step

(a) No field excitation	(a) Check circuit connections. Be sure field-applying contactor is operating. Check for open circuit in field or exciter. Check exciter output. Check rheostat. Set rheostat to give rated field current when field is applied. Check contacts of switches.
(b) Excessive load	(b) Reduce load. Check operation of unloading device (if any) on driven machine.
(c) Load inertia excessive	(c) May be a misapplication — consult manufacturer.

Motor Pulls Out of Step or Trips Breaker

(a) Low exciter voltage	(a) Increase excitation. Examine exciter. Check field ammeter and its shunt to be sure reading is not higher than actual current.
(b) Open circuit in field and exciter circuit.	(b) Test with magneto and repair break.
(c) Short circuit in field	(c) Check with low-voltage and polarity indicator and repair field.
(d) Reversed field spool	(d) Check with low-voltage and polarity indicator and reverse incorrect leads.
(e) Load fluctuates widely	(e) See "Motor Hunts."
(f) Excessive torque peak	(f) Check driven machine for bad adjustment, or consult motor manufacturer.
(g) Power fails	(g) Re-establish power circuit.
(h) Line voltage too low	(h) Increase if possible. Raise excitation.

Symptom and Possible Cause	*Possible Remedy*

Motor Hunts

(a) Fluctuating load

(a) Correct excessive torque peak at driven machine or consult motor manufacturer. If driven machine is a compressor, check valve operations. Increase or decrease flywheel size. Try decreasing or increasing motor field current.

Stator Overheats in Spots

(a) Rotor not centered

(a) Realign and shim stator or bearings.

(b) Open phase

(b) Check connections and correct.

(c) Unbalanced currents

(c) Check for loose connections and improper internal connections.

One or More Coils Overheat

(a) Short circuit

(a) Cut out coil as expedient (in motors up to 5 hp or 3.73 kW). Replace coil when the opportunity arises.

Field Overheats

(a) Short circuit in a field coil
(b) Excessive field current

(a) Replace or repair.
(b) Reduce excitation until stator current is at nameplate value.

Symptom and Possible Cause	*Possible Remedy*
All Parts Overheat	
(a) Overload	(a) Reduce load or increase motor size.
(b) Over- or under-excitation	(b) Adjust excitation to nameplate rating.
(c) No field excitation	(c) Check circuit and exciter.
(d) Reverse field coil	(d) Check polarity and, if wrong, change leads.
(e) Improper voltage	(e) See that nameplate voltage is applied.
(f) Improper ventilation	(f) Remove any obstruction and clean out dirt.
(g) Excessive room temperature	(g) Supply cooler air.

Summary

A synchronous motor is one that is in unison or in step with the phase of the alternating current that operates it. In construction, synchronous motors are nearly identical with the corresponding alternators, and consist essentially of the *stator* (armature) and the *rotor* (field).

When a balanced polyphase alternating current is supplied to the armature of a synchronous motor, it produces a rotating magnetic field that rotates in synchronization with that of the supply current. This rotating magnetic field acts on the damper or amortisseur squirrel-cage winding to produce a starting torque, causing the rotor to rotate.

The *speed* of a synchronous motor is determined by the frequency of the supply current and the number of poles of the motor. The equation for the determination of motor speed is:

$$\text{rpm} = \frac{\text{frequency} \times 120}{P}$$

in which

P = number of poles of the motor

The *torque* required to operate the driven machine at all moments occurring between initial breakaway and final shutdown is important in determining the motor characteristics. The various torques associated with synchronous motors are starting torque, running torque, pull-in torque, and pull-out torque. The running torque is determined by the horsepower and speed of the driven machine, and at any given point, the torque in lb.-ft. is:

$$T = \frac{5,250 \times hp}{\text{speed in rpm}}$$

To obtain metric measurement, take answer and multiply by 0.1383 to get *kilogram-meters*.

Since the metric system does have its place in both Canada and Europe, the United States will be meeting some of its terminology from time to time. It is best to have conversion tables handy so the conversational and text material can be interpreted as well as the mathematical.

Torque is measured in ounce-inches and pound-feet in the usual American terminology. However, in the SI or metric system it is referred to in units called *newtonmeters* (Nm).

Table 3-2. Conversion Table for Torque

SI Unit	Imperial/Metric to SI	SI to Imperial/Metric
newtonmeter (Nm)	1 lb.-ft. = 1.356 Nm	1 Nm = 0.738 lb.-ft.
	1 oz.-in. = 7.062 × 10⁻³ Nm	1 oz.-in. = 8.851 lb.-in.
	1 lb.-in. = 0.113 Nm	1 lb.-in. = 141.61 oz.-in.

The power in watts (W) delivered through a single-phase AC circuit is the product of the current, voltage, and power factor, as shown in the equation:

$$W = EI \cos \angle\theta$$

In a three-phase AC circuit the power in watts can be determined by the equation:

$$W = EI \cos \angle\theta\sqrt{3}$$

Methods most commonly used in starting synchronous motors are (1) across-the-line; (2) reduced voltage; (3) reactance; (4) resistance; and (5) Korndorfer.

The various classes of service for which synchronous motors may be used are classified as: (1) power-factor correction; (2) constant-speed, constant drives; and (3) voltage regulation.

Review Questions

1. Give the definition for a synchronous motor?
2. What is the basic operating principle of the synchronous motor?
3. How can the speed of a synchronous motor be determined?
4. List the various torques associated with synchronous motor operation.
5. What methods are commonly used to start synchronous motors?
6. What are some of the possible troubles with a synchronous motor if all parts overheat?
7. What is the difference between a stator and a rotor in a synchronous motor?
8. Why do synchronous motors have to be started with an auxiliary motor or some other method not common to other motors?
9. What is the *major* use for the synchronous motor?
10. What are the most common horsepowers available in this type of motor?
11. Why are small synchronous motors not used on electric clocks anymore?

CHAPTER 4

Squirrel-Cage Motors

The most common form of induction motor is the squirrel-cage type. This motor has derived its name from the fact that the rotor, or secondary, resembles the wheel of a squirrel cage. Its universal use lies in its mechanical simplicity, its ruggedness, and the fact that it can be manufactured with characteristics to suit most industrial requirements.

Construction

A squirrel-cage motor consists essentially of two units, namely:

1. Stator.
2. Rotor.

The stator (or primary) consists of a laminated sheet-steel core with slots in which the insulated coils are placed. The coils are so grouped and connected as to form a definite polar area and to produce a rotating magnetic field when connected to a polyphase alternating-current circuit.

The rotor (or secondary) is also constructed of steel laminations, but the windings consist of conductor bars placed approximately parallel to the shaft and close to the rotor surface. These windings are short-circuited, or connected at each end of the rotor, by a solid ring. The rotors of large motors have bars and rings of copper connected at each end by a conducting end ring made of copper or brass. The joints between the bars and end rings are usually electrically welded into one unit, with blowers mounted on each end of the rotor. In small squirrel-cage rotors, the bars, end rings, and blowers are of aluminum cast in one piece instead of welded together.

The air gap between the rotor and the stator must be very small in order for the best power factor to be obtained. The shaft must, therefore, be very rigid and furnished with the highest grade of bearings, usually of the sleeve or ball-bearing type. A cutaway view of a typical squirrel-cage induction motor is shown in Fig. 4-1.

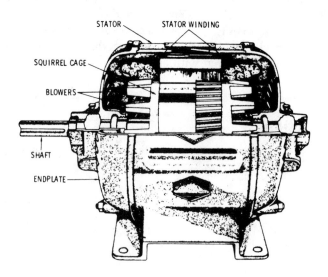

Fig. 4-1. Cutaway view of a typical squirrel-cage induction motor.

Principles of Operation

In a squirrel-cage motor, the secondary, or squirrel-cage, winding takes the place of the field winding in a synchronous motor. As in a synchronous motor, the currents in the stator set up a rotating magnetic field. This field is produced by the increasing and decreasing currents in the windings. When the current increases in the first phase, only the first winding produces a magnetic field. As the current decreases in this winding and increases in the second, the magnetic field shifts slightly, until it is all produced by the second winding. When the third winding has maximum current flowing in it, the field is shifted a little more. The windings are so distributed that this shifting is uniform and continuous. It is this action that produces a rotating magnetic field.

As this field rotates, it cuts the squirrel-cage conductors, and voltages are set up in these just as though the conductors were cutting the field in a DC generator. These voltages cause currents to flow in the squirrel-cage circuit — through the bars under the adjacent south poles into the other end ring, and back to the original bars under the north poles to complete the circuit.

The current flowing in the squirrel cage, down one group of bars and back in the adjacent group, makes a loop that establishes magnetic fields in the rotor core with north and south poles. This loop consists of one turn, but there are several conductors in parallel and the currents may be large. The poles in the rotor are attracted by the poles of the rotating field set up by the currents in the armature winding, and follow them around in a manner similar to the way in which the field poles follow the armature poles in a synchronous motor.

There is, however, one interesting and important difference between the synchronous motor and the induction motor — the rotor of the latter does not rotate as fast as the rotating field in the armature. If the squirrel cage were to go as fast as the rotating field, the conductors in it would be standing still with respect to the rotating field, rather than cutting across the field. Then there could be no voltage induced in the squirrel cage, no currents in it, no magnetic poles set up in the rotor, and no attraction between it and the rotating field in the stator.

Speed

The speed of a squirrel-cage induction motor is nearly constant under normal load and voltage conditions, but is dependent on the number of poles and the frequency of the AC source. This type of motor slows down, however, when loaded an amount that is just sufficient to produce the increased current needed to meet the required torque.

The difference in speed for any given load between synchronous and load speed is called the *slip* of the motor. Slip is usually expressed as a percentage of the synchronous speed. Since the amount of slip is dependent on the load, the greater the load, the greater the slip will be; that is, the slower the motor will run. This slowing of the motor, however, is very slight, even at full load, and amounts to from 1 to 4 percent of synchronous speed. Thus, the squirrel-cage type is considered a constant-speed motor.

The slip of an induction motor, as previously defined, is the difference between its synchronous speed and its operating speed and may be expressed in any of the following ways:

1. As a percentage of synchronous speed.
2. As a decimal fraction of synchronous speed.
3. Directly in revolutions per minute.

The most common method used for speed calculation of induction motors is by use of the formula:

$$\text{slip (\%)} = \frac{(\text{synchronous speed} \quad - \quad \text{operating speed}) \quad 100}{\text{synchronous speed}}$$

The synchronous speed of a motor is found by the following:

$$N_s = \frac{\text{frequency} \times 120}{\text{number of poles}}$$

Example — A three-phase squirrel-cage induction motor having four poles is operating on a 60-hertz AC circuit at a speed of 1,728 rpm. What is the slip of this motor?

Solution — By substituting the values in the previous formulas:

$$N_s = \frac{60 \times 120}{4} = 1,800 \text{ rpm}$$

$$\text{Slip} = \frac{1,800 - 1,728) 100}{1,800} = 4\%$$

From the foregoing it follows that this type of motor is not suitable in industrial applications where a great amount of speed regulation is required, because the speed can be controlled only by a change in frequency, number of poles, or slip. Speed is seldom changed by changing the frequency. The number of poles is sometimes changed either by using two or more distinct windings or by reconnecting the same winding for a different number of poles.

Torque

The starting torque of a squirrel-cage induction motor is the turning effort or torque the motor exerts when full voltage is applied to its terminals at the instant of starting. The amount of starting torque that a given motor develops depends, within certain limits, on the resistance of the rotor winding. Starting torque is usually expressed as a percent of full-load torque. An increase of rotor resistance gives an increase in slip and a decrease in efficiency. In fact, all the desirable characteristics are so interrelated that it is impossible to make one of them surpassingly good without adversely affecting the others. Fig. 4-2 shows the characteristic curves of one type of squirrel-cage motor.

Motors can be built for high efficiency alone, for high starting torque, or high power factor, but there is a commercial limit that prohibits a motor from excelling in all characteristics. In order to have a basis for normal starting torques, the National Electrical Manufacturers Association (NEMA) has established the minimum values for motors from 2 poles to 16 poles inclusive. With full rated voltage applied at the instant of starting, the starting torque will not be less than those listed in Table 4-1.

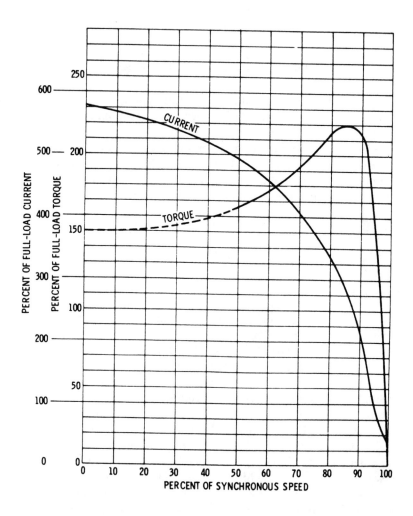

Fig. 4-2. Characteristic curves of a normal starting-torque, low starting-current, squirrel-cage motor.

Classification

Squirrel-cage motors are further classified by NEMA according to their electrical characteristics, as follows:

Class A — Normal torque, normal starting-current motors.
Class B — Normal torque, low starting-current motors.
Class C — High torque, low starting-current motors.
Class D — High-slip motors.
Class E — Low starting-torque, normal starting-current motors.
Class F — Low starting-torque, low starting-current motors.

**Table 4-1. Minimum Starting Torques for
Squirrel-Cage Motors**

No. of Poles	Percent of Full-Load Torque
2	150
4	150
6	135
8	125
10	120
12	115
14	110
16	105

Class A Motors

The Class A motor is the most popular, employing a squirrel-cage winding having relatively low resistance and reactance (Fig. 4-3). It has a normal starting torque and low slip at the rated load and may have sufficiently high starting current to require, in most cases, a compensator or resistance starter for motors above 7^1/$_2$ hp (5.595 kW).

Class B Motors

Class B motors are built (Fig. 4-4) to develop normal starting torque with relatively low starting current, and can be started at full voltage. The low starting current is obtained by the design of the motor to include inherently high reactance. The combined effect of induction and frequency on the current is termed *inductive reactance*. The slip at rated load is relatively low.

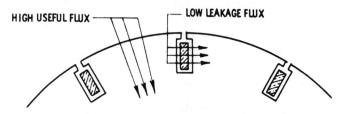

Fig. 4-3. Rotor construction of a Class A squirrel-cage motor to obtain normal torque with normal starting current. In this type, the rotor bars are placed close to the surface of the rotor, resulting in relatively low reactance.

Class C Motors

Class C motors are usually equipped with a double squirrel-cage winding, and combine high starting torque with a low starting current. These motors can be started at full voltage. The low starting current is obtained by design (Fig. 4-5) to include inherently high reactance. The slip at rated load is relatively low.

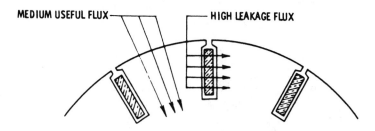

Fig. 4-4. To obtain normal torque with a low starting current (Class B motor), the rotor is constructed with deep and narrow rotor bars, resulting in high reactance during starting.

Class D Motors

Class D motors are provided with a high-resistance squirrel-cage winding (Fig. 4-6), giving the motor a high starting torque, low starting current, high slip (15 to 20 percent), and low efficiency.

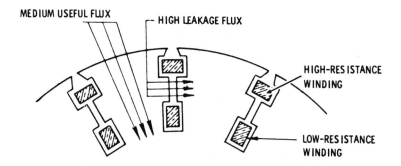

Fig. 4-5. **Rotor construction of a Class C motor. Two sets of bars are used, with the outer bars having a high resistance to produce a high starting torque with low starting current. At running speed, nearly all of the rotor current flows in the inner windings.**

Class E Motors

Class E motors have a low slip at rated load. The starting current may be sufficiently high to require a compensator or resistance started above 7¹/₂ hp (5.595 kW) (Fig. 4-7).

Class F Motors

Class F motors combine a low starting current with a low starting torque, and may be started on full voltage. The low starting current is obtained by design (Fig. 4-8) to include inherently high reactance.

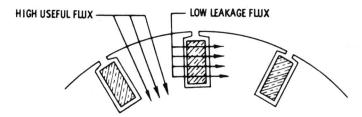

Fig. 4-6. **Rotor construction of a high-slip motor (Class D). To obtain low starting torque with low starting current, thin rotor bars are used to make the leakage flux in the rotor low and the useful flux high.**

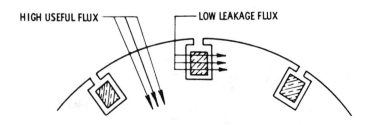

Fig. 4-7. Rotors of Class E motors are constructed with a low-resistance winding placed to offer a low reactance.

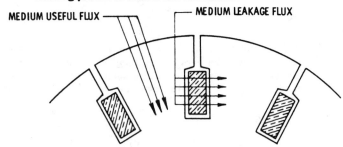

Fig. 4-8. Rotor construction of a low starting torque, low starting current, Class F motor. A low-resistance rotor winding is placed to offer high reactance during the starting period.

Double Squirrel-Cage Motor

Double squirrel-cage motors are so called because they are designed with two separate squirrel-cage windings, one within the other. Thus, a motor of this type is a combination having both a high- and a low-resistance squirrel-cage winding. The outer cage winding is made of high-resistance material, while the inner cage winding is made of a metal having a low resistance. In starting, the high-resistance winding gives the motor a high torque, while at full-load speed the low-resistance winding carries most of the current.

An example illustrating how the two squirrel cages operate is shown in Fig. 4-9. From the shape of the rotor slots, it is apparent that the bars of the inner cage are surrounded entirely by iron except for the constricted portion of the slot between the two cages, which constitutes an air gap. The bars of the outer cage are surrounded by iron at the sides only, and have two air gaps in the magnetic path around them; since this path is much less perfect magnetically than that around the inner conductors, its inductance is lower.

Fig. 4-9A shows the condition just as the motor starts. At this instant, the rotating field produced by the stator current is sweeping across both sets of rotor conductors at the full line frequency

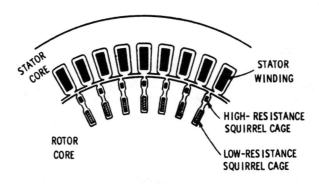

A. Motor at standstill.

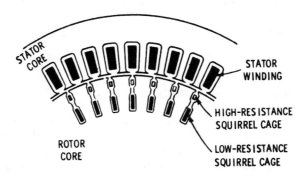

B. Motor at normal speed.

Fig. 4-9. Current relation in the rotor and stator windings of a double squirrel-cage induction motor. The high-resistance rotor winding carries most of the current induced in the rotor at the instant of start, thus giving high starting torque with low starting current. At full running speed, the low-resistance rotor winding carries most of the current, resulting in high efficiency.

and inducing currents in them. Since the outer conductors have a relatively low inductance, considerable current is set up, even at full line frequency. In the inner conductors, however, the current is greatly impeded by the combined action of the higher inductance of this winding and the high frequency of the current. In fact, the choking action of the self-induction at line frequency is so great that very little current can flow through this winding at the start. The relative density of the currents in the two sets of conductors is shown by their amount of shading.

As the rotor gains in speed, the frequency of the currents induced in it decreases, and the relation between the currents in the two cages gradually and automatically changes to that shown for normal speed in Fig. 4-9B. For this speed, the rotor currents are proportional to the slip, and they alternate at only about 1 hertz. At this low frequency, the higher inductance of the inner-cage winding is of relatively little importance and produces little choking effect.

The resistances of the two cages are now the chief limiting factor in the rotor currents. Consequently, the inner, low-resistance cage carries most of the total rotor current, with the advantageous results already mentioned. The starting torque of double squirrel-cage motors is greater than that of the ordinary squirrel-cage, but less than that of a motor with a single high-resistance squirrel-cage winding.

Multispeed Squirrel-Cage Motors

In certain industrial applications where two, three, or four different constant speeds are required, it is the usual practice to wind the stator with two or more separate independent windings, whose poles may be changed by changing the external connections, giving a limited number of different speeds. The speeds so obtained are always in a constant ratio, such as 1,800, 1,200, 900, 600, etc.

Multispeed motors can be classified into three groups according to output — constant torque, constant horsepower, and variable torque and horsepower. The relation between torque, horsepower, and speed is:

$$horsepower = \frac{torque \times speed}{5,252}$$

where torque is measured in pound-feet and speed in revolutions per minute.

The horsepower output of the constant-torque motor varies directly as the speed. With the constant-horsepower type, the torque is inversely proportional to the speed, the horsepower being the same at each speed. In the case of the variable-torque and horsepower type, both the horsepower and the torque decrease with a reduction in the speed. The latter motor is usually employed on loads that vary as the square or cube of the speed, such as fans and blowers, centrifugal pumps, etc. To convert horsepower to kilowatts (kW), multiply by 0.746.

Starting

The condition at the starting of a motor is similar to that of a transformer with a short-circuited secondary, since the rotor, by construction, is being short-circuited by means of heavy metal bars, as previously described.

When the rated voltage is applied to the stator windings of a squirrel-cage motor, a heavy current is drawn from the line. It ranges from four to seven times the full-load current, depending on the type of motor. This high current is momentary only, and falls off rapidly as the motor increases its speed. The magnitude of this current depends on the electrical design of the motor and is independent of the mechanical load. The duration of the starting current, however, depends on the time required for acceleration, which in turn depends on the nature of the driven load.

Squirrel-cage motors in industrial plants are usually started by one of the following methods:

1. Directly across the line.
2. By autotransformers (compensators).
3. By resistance in series with the stator winding.
4. By means of a step-down transformer.

Squirrel-cage motors of 5 hp (3.73 kW) or less are generally started with line switches, connecting them directly across the line. The switches are usually equipped with thermal devices that open the circuit when overloads are carried beyond predetermined limits. Large motors usually require various voltage-reducing methods of the type previously mentioned.

All types of squirrel-cage motors may be thrown directly across the line — provided, of course, that the starting currents do not cause voltage fluctuations that will interfere with power service elsewhere, and also that the starting currents do not exceed those permitted by local power regulations. When so started, a magnetic contactor is usually employed, operated from a start-stop pushbutton station, or automatically from a float switch, thermostat, pressure regulator, or other pilot-circuit control device.

Across-the-Line Starting

There are several across-the-line starting methods whose use depends on the duty of the motor and the method of control desired. It is customary to employ a straight single-throw switch or contactor for motors up to 5 hp (3.73 kW), with fuses rated for the starting current.

The methods of starting shown in Figs. 4-10 and 4-11 require two sets of fuses and a double-throw switch. The high-current fuses are used for starting only, while the running fuses have a lower current rating. These methods of starting are used for motors up to 250 volts.

With large motors having a voltage rating of from 440 to 600 volts, oil circuit breakers or magnetic switches are generally used. Both the oil circuit breakers and magnetic switches are provided with overload and undervoltage protection. Various types of across-the-line starters used with squirrel-cage motors are shown in Figs. 4-12 to 4-15.

Autotransformer Starting

The autotransformer, or compensator, method of starting employs two or three autotransformers for the reduction of voltage to the motor terminal. The autotransformer type of starter has the advantage of drawing less current from the line for a given

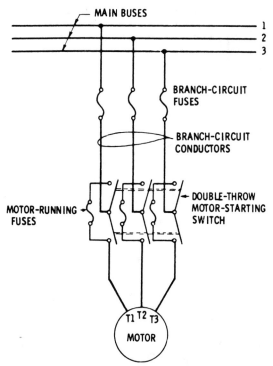

Fig. 4-10. Across-the-line starting method employing a double-throw switch. The switch is thrown to the lower position to start the motor, thus cutting out the running fuses. After running speed is reached, the switch is thrown to the upper position.

reduction in voltage to the motor terminals. This type of starter may employ oil circuit breakers or switches, as illustrated in Figs. 4-16 to 4-18.

With reference to Fig. 4-17, the compensator consists of an auto-transformer, a manually operated set of contacts, and a temperature overload relay. When the operating handle is pushed away from the operator, the autotransformer is connected to the power source, and, at the same time, the motor that is being controlled is connected to the taps on the autotransformer. These taps apply 50, 65, or 80 percent of the line voltage to the motor. This reduced voltage materially limits the torque on the motor in starting. In fact, the torque varies as the square of the applied voltage. If, for

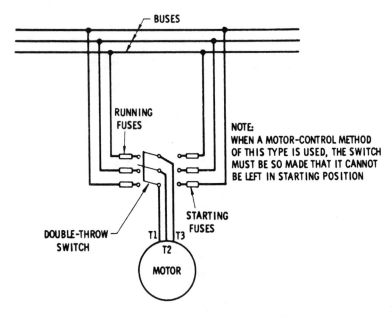

Fig. 4-11. An across-the-line starting arrangement with two sets of fuses, one set for starting and one set for normal running operation.

example, the 50-percent tap is used, the torque applied to the motor on starting will only be 25 percent of the full-load torque; if the 80-percent tap is used, the torque will be 64 percent of the full-voltage starting torque.

After the motor has accelerated, the operator quickly pulls the handle back to the running position, which connects the motor to full line voltage. The handle is held in the running position by an undervoltage trip coil. If the overload relay trips out, if the line voltage fails momentarily, or if the cover is removed, the under-voltage coil is de-energized, allowing the handle to return to the off position. To restart the motor, the overload relay must be reset by pushing a reset button on the outside of the case. Then, the operating handle must be thrown first to the starting and next to the running position. It is not possible to throw the handle from the off position directly to the running position (this would start the motor on full voltage), because the handle is so interlocked that it must first be pushed to the starting position before it can be pulled through the off position to the running position.

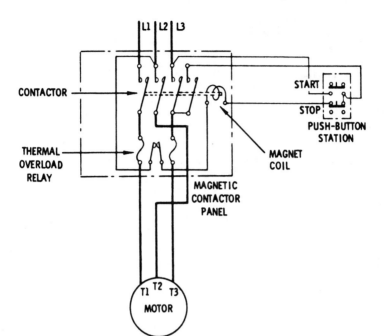

Fig. 4-12. An across-the-line starting method with remote control and thermal overload protection.

Resistance Starting

The resistance-type starter, Fig. 4-19, employs a heavy-duty resistance in series with the line conductors. The drop of voltage through this resistance produces a reduced voltage at the motor terminals. The starting current of the motor is reduced in direct proportion to the reduction of the voltage applied to the motor terminals.

It should be noted, however, that in the case of a resistance starter being used to reduce the voltage to half value, for example, the current drawn from the line will also be half of the full voltage value, whereas if an autotransformer starter is used, the current drawn from the line varies as the square of the voltage ratio of the transformer. Thus, if half voltage is applied to the motor terminal from the secondary of the autotransformer, the current drawn from the line will only be one-fourth the full-voltage value.

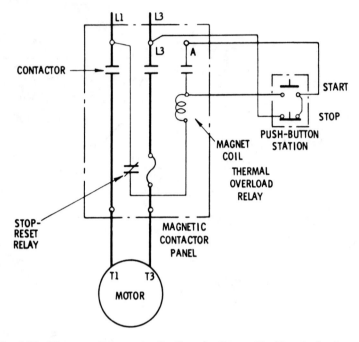

Fig. 4-13. Diagram of an across-the-line starting method for single-phase motors.

Step-Down Transformer Starting

The voltage step-down transformer method of starting is rarely used because it is more expensive than the autotransformer method. A step-down voltage transformer consists of two complete coils or windings, while an autotransformer requires but one tapped coil or winding. The comparison is shown in Fig. 4-20.

Control Equipment

Selecting the right squirrel-cage motor for the machine in question is but one step toward meeting the motor-application problem. Selection of the proper control apparatus is just as important as the selection of the motor itself. The intelligent choice of motor control involves a complete knowledge of the types of control apparatus available, as well as their functions.

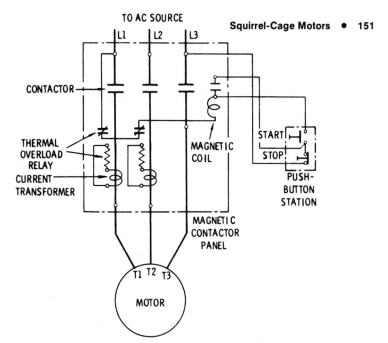

Fig. 4-14. Magnetic starter for a three-phase motor.

The functions of the control apparatus for squirrel-cage motors are:

1. Starting the motor on (a) full voltage, or (b) reduced voltage.
2. Stopping the motor.
3. Disconnecting the motor upon failure of voltage.
4. Limiting the motor load.
5. Changing the direction of rotation of the rotor.
6. Starting and stopping the motor (a) at fixed points in a given cycle of operation, (b) at the limit of travel of the load, or (c) when selected temperatures or pressures are reached.

Where multispeed motors are involved, the following functions may be added:

7. Changing the speed of rotation rpm.
8. Starting the motor with a definite speed sequence.

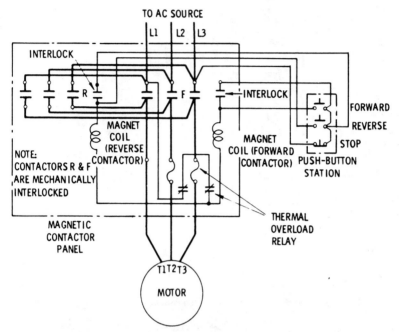

Fig. 4-15. An across-the-line reversing type starter.

Stopping the Motor

Control devices permit motors to be stopped as follows:

1. Under the direction of the operator.
2. Under the control of a pilot-circuit device, such as a thermostat, pressure regulator, float switch, limit switch, or cam switch.
3. Under the control of protective devices that will disconnect the motor under overload conditions detrimental to the motor, or upon failure of voltage or a lost phase.

Protecting the Motor

To protect motors against damage during severe momentary or sustained overloads, overcurrent devices are employed to disconnect them from the line. These devices are of two general types:

1. Thermal overload relay.
2. Dashpot overload relay.

The characteristics of the thermal-type overload relay are illustrated in Fig. 4-21, which indicates an inverse relationship between current and time — that is, the greater the current, the sooner the relay operates to disconnect the motor from the line. When the relay operates, it opens an independent, or pilot, circuit, which causes the main line contacts to drop out, thus disconnecting the motor from the line. Dashpot overload relays have a similar inverse time-current relationship. Thermal and dashpot overload relays are of two types:

1. Hand reset.
2. Automatic reset.

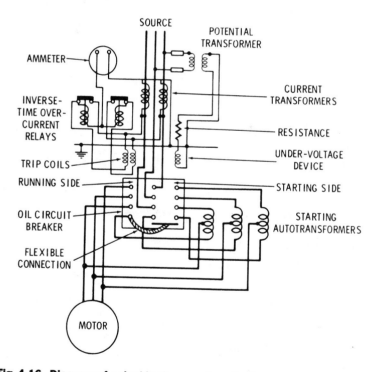

Fig. 4-16. Diagram of a double-throw motor-starting oil circuit breaker with overcurrent protection, used in starting large squirrel-cage motors.

As the names denote, the hand-reset type must be reset by hand after having tripped (usually by pressing a button projecting through the enclosing case), whereas the automatic reset types reset themselves automatically. Dashpot overload relays are generally of the automatic-reset type, but they can be made of the hand-reset type by providing a hand-reset attachment that will prevent automatic resetting.

Standard polyphase motor control devices, almost without exception, are equipped with two thermal-overload relays of the hand-reset type. Thermal-overload relays of the automatic-reset

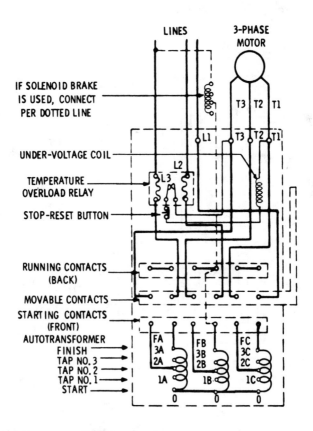

Fig. 4-17. Diagram of an autotransformer type, manual reduced-voltage starter for a three-phase squirrel-cage motor.

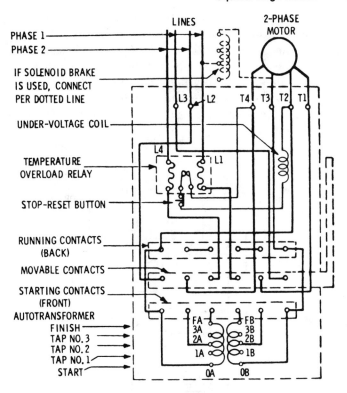

PHASE 1
PHASE 2

IF SOLENOID BRAKE
IS USED, CONNECT
PER DOTTED LINE

UNDER-VOLTAGE COIL

TEMPERATURE
OVERLOAD RELAY

STOP-RESET BUTTON

RUNNING CONTACTS
(BACK)

MOVABLE CONTACTS

STARTING CONTACTS
(FRONT)
AUTOTRANSFORMER
FINISH
TAP NO. 3
TAP NO. 2
TAP NO. 1
START

LINES

2-PHASE
MOTOR

Fig. 4-18. Diagram of an autotransformer type, manual reduced-voltage starter for a two-phase squirrel-cage motor.

type are also used. Their purpose is to prevent burnout of the motor windings; hence, they should be selected carefully to accomplish that end. Thermal relays must be of the proper rating, and dashpot relays must be properly set to disconnect motors from the line before operating conditions tax the motor windings to the point of breakdown.

Overload relays protect polyphase motors against phase failure due to a blown fuse or some other power interruption in one line of the supply circuit. If phase failure occurs while the motor is at rest, the motor will trip the overload relay, thereby disconnecting the motor from the line. If phase failure occurs while the motor is running, the motor will continue to run, provided the load is not too great. The current taken by the motor when running single-phase will be from two to three times the normal three-phase value, and unless the motor is very lightly loaded, will be sufficient to trip the overload relays before the motor windings can be damaged. If the load is very light, the overload relays will not trip, and the motor will continue to operate without damage until shut down, but will refuse to start the next time an attempt is made to operate it.

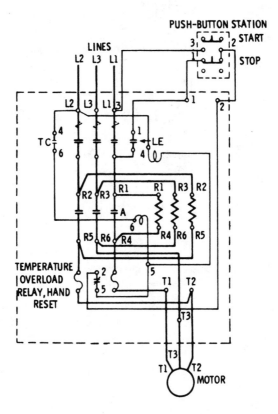

Fig. 4-19. Diagram of a resistance-type starter for a three-phase squirrel-cage motor.

After hand-reset overload relays have been tripped, they must be manually reset by an operator before the control will again connect the motor to the line. The operator, noting the failure of the motor to start, will presumably investigate and remove the trouble.

Automatic-reset overload relays in magnetic starters controlled by two-wire pilot-circuit devices, such as float switches, pressure regulators, thermostats, snap switches, etc., will cause the motor to be connected to the line again and again; each time a heavy load current will be drawn from the line, which will reheat the motor windings until they burn out.

Voltage failure also requires protection to operators and motors. Control apparatus designed so that the main-line contacts are held in by a magnet coil operated across one phase of the supply circuit can have their control or pilot circuits so arranged that, upon failure of the voltage, the motor will not restart until the operator goes through the necessary starting operations. Three-wire pilot-circuit control is required for such control devices to cause and maintain the interruption of power to the main circuit. Motor applications involving control apparatus that starts and stops the motor automatically by means of two-wire pilot-circuit devices, such as thermostats, float switches, and pressure regulators, cannot be arranged for low-voltage protection.

To reverse squirrel-cage motors, it is only necessary to interchange two leads — (a) any two leads to three-phase motors; (b) two leads of either phase to two-phase four-wire motors, leaving one phase intact; (c) the two outside leads to two-phase three-wire motors, leaving the common lead intact. Reversal is accomplished by means of either a small drum controller for small motors, or a

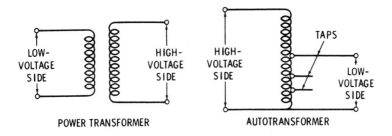

POWER TRANSFORMER   AUTOTRANSFORMER

Fig. 4-20. Comparison of the copper requirements of a single-phase power transformer and an autotransformer for motor starting.

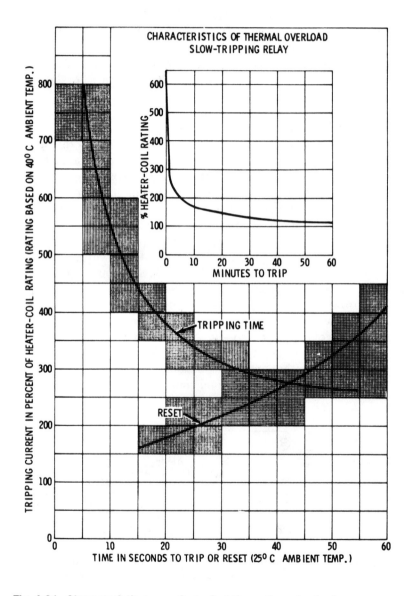

Fig. 4-21. Characteristic curve of a typical thermal overload relay.

reversing magnetic contactor for larger motors like the one shown previously in Fig. 4-15.

Control for Multispeed Motors

Multispeed motors involve all the devices heretofore mentioned to effect starting and stopping under the six conditions listed. Additional control devices are required, however, for changing their speeds, which is accomplished by regrouping the motor windings. Two types of controls used are:

1. Manually operated speed-setting drums.
2. Combinations of magnetic contactors.

Magnetically operated controllers can be so arranged that the multispeed motors will always start up through low speed, subsequently being transferred to a higher operating speed. The auxiliary control device accomplishing this is known as a *compelling relay*. It may be operated either manually or automatically by means of a time-delay device.

Application

As pointed out in the beginning of this chapter, the squirrel-cage motor, because of its simplicity of construction and because it can be built with electrical characteristics to suit almost any industrial requirement, is one of the most widely used machines. As noted from the characteristics and classification, squirrel-cage motors as a rule are not suitable where a high starting torque is required, but are most suitable where the starting-torque requirements are of a medium or low value.

In selecting a motor for a certain application, there are, in addition to torque and starting-current considerations, a large number of factors affecting an economical and efficient operation. The selection is determined in whole or in part by the user on the basis of available data, such as speed range and regulation, mechanical arrangement, available voltage, direction of rotation and reversing, operating schedule, method of control, surrounding atmosphere, etc.

Since the power factor and efficiency of any induction motor are lower at light loads than at heavy loads, it is obvious that in selecting a motor for a definite load, the size should be such as to permit the operation of the motor as nearly as possible at full load. Also, low-speed motors have a lower power factor and weigh more per horsepower than high-speed motors. Therefore, in choosing a motor for a certain duty, size, and speed should be carefully considered so as to give the most economical and satisfactory service.

For the largest group of applications, such as for fans, pumps, compressors, conveyors, etc., which are started and stopped infrequently and have low inertia loads so that the motor can accelerate in a few seconds, the conventional general-purpose NEMA Class A motor can be used.

If a motor is to be installed in a location where there is a limitation on the starting current, the modified general-purpose NEMA Class B motor can be used. If the starting current is still in excess of what can be permitted, then reduced-voltage starting is employed.

On those applications where reduced-voltage starting does not give sufficient torque to start the load with either NEMA Class A or B motors, Class C motors with their high inherent torque, reduced starting current, and reduced-voltage starting may be used.

Conveyors and Compressors

For applications such as conveyors and compressors, which sometimes require a starting torque of at least twice full-load torque, NEMA Class C motors may be used with full-load voltage starting.

Large Fans

This type of drive is one which requires special consideration. These drives, once they are accelerated, run continuously at full load; therefore, it is desirable to have the best possible efficiency and power factor. Some of the fans, however, have extremely high values of moment of inertia (WR^2) and motors with normal starting-torque characteristics may require from 30 seconds to 1 minute to accelerate. Starting current flowing in both the rotor and stator for this long period of time may generate sufficient heat to damage the windings.

To meet this application with a squirrel-cage motor, special NEMA Class B motors are used to reduce the starting current to a minimum. The rotors of these special motors are designed with a large mass of material, especially in the end rings, so that it is possible for the motor rotor to absorb the tremendous losses during the accelerating period without reaching excessive temperatures. Once the motor reaches its full-load speed, the losses return to normal and the rotor rapidly cools down to its normal operating temperatures.

Flywheels, Presses, and Bolt Headers

Another continuous-running motor application that requires special consideration is on drives where there is high external inertia, quite often in the form of a flywheel, and where the load, instead of being a continuous full-load torque, is pulsating in nature. Typical examples of this type of application are presses and bolt headers. In this type of application, no work is being done most of the time; then a peak load occurs that may require torques of many times the full-load torque of the motor.

Under these conditions, the running efficiency of the motor at full load is not important, because the motor never operates at that point. Therefore, it is deliberately designed with more than normal secondary resistance so that it has a tendency to slow down as the load comes on the drive. This tendency for the motor to slow down permits the flywheel to give up energy to absorb the peak load, with the motor exerting no more than full-load torque. However, the energy that is taken out of the flywheel during the work stroke must be returned to it by the motor. Thus, the motor must have sufficient torque to accelerate the flywheel before the next work stroke is made.

Since presses work at widely varying rates, say from 2 or 3 to as high as 100 to 150 work strokes per minute, the correct amount of rotor resistance will vary, depending on the number of strokes per minute, there is ample time for the flywheel to slow down 10 to

15 percent during the work stroke and give up a considerable part of its kinetic energy. Therefore, motors used on this type of press should have high slip. One of the standard ratings listed by the motor manufacturers is a motor having 8 to 13 percent slip.

As the number of strokes per minute increases, the length of time of the working stroke decreases, and there is not much time available for the flywheel to slow down. Neither is there much time for the motor to reaccelerate the flywheel between strokes, so that the amount of slowdown of the flywheel is usually between 5 and 10 percent. The standard line of press motors with 5 to 8 percent slip has been designed for this application. These presses usually make 10 to 40 strokes per minute.

On the smaller presses, making 100 to 150 strokes per minute, there is very little time for the flywheel to accelerate or decelerate. For these, a standard motor that has approximately 3 percent slip is entirely adequate; there is no object in supplying a high-slip motor. Quite often, someone tries to use NEMA Class C motors on these high-torque applications, and if there is not trouble due to overheating, trouble may develop in the form of mechanical failure of the rotor due to the unequal heating in the double-deck winding, as previously explained.

Squirrel-Cage Motor Troubleshooting Chart

Symptom and Possible Cause   *Possible Remedy*

Motor Will Not Start

(a) Overload control tripped

(a) Wait for overload to cool. Try starting again. If motor still does not start, check all the causes as outlined in the following.

(b) Power not connected

(b) Connect power to control and control to motor. Check clip contacts.

Symptom and Possible Cause	Possible Remedy
(c) Faulty (open) fuses	(c) Test fuses.
(d) Low voltage	(d) Check motor nameplate values with power supply. Also check voltage at motor terminals with motor under load to be sure wire size is adequate.
(e) Wrong control connections	(e) Check connections with control wiring diagram.
(f) Loose terminal lead connection	(f) Tighten connections.
(g) Driven machine locked	(g) Disconnect motor from load. If motor starts satisfactorily, check driven machine.
(h) Open circuit in stator or rotor winding	(h) Check for open circuits.
(i) Short circuit in stator winding	(i) Check for shorted coil.
(j) Winding grounded	(j) Test for grounded winding.
(k) Bearings stiff	(k) Free bearings or replace.
(l) Grease too stiff	(l) Use special lubricant for special conditions.
(m) Faulty control	(m) Check control wiring.
(n) Overload	(n) Reduce load.

Motor Noisy

(a) Motor running single phase	(a) Stop motor, then try to start. (It will not start on single phase.) Check for "open" in one of the lines or circuits.

Symptom and Possible Cause	*Possible Remedy*
(b) Electrical load unbalanced	(b) Check current balance.
(c) Shaft bumping (sleeve-bearing motors)	(c) Check alignment and condition of belt. On pedestal-mounted bearing, check end play and axial centering of rotor.
(d) Vibration	(d) Driven machine may be unbalanced. Remove motor from load. If motor is still noisy, rebalance rotor.
(e) Air gap not uniform	(e) Center the rotor and, if necessary, replace bearings.
(f) Noisy ball bearings	(f) Check lubrication. Replace bearings if noise is persistent and excessive.
(g) Loose punchings or loose rotor on shaft	(g) Tighten all holding bolts.
(h) Rotor rubbing on stator	(h) Center the rotor and replace bearings if necessary.
(i) Object caught between fan and end shields	(i) Disassemble motor and clean. Any rubbish around motor should be removed.
(j) Motor loose on foundation	(j) Tighten hold-down bolts. Motor may possibly have to be realigned.
(k) Coupling loose	(k) Insert feelers at four places in coupling joint before pulling up bolts to check alignment. Tighten coupling bolts securely.

Symptom and Possible Cause   *Possible Remedy*

Motor Running Temperature Too High

(a) Overload

(a) Measure motor loading with ammeter. Reduce load.

(b) Electrical load unbalance (fuse blown, faulty control, etc.)

(b) Check for voltage unbalance or single phasing. Check for "open" in one of the lines or circuits.

(c) Restricted ventilation

(c) Clean air passages and windings.

(d) Incorrect voltage and frequency

(d) Check motor nameplate values with power supply. Also check voltage at motor terminals with motor under full load.

(e) Motor stalled by driven machine or by tight bearings

(e) Remove power from motor. Check machine for cause of stalling.

(f) Stator winding shorted or grounded

(f) Test winding for short circuit or ground.

(g) Rotor winding with loose connections

(g) Tighten, if possible, or replace with another rotor.

(h) Motor used for rapid reversing service

(h) Replace with motor designed for this service.

(i) Belt too tight

(i) Remove excessive pressure on bearings.

Symptom and Possible Cause	*Possible Remedy*

Bearings Hot

(a) End shields loose or not replaced properly

(a) Make sure end shields fit squarely and are properly tightened.

(b) Excessive belt tension or excessive gear slide thrust

(b) Reduce belt tension or gear pressure and realign shafts. See that thrust is not being transferred to motor bearings.

(c) Bent shaft

(c) Straighten shaft or replace.

Sleeve Bearings Hot

(a) Insufficient oil

(a) Add oil. If oil supply is very low, drain, flush, and refill.

(b) Foreign material in oil, or poor grade of oil

(b) Drain oil, flush, and re-lubricate, using industrial lubricant recommended by a reliable oil company.

(c) Oil rings rotating slowly or not rotating at all

(c) Oil too heavy; drain and replace.

(d) Motor tilted too far

(d) Level motor or reduce tilt and realign, if necessary.

(e) Oil rings bent or other-wise damaged in reassembling

(e) Replace oil rings.

(f) Oil ring out of slot

(f) Adjust or replace retaining clip.

Symptom and Possible Cause	*Possible Remedy*
(g) Motor tilted, causing end thrust	(g) Level motor, reduce thrust, or use motor designed for thrust.
(h) Defective bearings or rough shaft	(h) Replace bearings. Resurface shaft.

Ball Bearings Hot

(a) Too much grease	(a) Remove relief plug and let motor run. If excess grease does not come out, flush and relubricate.
(b) Wrong grade of grease	(b) Add proper grease.
(c) Insufficient grease	(c) Remove relief plug and re-grease bearing.
(d) Foreign material in grease	(d) Flush bearings, relubricate; make sure that grease supply is clean. Keep can covered when not in use.
(e) Bearings misaligned	(e) Align motor and check bearing housing assembly. See that the bearings races are exactly 90° with shaft.
(f) Bearings damaged	(f) Replace bearings.

Summary

The squirrel-cage motor is the most common form of induction motor. The motor derives its name from the fact that the rotor, or secondary, resembles the wheel of a squirrel cage.

A squirrel-cage motor consists essentially of a *stator* (or primary) and a *rotor* (or secondary). The stator consists of a laminated sheet-steel core with slots containing the insulated coils. The coils are so grouped and connected to a polyphase alternating-current circuit.

The rotor is also constructed of steel laminations, but the windings consist of conductor bars placed approximately parallel to the shaft and close to the rotor surface. These windings are short-circuited, or connected together at each end of the rotor, by a solid ring.

The squirrel-cage (or secondary) winding takes the place of the field winding in the synchronous motor. As in a synchronous motor, the currents in the stator set up a rotating magnetic field that is produced by increasing and decreasing currents in the windings. When the current increases in the first phase, only the first winding produces a magnetic field. As the current decreases in the first winding and increases in the second, the magnetic field shifts slightly, until it is all produced by the second winding. When maximum current is flowing in the third winding, the field is shifted a little more. The windings are so distributed that the shifting is uniform and continuous, which produces a rotating magnetic field.

The speed of a squirrel-cage motor is nearly constant under normal load and voltage conditions, but is dependent on the number of poles and frequency of the AC source. This type of motor slows, however, when loaded barely enough to produce the increased current for the required torque.

The starting torque of a squirrel-cage induction motor is the turning effort or torque that the motor exerts when full voltage is applied to its terminals at the instant of starting. The amount of starting torque depends, within limits, on the resistance of the rotor winding, and it is usually expressed as a percentage of full-load torque.

The squirrel-cage motor is simple in construction and can be built to suit almost any industrial requirement; therefore, it is one of the most widely used machines. As a rule, squirrel-cage motors are not suitable where high starting torque is required, but they are most suitable for medium- or low-starting-torque requirements.

Review Questions

1. What are the two basic units in a squirrel-cage motor?
2. What is the basic operating principle of the squirrel-cage motor?
3. What is the basic difference between the synchronous motor and the squirrel-cage motor?
4. How is starting torque expressed?
5. Why is the squirrel-cage motor used widely in industry?
6. List five industrial applications for the squirrel-cage motor.

CHAPTER 5

Wound-Rotor Motors

As far as the stator is concerned, the squirrel-cage motor and the wound-rotor motor are identical. The outstanding difference between the two motors lies in the rotor winding.

In the squirrel-cage motors previously described, the rotor winding is practically self-contained; it is not connected either mechanically or electrically to the outside power-supply or control circuit. It consists of a number of straight bars uniformly distributed around the periphery of the rotor and short-circuited at both ends by end rings to which the bars are integrally joined. Since the rotor bars and end rings have fixed resistances, such characteristics as starting and pull-out torques, rate of acceleration, and full-load operating speed cannot be changed for a given motor installation.

In wound-rotor motors, however, the rotor winding consists of insulated coils of wire that are not permanently short-circuited, but are connected in regular succession to form a definite polar area having the same number of poles as the stator. The ends of these rotor windings are brought out to collector rings, or slip rings, as they are commonly termed.

The currents induced in the rotor are carried by means of slip rings (and a number of carbon brushes riding on the slip rings) to an externally mounted resistance, which can be regulated by a special control, as indicated in Fig. 5-1. By varying the amount of

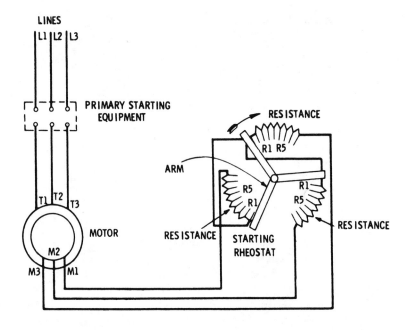

Fig. 5-1A. Diagram of a starter or controller for a wound-rotor induction motor. The resistances in each phase of the rotor are gradually cut out to increase the motor speed.

resistance in the rotor circuit, a corresponding variation in the motor characteristics can be obtained. Thus, by inserting a high external resistance in the rotor circuit at starting, a high starting torque can be developed with a low starting current. As the motor accelerates to full speed, the resistance is gradually cut out until, at full speed, the resistance is entirely cut out and the rotor windings are short-circuited.

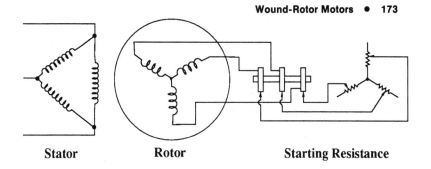

Stator   Rotor   Starting Resistance

Fig. 5-1B. Schematic representation for starting resistance in each phase of a wound-rotor 3φ motor.

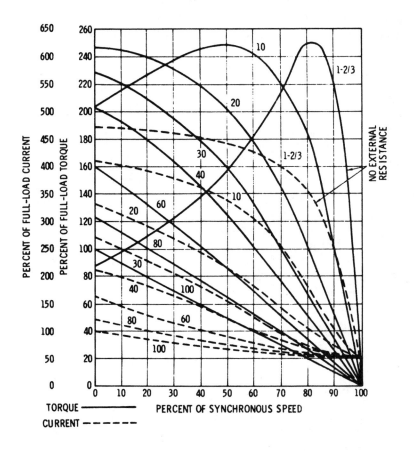

Fig. 5-2. Speed-torque and speed-current curves of a typical wound-rotor induction motor.

Characteristics

Varying the values of resistance in the rotor circuit can affect the characteristics of the motor as follows:

1. Variation in the starting torque and current.
2. Smooth acceleration.
3. Variation in operating speed.

The operating speed, of course, depends entirely on the amount of resistance incorporated in the control equipment.

Fig. 5-2 shows the speed-torque and corresponding speed-current curves obtainable for a typical wound-rotor induction motor using different values of external rotor resistance. The numbers given on the curves indicate the rotor-circuit resistance in percent of the value required to give full-load torque at standstill.

The curves in Fig. 5-2 illustrate that external resistance in the rotor circuit reduces the speed at which the rotor will operate with a given load torque. If the rotor resistor is designed for continuous duty, a portion of it may be allowed to remain in the circuit, thus obtaining reduced-speed operation. Therefore, the motor has a varying speed characteristic — that is, any change in load results in a considerable change in speed. It should be borne in mind, however, that the efficiency of a wound-rotor motor, including the I²R losses in the rotor resistance, is reduced in direct proportion to the speed reduction obtained.

Thus, the wound-rotor motor has found its use in cranes, hoists, and elevators. These are operated intermittently and for short periods, where exact speed regulation and loss in efficiency are of little consequence. If, however, lower speed is required over longer periods, poor speed regulation and loss in efficiency may become prohibitive.

Control Equipment

The foregoing has noted that wound-rotor motors have certain inherent speed-torque characteristics, and that these can be considerably altered by the secondary control equipment, which introduces varying values of resistance into the rotor circuit. Since the control equipment has this ability, a further study of the most common types of controllers and associated equipment is appropriate.

The functions of controllers used with wound-rotor motors are:

1. To start the motor without damage or undue disturbance to the motor, driven machine, or power supply.
2. To stop the motor in a satisfactory manner.
3. To reverse the motor.
4. To run the motor at one or more predetermined speeds below synchronous speed.
5. To handle an overhauling load satisfactorily.
6. To protect the motor.

The various types of controllers utilized may be divided into the following groups, depending on the size and function of the motor:

1. Faceplate starters.
2. Faceplate speed regulators.
3. Multiswitch starters.
4. Drum controllers.
5. Motor-driven drum controllers.
6. Magnetic starters.

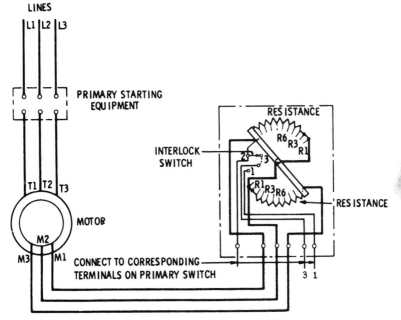

Fig. 5-3. Diagram of a faceplate starter for a wound-rotor induction motor.

Faceplate Starters

A typical wiring diagram of a combined faceplate starter and speed regulator is shown in Fig. 5-3. This particular type requires a separate switch for control of the primary circuit. Faceplate starters are usually adaptable for motors up to 50 hp (37.3 kW) where the full-load current in the rotor circuit is less than 150 amperes.

As shown in the diagram, the resistance is made up of three delta-connected sections, but is connected in two phases only, the third being a fixed step. The starting lever is equipped with a spring similar to that used in DC motor starters. The spring returns the lever to the starting (all-resistance) position when it is released. As an additional safety feature, the contact arm or lever cannot be left in the full-in position unless the primary (stator) contactor is closed; nor can it be left in an intermediate position at any time.

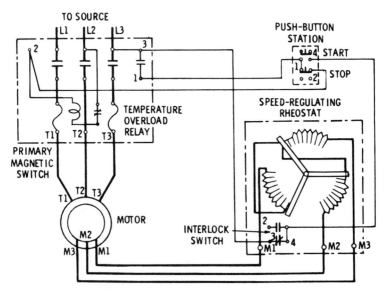

Fig. 5-4. Diagram of a secondary speed-regulating rheostat and a primary magnetic switch for a wound-rotor motor.

Faceplate Speed Regulators

A typical wiring diagram of a combined faceplate starter and speed regulator for use in the rotor circuit of wound-rotor induction motors is shown in Fig. 5-4. Starters of this type usually provide for a 50-percent speed reduction when the motor operates under full load at normal speed. They are usually built for motors in sizes up to 40 hp (29.84 kW) and for rotor currents up to 100 amperes. Since the starter is not connected with the primary circuit of the motor, a magnetic switch, an oil circuit breaker or similar device must be installed to control the primary circuit.

Multiswitch Starters

Multiswitch starters are used in the secondary circuits of large wound-rotor induction motors up to 2,000 hp (1,492 KW) with rotor currents up to 1,000 amperes. A typical wiring diagram of a multiswitch starter is shown in Fig. 5-5.

The type of switch to be used in the primary (stator) circuit depends on the voltage of the supply source. The resistor units in the secondary (rotor) circuit are balanced on all steps of the controller. The contact levers are of the double-pole type and are mechanically arranged in such a manner that they must be closed in a predetermined sequence, and only one at a time. Since

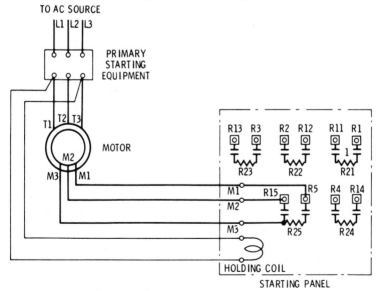

Fig. 5-5. Diagram of a typical multiswitch starter for a wound-rotor motor.

the switches are designed for hand-over-hand operation, a desirable time element is introduced that prevents too rapid acceleration of the motor. When the final switch has been closed, it is held in place by a magnetic coil, and because of the mechanical interlocking feature, all other switches remain closed.

A failure of the motor supply source will cause the magnet coil to release the switches, returning the starter to the starting position with all the resistance in the secondary circuit. This type of starter is designed for starting duty only and is not suitable where speed regulation is required.

Drum Controllers

Drum controllers (Fig. 5-6) are frequently employed for starting and for speed control of wound-rotor induction motors. Because of their construction features, they offer a compact, sturdy, and dependable control means and also add to the safety of the operator. Drum controllers are built to handle both stator and rotor circuits, the cylinder mounting the contact segments being built in two insulated sections. They are also built to handle the rotor circuit only, in which case the stator circuit is controlled by a circuit breaker or line starter. Other types are being built for operation by means of an independent motor. In addition to starting and regulating, speed-regulating drum collectors are commonly used for speed-reversing duty as well.

On small- and medium-sized motors, the faceplate starters described earlier are commonly used, but with large motors, drum controllers are generally preferred because their contacts are heavier and better able to handle the heavy currents required.

Fig. 5-7 shows the connections of a typical drum controller for reversing and speed-regulation duty with a wound-rotor motor. The resistor material is mounted separately from the drum, and connections from the resistor are brought to the drum fingers.

The resistor connections shown in Fig. 5-7 are for starting the motor with one phase of its secondary open on the first point of the controller. When higher starting torque is desired, or when connecting a motor that is rated above 80 hp (59.68 kW) connect R_1 to R_{11} at the resistor and the finger marked R_1 on the controller to terminal R_3 on the resistors. Resistor steps R_5 to R_6, R_{15} to R_{16}, and R_{25} to R_{26} are for resistances that remain permanently in the circuit. When these are not furnished, connect M_1, M_2, and M_3 to R_5, R_{15}, and R_{25} as indicated by the dotted lines.

Motor-driven drum controllers are used in certain drives requiring close automatic speed regulation such as in large air-

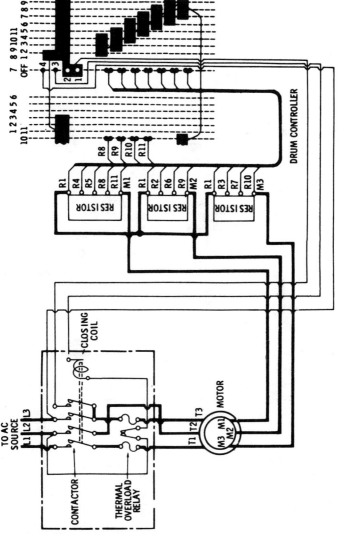

Fig. 5-6. Diagram of a nonreversing drum controller for a wound-rotor motor having a three-phase secondary.

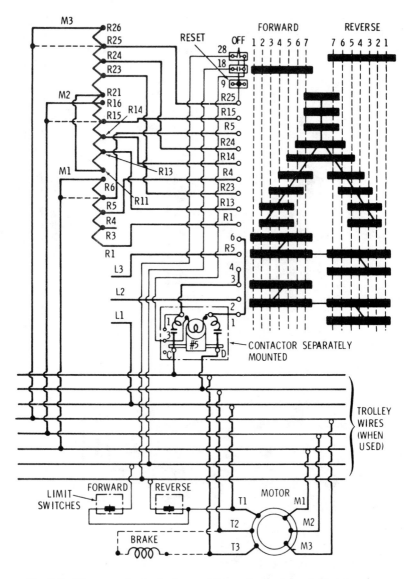

Fig. 5-7. Diagram of a typical reversing and speed-regulating drum controller for a wound-rotor induction motor.

conditioning plants, blowers, stokers, etc. In construction, these controllers differ from those previously described in that cam-operated switches are used in place of the segments and fingers.

These drums are built with either 13 or 20 balanced speed-regulating points, and the cam-switch construction saves space when so many balanced points are needed. Each switch carries two contacts, both of which are connected to the middle leg of the resistor. They make contact simultaneously with stationary contacts, one connected to each of the outside legs of the resistor. In this way, the closing of any cam-operated switch short-circuits the resistor at that point. The drum is driven by a small motor connected to the drum shaft through suitable gearing. The pilot motor is energized by some automatic device, such as a thermostat or a regulator.

A typical motor-driven drum may have 20 balanced speed points and be rated at 600 amperes and 1,000 volts. It may have a positioning device that ensures that the pilot motor stops only at positions in which the switches are fully closed. It may also have self-contained limit switches that open to stop the motor at each end of the drum travel.

Magnetic Starters

When used for starting and control of wound-rotor motors, magnetic starters are built in three different forms, depending on the duty of the motor.

1. Plain starting.
2. Speed regulating.
3. Speed setting.

Magnetic starters consist of a magnetic contactor for connecting the stator circuit to the line, and one or more accelerating contactors to commutate the resistance in the rotor circuit. The number of secondary accelerating contactors varies with the rating, a sufficient number being employed to assure smooth acceleration and to keep the inrush current within practical limits. The operation of the accelerating contactors is controlled by a timing device, which provides definite time acceleration. For high-voltage service, the primary contactor is usually of the oil-immersed type. The diagram of a typical magnetic starter for use with a wound-rotor induction motor is shown in Fig. 5-8.

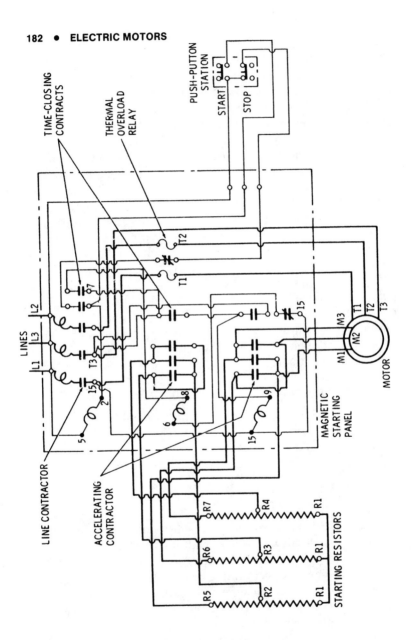

Fig. 5-8. Diagram of a magnetic contactor for use with a wound-rotor induction motor.

Magnetic-starting speed regulators are very similar to magnetic starters except that the secondary circuit is controlled by a series of magnetic contactors operated from a dial switch in the control station or by a faceplate type of rheostat mounted in the controller itself. Where magnetic contactors are used, they also act as accelerating contactors on starting. If a rheostat is used, accelerating contactors on the panel accelerate the motor to the speed for which the rheostat is set.

Controller Resistors

All standard wound-rotor phase circuits, whether for two- or three-phase circuits, have their secondaries wound for three-phase. The controller resistors used for each phase are identical with the exception of the terminal marking. The resistor for the first phase has its terminals marked consecutively (R_1, R_2, R_3, etc.), the second phase R_{11}, R_{12}, R_{13}, etc., the third phase R_{21}, R_{22}, R_{23}, etc., as shown in Fig. 5-9. The actual resistor will consist of one, two, three, or multiples of these frames of tubes or grids.

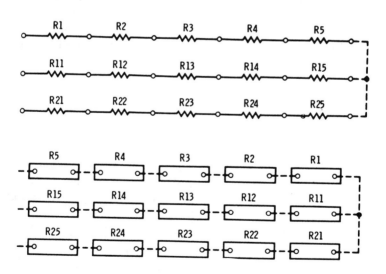

Fig. 5-9. Typical connections of secondary resistor units.

Secondary resistors for wound-rotor induction motors are, as a rule, designed for star connection. Resistors for most manual controllers may be connected with all three secondary phases closed or with one secondary phase open on the first point of the controller. Resistors for magnetic controllers are connected with all three phases closed in the secondary on the first point. The torque obtained with a resistor of a given class number varies with the connection used on the first point of the controller.

The NEMA resistor classifications for wound-rotor induction motors are given in Table 5-1. For example, Class 114 is for starting duty only, and for a motor that will be started and brought up to speed in approximately five seconds with a minimum of 75 seconds in between successive starts.

The capacity of resistors depends largely on the ventilating space. As a rule, the resistor frames should never be stacked more than four in height, and when space is available, each frame should be separated from the next by approximately the width of the end frame. Frames may be mounted on the floor, platform, or wall, but in such a way as to obtain free ventilation.

It must also be emphasized that a close periodic inspection of resistors and associated connections should be made. This inspection should include the tightening of loose lock nuts, connections, etc. The collection of dirt and dust should be blown out from between the resistor units.

Application

Wound-rotor motors have the ability to start extremely heavy loads. Hence they are suitable for:

1. Driving various types of machinery that require development of considerable starting torque to overcome friction.
2. Accelerating extremely heavy loads that have a flywheel effect or inertia.
3. Overcoming back pressures set up by fluids and gases in the case of reciprocating pumps and compressors.

Double squirrel-cage motors are also applicable on many of the heavier machines involving the problems mentioned above. However, if a considerable length of time is required to accelerate the load to full speed, double squirrel-cage rotors may burn out

Table 5-1. NEMA Resistor Classification for Wound-Rotor Motors

Per Cent Full-Load Current on First Point	Starting Torque % of Full Load					Resistor Class Number						
	Series Motors	Compound Motors	Shunt Motors	Wound-Rotor Induction Motors		5 Sec. on Out of 80 Sec.	10 Sec. on Out of 80 Sec.	15 Sec. on Out of 90 Sec.	15 Sec. on Out of 60 Sec.	15 Sec. on Out of 45 Sec.	15 Sec. on Out of 30 Sec.	Cont.
				1 Ph. Stg.	3 Ph. Stg.							
25	8	12	25	15	25	111	131	141	151	161	171	91
30	30	40	50	30	50	112	132	142	152	162	172	92
70	50	60	70	40	70	113	133	143	153	163	173	93
100	100	100	100	55	100	114	134	144	154	164	174	94
150	170	160	150	85	150	115	135	145	155	165	175	95
200	250	230	200		200	116	136	146	156	166	176	96

before full speed is attained. For that reason, wound-rotor motors should be used instead. Where high starting torque alone is involved, double squirrel-cage motors will qualify, but the fact that all the heat developed in the secondary circuit is confined to the rotor prevents their use if the starting period is too long. Frequent starting has the same effect of overheating double squirrel-cage motors, for which reason wound-rotor motors should be used on machinery started frequently.

Operating Speed Variation

Variations in operating speed are essential on many applications. It is often desirable to vary the operating speed of conveyors, compressors, pulverizers, stokers, etc., in order to meet varying production requirements. Wound-rotor motors, because of their adjustable speeds, are ideally suited for such applications. However, if the torque required does not remain constant, the speed of the wound-rotor motor will vary over wide limits, a characteristic constituting one of the serious objections to the use of wound-rotor motors for obtaining reduced speeds. Another factor that must be taken into consideration when wound-rotor motors are selected for reduced-speed operations is that of lowered motor efficiency.

This type of motor is suitable where the speed range required is small, where the speeds desired do not coincide with the synchronous speed of the line frequency, and where the speed must be gradually or frequently changed from one value to another.

The wound-rotor motor also gives high starting torque with a low current demand from the line, but it is not efficient when used a large proportion of the time at reduced speed, since power corresponding to the percent of drop in speed is consumed in the external resistance without doing any special work.

The wound-rotor motor can operate at any speed from its maximum full-load speed down to almost standstill. Speed reduction below one-half is not recommended because of poor speed regulation (the no-load speed is always synchronous speed) and because of increased heating of the motor due to the decreased ventilation.

When wound-rotor motors are used for cranes, hoists, and elevators — machinery operated intermittently and for short periods — poor speed regulation and loss in efficiency are of little consequence. However, if lowered speed is required over longer periods, poor speed regulation and loss in efficiency may become prohibitive.

Smooth Starting

By the use of external resistors in the slip-ring rotor windings, a wide variation in rotor resistance can be obtained with a resultant variation in acceleration characteristics. Thus, a heavy load can be started as slowly as desired, without jerk, and can be accelerated smoothly and uniformly to full speed. It is merely a matter of supplying the necessary auxiliary control equipment to insert sufficient high resistance at the start, and to gradually reduce this resistance as the motor picks up speed.

Low Starting Current

Many power companies have established limitations on the amount of current that motors may draw at starting. The purpose is to reduce voltage fluctuations and prevent flickering of lights. Because of such limitations, the question of starting current is often the deciding factor in the choice of wound-rotor motors instead of squirrel-cage motors.

Wound-rotor motors, with proper starting equipment, develop a starting torque equal to 150 percent of full-load torque with a starting current of approximately 150 percent of full-load current. This compares very favorably with squirrel-cage motors, one type of which requires a starting current of as much as 600 percent of full-load current to develop the same starting torque of 150 percent.

Wound-Rotor Motor Troubleshooting Chart

Symptom and Possible Cause	*Possible Remedy*
Motor Runs at Low Speed with External Resistance Cut Out	
(a) Wires to the control unit too small	(a) Use larger cable to the control unit.
(b) Control unit too far from motor	(b) Bring control unit nearer motor.
(c) Open circuit in rotor circuit (including cable to the control unit)	(c) Test by ringing out circuit and repair.
(d) Dirt between brush and ring	(d) Clean rings and insulation assembly.
(e) Brushes stuck in holders	(e) Use right size brush.
(f) Incorrect brush tension	(f) Check brush tension and correct.
(g) Rough collector rings	(g) File, sand, and polish.
(h) Eccentric rings	(h) Turn down on lathe, or use portable tool to true-up rings without disassembling motor.
(i) Excessive vibration	(i) Balance motor.
(j) Current density of brushes too high (overload)	(j) Reduce load. If brushes have been replaced, make sure they are of the same grade as originally furnished.

Summary

The stator in the wound-rotor motor is identical to the stator in the squirrel-cage motor. The basic difference in the two motors lies in the rotor winding.

In the squirrel-cage motor, the rotor winding is nearly always self-contained; it is not connected either mechanically or electrically to the outside power-supply or control circuit. However, in wound-rotor motors, the rotor winding consists of insulated coils of wire that are not permanently short-circuited, but are connected in regular succession to form a definite polar area having the same number of poles as the stator. The ends of these rotor windings are brought out to collector rings, or slip rings.

External resistance in the rotor circuit reduces the speed at which the rotor will operate with a given load torque. If the rotor resistor is designed for continuous duty, a portion may be permitted to remain in the circuit to obtain reduced-speed operation. Therefore, the motor has a varying speed characteristic — any change in load results in a considerable change in speed.

The wound-rotor motor is often used in cranes, hoists, and elevators. These devices are operated intermittently and for short periods of time, where exact speed regulation and loss in efficiency are of little consequence.

Wound-rotor motors can be used to start extremely heavy loads. Hence, they are suitable for: (1) driving various types of machinery that require development of considerable starting torque to overcome friction; (2) accelerating extremely heavy loads that have a flywheel effect or inertia; and (3) overcoming back pressures set up by fluids and gases, as in reciprocating pumps and compressors.

Review Questions

1. What is the chief difference in the squirrel-cage motor and the wound-rotor motor?
2. How is the wound-rotor motor provided with varying speed characteristic?

3. List five functions of controllers used with wound-rotor motors.

4. List five types of controllers.

5. List three types of equipment that are suitable for wound-rotor motors.

6. Draw the schematic of a wound-rotor motor with resistors in series with the rotors.

CHAPTER 6

DC Motors

Direct-current motors are classified according to the type of winding employed, as:

1. Series-wound.
2. Shunt-wound.
3. Compound-wound.
4. Permanent-magnet (PM).

The series-wound motor (Fig. 6-1) is one in which the field coils and armature are connected in series, and the entire current flows through the field coils. The shunt-wound motor (Fig. 6-2) derives its name from the fact that the field coils and the armature are connected in shunt (parallel), and because of this, the field current is only a small portion of the total or line current.

The compound-wound motor (Fig. 6-3) incorporates both the series-wound and the shunt-wound windings. In other words, it has both the series and the shunt windings previously illustrated.

The PM motor has an armature, brushes, and permanent magnets that replace the previous field coils. There are no field coils in a PM motor.

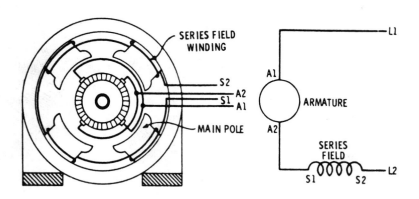

Fig. 6-1. Diagrams of the connections for a series-wound DC motor.

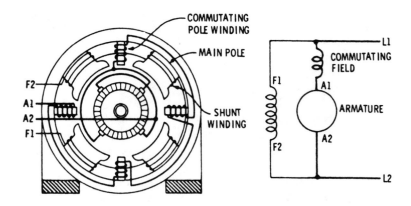

Fig. 6-2. Wiring diagrams of a shunt-wound DC motor with interpoles. The interpoles (commutating poles) are in series with the armature so that their field strength will be proportional to the load on the motor.

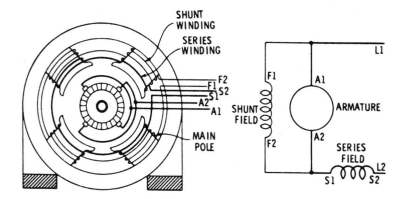

Fig. 6-3. Wiring diagrams of a cumulatively compound-wound DC motor.

Construction

All DC motors, regardless of size, have a stationary field member (usually called a *frame* or *yoke*) and a rotating armature member. The frame, which is made of cast or fabricated steel, serves as a means of support for the motor and forms a part of the magnetic circuit connecting the field poles and commutating poles. The field poles upon which the field coils are wound are made of cast steel, forced steel, or steel laminations. When cast or forged steel is used, the cores are usually made with a circular cross section. Laminated poles (Fig. 6-4), which are most commonly used (except for very small motors), have a rectangular cross section and are fastened to the frame by bolts.

The armature is made of machine-wound coils embedded in parallel slots on the surface of the armature core. The core is made of thin wrought-iron or mild-steel stampings of from 18 to 25 mils in thickness.

The brushes are constructed as a means of carrying the current from the external to the internal circuit. They are usually made of carbon and are carried in brush holders (Fig. 6-5), which are mounted on brush-holder studs or brackets.

The commutator (Fig. 6-6) is built up of segments of hard-drawn copper insulated from the supporting rings by "built-up"

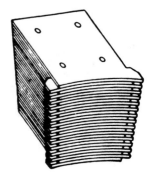

Fig. 6-4. Construction details of a typical laminated field-pole piece.

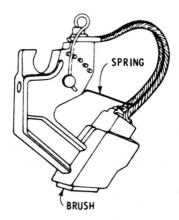

Fig. 6-5. Brush and holder for a DC motor.

mica to form a cylinder. These segments are tightly clamped together by means of a heavy external ring. In order to improve commutation, modern motors are equipped with *auxiliary poles,* variously termed *interpoles* or *commutating poles.* These are small auxiliary poles placed between the regular field poles. Their purpose is to assist commutation and prevent sparking at the brushes for different loads.

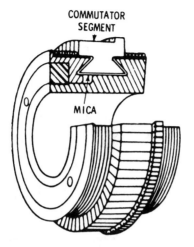

COMMUTATOR
SEGMENT

MICA

Fig. 6-6. Cutaway view of a DC-motor commutator.

Operating Characteristics

Series Motors

The load characteristic curves of a series motor are given in Fig. 6-7. A study of the current-torque curves indicates that the torque varies approximately as the square of the armature current. From this, it follows that this type of motor is suitable in load applications where it is necessary to supply a large torque with a moderate increase in current, such as in traction work, crane operation, etc.

The speed of a series motor, as indicated, varies greatly with the change in load. Because of this speed characteristic, and the resultant possibility of dangerously high speed at light loads, this motor is not suitable for belt drive or for use on any load where the torque might drop below 15 percent of full-load torque.

Shunt Motors

From the characteristic curves of a shunt motor (Fig. 6-8) it will be noticed that this motor will run at very nearly the same speed at any load within its capacity, and will not slow very much even when greatly overloaded. There is but a slight drop in speed

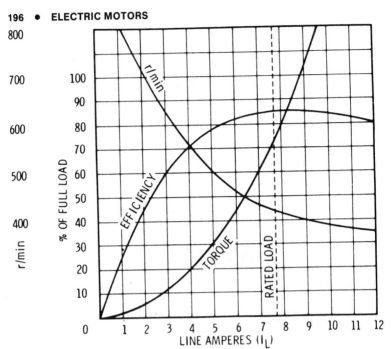

Fig. 6-7. Operating characteristic curves of a series DC motor.

from no load to full load. This drop may vary from 5 to 15 percent of the full-load speed, being dependent on saturation, armature reaction, and brush position.

Because of commutation limitations, shunt motors in integral horsepower sizes are not suitable for across-the-line starting. Shunt motors designed for operation over a given speed range by field control are not technically shunt motors, in that a stabilizing series field is added to assure stable speed under weak field conditions. When this winding is included, the possibility of armature reaction demagnetizing the weakened shunt field with a change in load is eliminated. Shunt motors without speed control are used to drive machinery that is designed to run continuously at a constant speed.

Compound Motors

The addition of a cumulative series-field winding to the shunt field produces a compound motor. The addition of this series-field

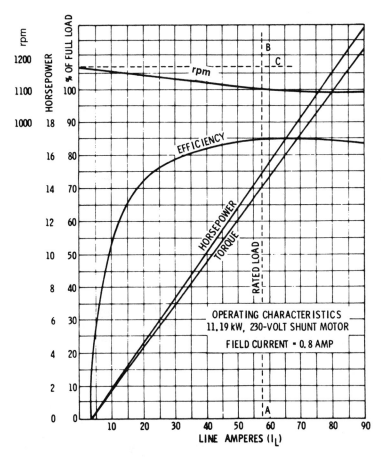

Fig. 6-8. Operating characteristics curves of a typical 15 hp (11.19 kW) 230-volt, shunt-wound DC motor.

winding gives the motor a characteristic that is a combination of the series and shunt motor.

The speed changes with the load, but it does not change as much as in a series motor and it usually changes a great deal more than in a shunt motor. Like the series motor, the compound motor has excellent characteristics for starting heavy loads and yet is in no danger of "running away" at light loads when used without a speed controller.

Compound motors are used for loads requiring high starting torque, or for loads subject to torque pulsations. They are not practical for applications requiring adjustable speed by field control. With a weakened shunt field, the series-field flux becomes a greater portion of the total flux; hence the changes in load may produce unstable speeds.

Compound motors are commonly employed for elevators, air compressors, ice machines, certain kinds of hoisting and conveying machinery, printing presses, paper cutters, pumps, and other machinery where the load fluctuates suddenly or periodically, and where constant speed is not essential. They are usually designed so that there will be a drop in speed of about 20 percent between no load and full load. Compound motors may be wound for greater or less speed change to meet special conditions and are usually wired for across-the-line starting in sizes up to 5 hp (3.73 kW). Comparisons of the speed-torque characteristic curves of series, shunt, and compound DC motors are given in Fig. 6-9.

Permanent-Magnet (PM) Motor

The permanent-magnet motor is another DC type being used in the fractional horsepower sizes to do a number of jobs, such as in power seats, power windows, and windshield wipers for automobiles. It has many other uses that are just being discovered (Fig. 6-10).

The field of the motor is replaced with a couple of ceramic magnets. For a given field strength, the permanent-magnet ring and magnet assembly are considerably smaller in diameter than in the wound-field counterpart, the shunt DC motor. This can make it useful in places where a larger motor would not fit. When weight is a factor, it also has the advantage over the shunt motor. Another plus for this type of motor is that its field is not affected by the armature. There is no armature reaction with the field since the field is distinctly different and does not rely upon the power source to produce its magnetic field.

Armature reaction does weaken the field strength of a regular DC shunt-wound motor. This is particularly noticeable at loads of 200 percent of rated load. This particular characteristic is responsible for the dropoff in torque associated with the shunt motor design. Permanent-magnet motors produce relatively high torques at low speeds. This means they can be used in places where an expensive gearmotor would ordinarily be needed. Gearmotors have inherent mechanical characteristics that cannot be tolerated in some applications (Fig. 6-11). The *backlash* and *windup* of gearing in gearmotors can be detrimental in some applications. This gives the PM motor an advantage since it does not suffer from either of these mechanical conditions.

Keep in mind: Permanent-magnet motors cannot be continuously operated at the high torques they are able to generate. They

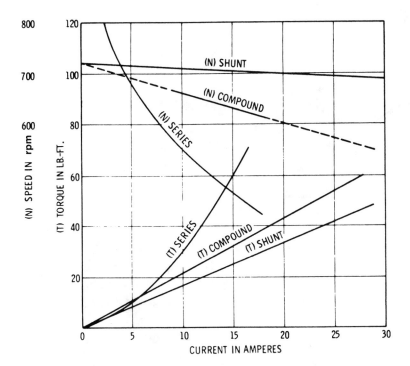

Fig. 6-9. Speed torque characteristic curves for compound, shunt, and series DC motors of equal size.

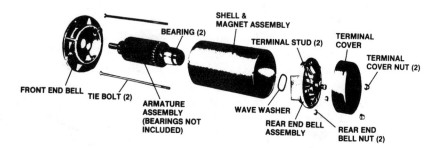

Fig. 6-10. Permanent magnet motor.

Fig. 6-11. Parallel shaft gearmotor. (Courtesy Bodine)

will overheat rapidly. Overheating can destroy the magnetic field of the permanent magnet.

The linear speed and torque of PM motors, coupled with their ability to be easily controlled electronically, make them ideal for multiple speed and servometer uses.

The PM motor can generate very high starting torque. This high starting torque can be a valuable asset in many motor applications. It functions well for actuator drives and other types of intermittent duty applications (Fig. 6-12).

The permanent magnets also provide some self-braking when the power is removed. There is, of course, some shaft coast. They require only two leads instead of four as the shunt motor does. This means reversing is simpler with a switch that will switch the two leads to reverse the polarity. Dynamic braking can be accomplished by simply shunting the two leads once they are removed from the power source.

Disadvantages—While the PM motor has many advantages, it also has some disadvantages that should be considered before it is used as a replacement for a shunt-type motor.

At the lower end of the temperature spectrum, generally below 0°C and lower, the ceramic magnets become increasingly susceptible to demagnetizing forces. Armature reaction, which is capable of producing the threshold limit for demagnetization to occur, takes on greater importance at lower temperature use. Certain improvements in temperature corrections may be available from a particular manufacturer.

At the higher end of the temperature spectrum, as temperature increases, residual or working flux of PM motors decreases at a moderate rate. This flux decrease is much like the decrease of field flux strength in wound-field motors as copper resistance increases with temperature.

Some permanent-magnet motors cannot be disassembled without having a *circuit keeper* inserted to retain the magnetism of the permanent magnets. Although many of the newer ceramic materials permit disassembly without loss of magnetic field strength, the user should inquire before attempting to disassemble the motor.

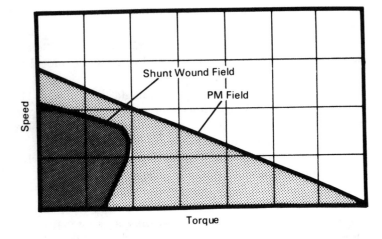

Fig. 6-12A. Comparison of shunt and PM motor curve shapes. *(Courtesy Bodine)*

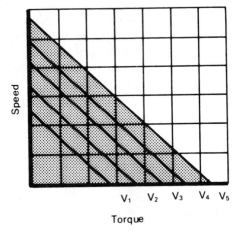

Fig. 6-12B. A typical family of speed torque curves for a PM motor at different voltage inputs, with V5 > V4 > V3 > V2 > V1. *(Courtesy Bodine)*

Cautions—Because of their high starting torque characteristic, care must be exercised in applying permanent-magnet gear-motors. A PM gearmotor application should be carefully checked for any high-inertia loads or high starting-torque loads that could cause it to transmit excessive torque into the gearhead. SCR controls having current-limiting circuits or overload-slip clutches are often used to protect gearing used with PM motors.

SCRs—A silicon controlled rectifier (SCR) is a semiconductor device. The SCR is also called a thyristor. The SCR term was first used by General Electric and it has replaced thyristor in the literature.

The SCR is a device that can have its current flow controlled by very small voltages placed between its cathode and gate terminals. It has an anode, cathode, and gate. The current flow through the SCR is from the cathode to the anode. The current is controlled by a small voltage between cathode and gate. The junction between the semiconductor material making up the anode and the semiconductor comprising the cathode can have its resultant resistance changed by a slight voltage differential between gate and cathode. The low resistance between cathode and anode will allow more current to flow than where there is high resistance between the anode and cathode. The gate to cathode voltage has the effect of lowering or raising the ability of the SCR to conduct current. This ability to increase or decrease current flow directly affects the speed and performance of the motor whose circuit is controlled by the SCR.

Speed Control

The factors that determine the speed of a motor are the voltage across its armature and the strength of the field. If the voltage supplied to the armature is reduced to one-half its original value, the speed of the motor will be reduced in the same proportion. Similarly, if the voltage across the armature is doubled, the speed will be doubled, provided the field strength remains the same. In the same manner, if the magnetic field is weakened, the motor speed will increase, and if the field is strengthened, the motor speed will be reduced. There are four principal methods whereby the speed of a DC motor can be regulated. They are:

1. Shunt-field control.
2. Armature-resistance control.
3. Variable-voltage control.
4. Multivoltage control.

Shunt-Field Control

It has been previously pointed out that weakening the field increases the speed of a motor, and strengthening the field decreases the speed. This method of varying the speed can be accomplished by means of a variable resistance inserted in series with the shunt field, and is the most commonly used means of speed control for DC motors. By this method, the speed of a shunt motor may be varied from normal to four times normal, depending largely upon the mechanical construction of the motor.

It is generally impractical to build adjustable-speed motors for a range exceeding four-to-one, because the shunt field becomes so weakened that the characteristics of the motor assume more the shape of a heavily compounded motor curve than shunt curves. A schematic diagram showing this method of speed control is shown in Fig. 6-13.

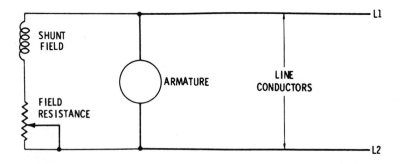

Fig. 6-13. Diagram showing a method of speed control for a DC motor by connecting a variable resistance in series with the shunt field.

The variable field resistance may be a simple field rheostat or it may be combined with a starting rheostat. Shunt-motor speed control by means of variation in field resistance is most satisfactory and efficient. The wattage loss (I^2R) in the field-circuit resistance is small since the field current is small compared to the armature current.

This method of speed control is also used for compound motors, although for any speed setting, there is some variation in the speed with load, in contrast to the shunt motor, where the speed is but slightly affected by the load.

For speed control of series motors, this method is often applied, although in a modified form. Here, it consists in shunting part of the current around the field coils, thus weakening the magnetic field and increasing the speed of the motor.

Armature-Resistance Control

The speed of a DC motor may be decreased by inserting a resistance in series with its armature, as shown in Fig. 6-14. The speed at full load may be decreased to any desired value, depending on the amount of this resistance. With this method, the armature voltage drops as the current passes through the series resistance, and the remaining voltage applied to the armature is lower than the line voltage. Thus, the speed is reduced in direct proportion to this voltage drop at the armature terminals. For example, if at full load an amount of resistance is inserted so that one-half of the voltage is consumed in the resistance, the motor speed will be reduced by one-half.

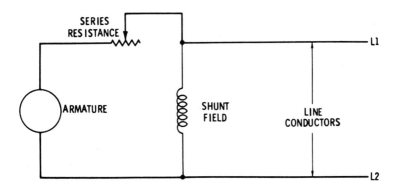

Fig. 6-14. Diagram showing a method of speed control for a DC motor by connecting a variable resistance in series with the armature.

In the case of a 220-volt motor with an armature current of say 50 amperes and a series resistance of 2 ohms inserted in the armature circuit, the voltage drop will be 2 $\times$ 50, or 100 volts. In this case, 120 volts will be applied to the armature and the resulting speed reduction will be 120/220, or 55 percent of normal speed. The wattage loss (I^2R) in this case will be 50^2 $\times$ 2, or 5,000 watts.

If, on the other hand, the armature current were only 25 amperes and all other factors remained the same, then the voltage drop would be 25 $\times$ 2, or 50 volts. The voltage applied to the armature would then be 220 - 50, or 170 volts, and the speed of the motor would be 170/220, or 77 percent of normal. The wattage loss in this latter case would be 1,250.

Thus, it will readily be observed from the examples shown that this method of speed control is inefficient and costly, since a large proportion of the energy to the motor is dissipated in heat. It is sometimes difficult to dissipate this heat, and the regulation is very poor.

Standard speed controllers are commonly made to reduce the speed 50 percent when full-load current is flowing through the armature. If only half the full-load current is flowing through the armature, the speed will be reduced only about 25 percent.

Variable-Voltage Control

This system of speed control requires a generator for each motor, the motor field being maintained constant while its armature voltage is varied by varying the generator field strength. This method gives a large speed range with any desired number of speed points. It is essentially a constant-torque system, because the horsepower delivered by the motor decreases with a decrease in applied voltage and a corresponding decrease in speed.

This particular system has a further advantage that can be used to provide excellent starting characteristics by bringing the generator voltage gradually up from zero, starting and bringing the motor up to speed with a comparatively slowly increasing voltage. Because of the excellent starting characteristics, this system is used largely for modern high-speed elevators, and on account of the combination of excellent starting characteristics and the wide speed range available, it is used to some extent for reversing-planer installations. It is not applied to any great extent, generally on account of the expense of the generating equipment.

Multivoltage Control

The multivoltage system consists of a generating plant or system so arranged that a number of different voltages may be applied to the armature of a motor whose field is connected across a constant voltage, as illustrated in Fig. 6-15. Six different voltages may be obtained with this particular circuit. By means of a suitable controller, any one of the voltages may be supplied to the motor armature terminals, while the shunt field is connected directly to the outside line wires. Since the speed of the motor is directly proportional to the voltage impressed on the armature terminals, a variety of motor speeds may thus be obtained.

Because of the high cost of the power-generating machinery and the current-carrying conductors, however, this type of motor speed control is seldom used. With modern commutating-pole motors, a more economical speed range may be obtained by the common field-resistance control previously described.

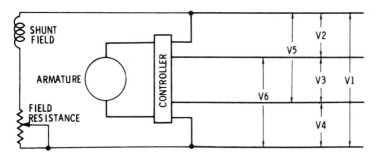

Fig. 6-15. Multivoltage type of speed control for a DC motor.

Ward-Leonard Speed Control

This system of speed control (Fig. 6-16) utilizes a motor-generator set comprised of a DC generator supplying current to a shunt motor connected as indicated. Here, the field circuit of the generator is supplied from the main circuit and has a field rheostat with sufficient resistance to vary the field current from its full value to nearly zero. The drive motor whose speed it is desired to regulate has its armature connected to the variable-voltage generator, while its field current is supplied from the main line. With this arrangement, any speed of the drive motor, from full value to nearly zero, may be obtained by adjustment of the generator field current.

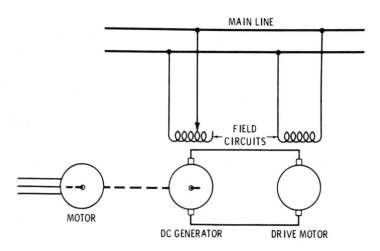

Fig. 6-16. Ward-Leonard method of speed control for a DC motor.

Due to the cost of the machinery involved (three full-sized machines required instead of one), this method is used only when the required range and precision of speed adjustment cannot be obtained by a more economical method. It is used to some extent in the operation of large printing presses, in gun turrets aboard battleships, etc.

Special Control Methods

Another method of motor speed control is obtained by changing the air gap (field flux) between the poles and the motor armature (Fig. 6-17). In some motors of this type, the field cores are geared together at their outer ends so that they may all be moved simultaneously in or out by means of a gear mechanism and handwheel.

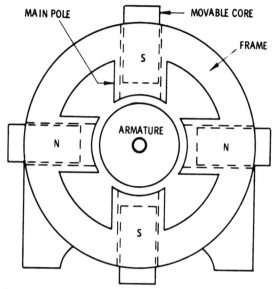

Fig. 6-17. A method of speed control of a DC motor by using adjustable air gaps between the field core and armature.

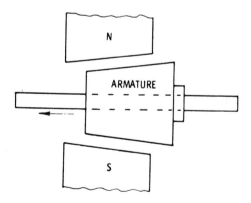

Fig. 6-18. A method of speed control of a DC motor by moving its wedge-shaped armature axially, as indicated.

In others, the armature and the field poles are a slightly tapered (Fig. 6-18), and a gear mechanism shifts the armature along the shaft. In both methods, the speed is varied by adjusting the air gap.

The speed is increased by lengthening the air gap (weakening the field) and is decreased by shortening the gap (strengthening the field). This method (which is but another adaptation of speed control by field manipulation) is seldom used, mainly because of the high construction cost.

Dynamic Braking

With reference to Fig. 6-19, a dynamic braking of a shunt motor is accomplished by inserting a resistance in parallel with the armature, as shown. When dynamic braking is desired, the armature is disconnected and immediately connected to the *off* or *brake* position, placing the resistance bank directly across the armature terminals. The rotation of the armature under the influence of the excited shunt field causes a voltage to be generated, sending current through the resistance. As a result, the motor functions as a loaded generator and develops a retarding torque that rapidly stops the motor.

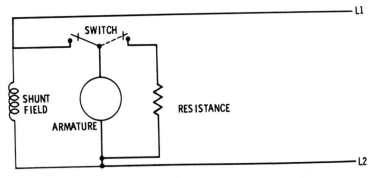

Fig. 6-19. Diagram showing the connections of a dynamic brake to a DC motor.

Starting and Reversing

In order to prevent damage to a motor when starting, it is necessary to use a variable starting resistance (starting rheostat) in series with the armature. The need for such a starting resistance becomes apparent if it is remembered that when a motor whose armature is at rest is connected directly to the line, the initial inrush of current will be limited only by the resistance and self-inductance of the armature and series-field winding (when used). Since these resistances are very small, this current would reach a value several times the full-load current and, as a consequence, the motor would be severely damaged if not completely burned out. Thus, the need for a current-limiting device becomes apparent.

This current-limiting device is known as the starting rheostat, and consists of a number of resistance steps in which a step is cut out each time the current falls to a predetermined value. The starting resistance is generally designed so that the motor will start a full load with a starting current not to exceed 150 percent of the full-load current.

To reverse the direction of rotation in any DC motor, the connections to the armature terminals alone, or the connections to the field winding (or windings) alone, must be reversed. If both (or all) are reversed, the direction remains unchanged. The foregoing will be apparent from a study of the equation for the torque, which is:

$$T_t = K_t \phi I_a$$

where

T_t = torque
K_t = constant
I_a = armature current
ϕ = flux entering the armature from one N pole

A negative torque (which will result in a reversal of direction) will be obtained by changing the polarity of either the current or the flux. If both are changed, the result will be a positive or unchanged torque, since a negative value multiplied by another negative value will give a positive product.

Controllers

DC motor controllers are divided into two principal classes, depending on their method of operation:

1. Manual controllers.
2. Magnetic controllers.

In the former, the acceleration of the motor armature is performed entirely by manipulation of the rheostat arm (or arms) by hand, whereas in the latter, the starter is moved into position by the magnetic action of a coil (or coils), which are usually energized by the closing of a light auxiliary switch, or switches, usually called *starting buttons*.

Manual Controllers

Manual controllers are supplied in three principal forms, termed as:

1. Faceplate controllers.
2. Multiswitch controllers.
3. Drum controllers.

Faceplate Controllers — Faceplate controllers are divided into various classifications, depending on their duty, as follows:

1. For starting duty only.
2. For starting and regulating duty, and for providing the speed increases of from one to four times full-speed value by field control only.
3. For starting and regulating duty providing a minimum of 50-percent speed reduction by armature control only.
4. For starting and regulating duty providing a 50-percent speed reduction by armature control and a 25-percent increase by field control.

From speed regulation by armature control, two classes of service are obtained — constant torque and varying torque. Constant torque applications include machine tools, plunger-type pumps, and similar loads where the horsepower output of the motor decreases directly with the motor speed, while the torque remains constant. Varying-torque applications include blowers, centrifugal pumps, and similar loads where the torque varies approximately as the square of the motor speed.

Operation — Fig. 6-20 shows the wiring diagram of a Class 1 (for starting duty only) faceplate controller. The operation is as follows: connection is made from one of the line wires to a contact arm whose function it is to carry the current over a series of stationary contact buttons provided with graduated resistance steps to the motor winding. When the contact arm is moved toward the

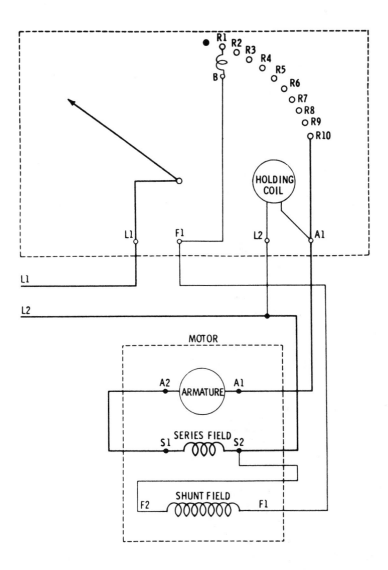

Fig. 6-20. Diagram of the connections between a compound DC motor and a faceplate starter.

right, the first connection serves to put the shunt-field circuit directly across the line. When the contact arm is moved over all the way to the right (which is the running position), it makes contact with a magnet coil (variously termed *holding coil* or *no-voltage release*) whose purpose it is to hold the contact arm in position during normal conditions of the circuit. If the voltage fails, the magnet coil will be de-energized and spring action will carry the contact arm back to its original position. This action prevents the motor from being connected directly across the line when voltage is restored to the line.

Faceplate controllers in Class 2 (designed for both starting and speed-regulating duty, Fig. 6-21) combine the functions of starting by means of armature resistance, and speed regulation by means of field-circuit resistance. Thus, the rheostats have a compound contact arm, one for starting and one for regulating speed. The starting arm is spring-retained. In starting, the starting arm is moved to the running position (by the regulating arm), where it is held by the low-voltage release magnet, while the regulating arm is carried back to the correct point for the desired speed.

Faceplate controllers in Class 3 (designed for speed regulation below normal speed by armature resistance only) are very similar to rheostats of Class 1, except that all the starting buttons or segments must be dimensioned to carry the full-load current continuously. In this type of rheostat, the no-voltage release is arranged to hold the movable contact arm at any desired point. This is accomplished by equipping the contact arm with a pawl, and the magnet operates to hold the contact arm in any operating position.

Class 4 faceplate controllers are designed to reduce the speed of the motor 50 percent, when full-loaded in normal speed, by inserting resistance in the armature circuits, and to increase the speed 25 percent by regulating the resistance in the field circuit. A typical wiring diagram for a controller of this class is shown in Fig. 6-22. These controllers are designed in two types, one having a small number of field-regulating steps, and the other a large number.

Controllers equipped with a relatively small number of buttons (resistance steps) are used for standard constant-speed motors where it is desired to increase the speed by 25 percent by weakening the field current. A large number of buttons (resistance steps)

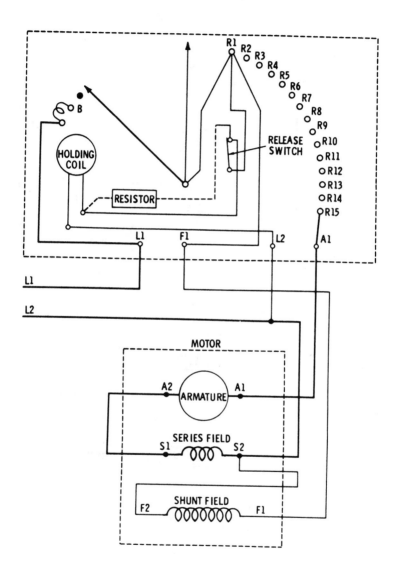

Fig. 6-21. Diagram of the connections between a compound DC motor and a speed-regulating faceplate starter.

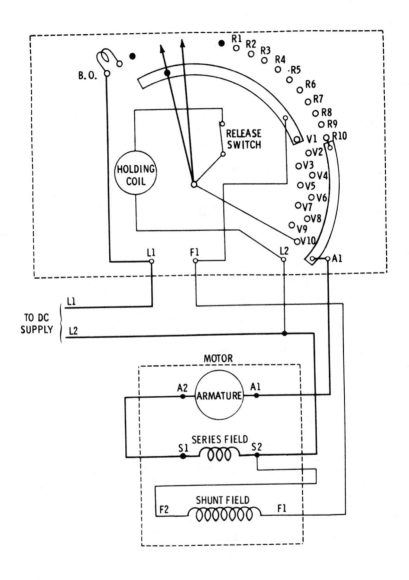

Fig. 6-22. Wiring diagram of the connections between a compound DC motor and a faceplate starter used also for speed control.

signifies starting rheostats where a wide range of speed is desired through field-control regulation.

Multiple Switch Controllers — These are designed with a series of switches that must be closed manually in a left-to-right sequence, and with a hand-over-hand motion. This arrangement provides a certain element of safety because of the time element required, compared to the manual starter where the operator often gives the controller rough treatment by cutting out the resistance too rapidly.

Successive portions of the starting resistance are cut out as each switch is closed. The switches are so mechanically interlocked as to prevent closing in any other but the proper sequence. Each successive switch mechanically locks the preceding one. The last switch is locked in by a catch held by a low-voltage protection magnet. Ratings run to 150 hp (111.9 kW) and are especially adaptable to large-motor manual starting because of the time element introduced by the hand-over-hand closing procedure necessary. Multiple-switch units are for starting duty only, without dynamic braking, and for nonreversing service of either shunt, series, or compound motors.

Drum Controllers — These are used for starting and control of adjustable-speed motors having speed ranges of four-to-one or less. Drum controllers are used for control of both AC and DC motors in various services, such as in the machine-tool industry, hoisting work, for control of trolley-car motors, railroad motors, etc. When used in connection with DC motors, drum controllers are available for both reversing and nonreversing service and for armature-starting, field-regulating applications.

With reference to Fig. 6-23, the drum controller consists essentially of a drum cylinder insulated from its central shaft to which the operating handle is attached. To facilitate the operation, copper segments are attached to the drum. These segments are connected to and insulated from one another, as shown in the diagram of connections.

A series of stationary fingers is arranged to make contact with the segments. These fingers are insulated from one another but interconnected to the starting resistance and the motor circuit. The drum assembly has a notched wheel keyed to the central shaft, the function of which is to indicate to the operator when complete contact at each step is made. When controller is moved forward one notch, the fingers are in position *1*. The current then flows from L_1 through all the series resistance to L_2, and the motor starts rotating. When the handle is moved further, the resistance is gradually being cut out of the armature circuit and inserted into the field circuit. Finally, when the handle is turned to notch *4*, all of the resistance has been transferred from the armature to the field circuit, and the motor is running at full speed.

Drum control units may be obtained with dynamic braking resistors and, in some instances, a magnetic contactor. A protective panel can also be provided to obtain overload and low-voltage protection. Drum controllers are generally provided with the reset contact that requires the drum to be returned to the *off* position before the protective panel can be reset after the overload or voltage failure.

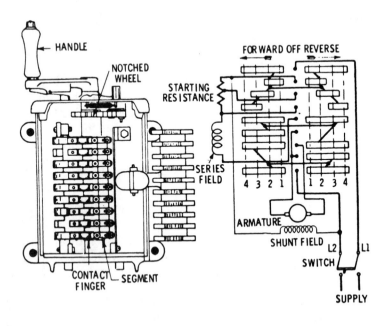

Fig. 6-23. Interior view and wiring diagram of a typical drum controller.

Overload protection is usually obtained by instantaneous thermal relays that have an inverse time characteristic for normal overloads, but that trip instantaneously on abnormal overloads (about four times the normal load or more). Reset can be manual (after returning controller to *off*) or can be made automatic.

The function of the blowout magnet sometimes used on drum controllers is to extinguish the electric arc formed as the contacts open, thus preventing the fingers and corresponding contacts from damage. This extinguishing effect is provided by means of a magnetic blowout coil, which is shown in Fig. 6-24.

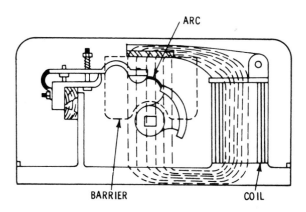

Fig. 6-24. Blowout coil and its magnetic field sometimes used on drum controllers to extinguish the electric arc between the contacts.

An asbestos plate is hinged to the core of the blowout coil and extends the full length of the drum, and from it arc barriers project down between the fingers. The current through the blowout coil sets up a strong magnetic flux, which passes across the contacts. As the contact opens, the current forms an electric arc, which is deflected upward or downward, depending on the direction of the current in the arc. The arc, in deflecting, will strike the asbestos barrier above or below the finger and will draw out to such an extent that it will be dissipated.

In other controllers, the magnet blowout coil is of a long narrow shape surrounding the pole piece and carried just over the line of the contact fingers. In all cases, however, the winding of the blowout coil is connected in series with the armature of the motor, so that whatever current passes through the blowout coil provides the necessary magnetic flux, which is approximately proportional to the amount of current supplied to the motor. The blowout coil is made of copper wire or copper bars, the size and turns being determined by the motor current in each instance.

Magnetic Controllers

Magnetic controllers or starters are classified according to their control features as:

1. Semiautomatic.
2. Automatic.

Semiautomatic control requires manual actuation, such as an operator pushing a button for *start, stop, inch, forward,* or *reverse.* An automatic starter, on the other hand, functions without human actuation and exists wholly due to special conditions and design features. Examples include a motor-operated float switch, a pressure-operated switch, a time-limit switch, etc.

1. Definite time limit.
2. Current limit.
3. Counter-emf.

Definite Time-Limit Starters — Time-limit starters (Fig. 6-25) are simple in construction, accelerate the motor with low current peaks, use less power during acceleration, and always accelerate the motor in the same time regardless of variations in load. Definite time delay may be obtained in one of several ways. A pawl-and-ratchet mechanism attached to the gears of a timing device permits the solenoid to operate at a predetermined moment. An escapement mechanism is another way of obtaining time delay.

Inductive time-limit acceleration makes use of the inductive time-lag effect of the decay of current a heavily inductive circuit. Sometimes called *magnetic timing,* the inductive time-limit contactors may be of the normally open hold-out type, or of the normally closed type. In a similar manner, capacitor time-limit

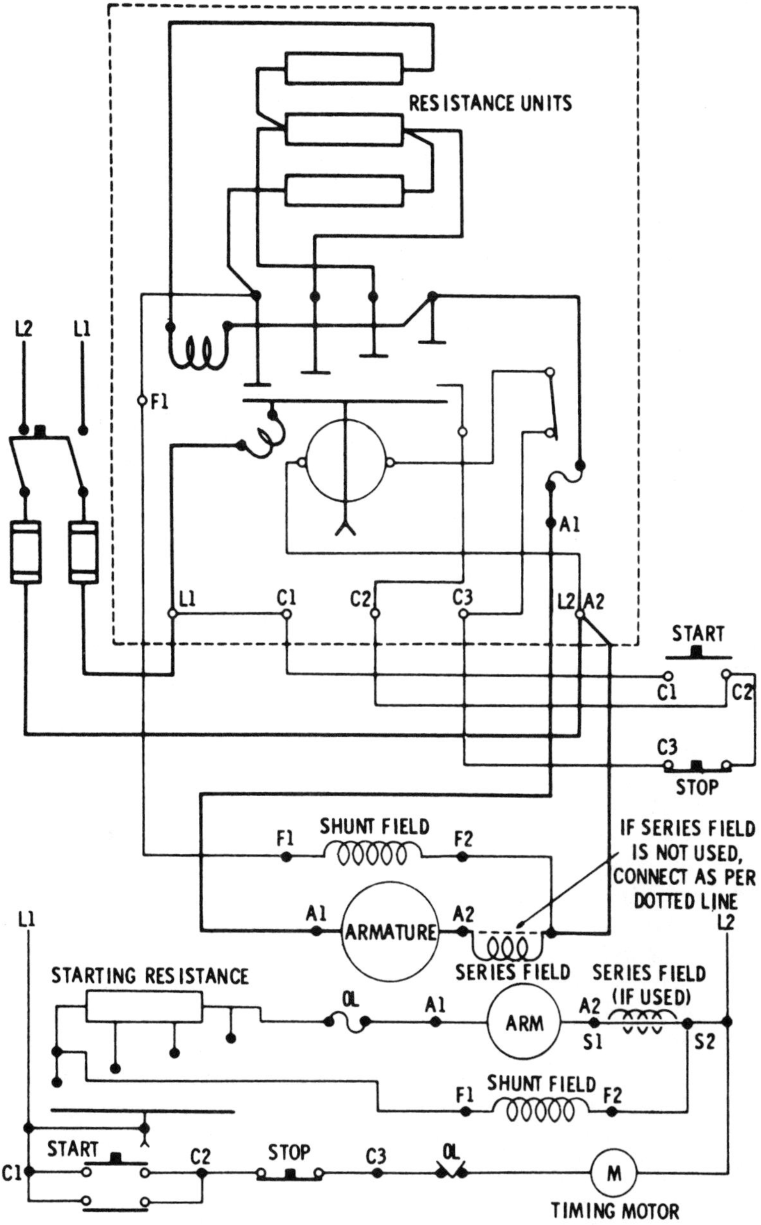

Fig. 6-25. Diagram of a typical DC time-limit acceleration starter.

acceleration makes use of the constant discharge time of a capacitor in the sequence operation of accelerating contactors.

Standard magnetic accelerating equipment is available in ratings up to 150 hp (111.9 kW) for reversing or nonreversing service, and with or without dynamic braking. Braking can be obtained by several methods. Dynamic braking makes use of a discharge resistor placed across the motor terminals after the main contactor has disconnected the motor from the line. Plugging is obtained by reversing the armature connections while running, and accomplishes a very rapid reversal of the armature. Magnetic brakes are used to obtain quick, accurate stopping and to hold the load after stopping. Most brakes are electrically released and spring-set, so that braking will be obtained even though an electrical failure should occur.

In addition to overload and low-voltage protection, other types of protection are often required. Overspeed or underspeed protection can be obtained by means of centrifugal governors or tachometers. Control functions may thus be initiated at any predetermined speed value to accomplish any desired reaction.

Field-failure protection is usually provided to disconnect the motor from the line in the event of an open shunt field in either shunt or compound motors. A relay coil is connected in series with the shunt field to give the signal on an open-field condition.

Field-protective relays are often used to insert resistance in series with the shunt field when the motor is not in operation, thus preventing overheating of the shunt field while the motor is at a standstill. When shunt-field circuits must be opened, field-discharge resistances should be provided to limit induced voltage to a value that will not damage the field-winding insulation when the field circuit is open.

Current-Limit Starters — Acceleration by current limit depends wholly on values of the armature current, and contactor operation may be regulated to any predetermined value. For example, when the *start* button is pressed and the main-line contactor closes with all the starting resistance in the circuit, a certain value of inrush current (say 150 percent) is encountered, which finally decreases to some other value (say 115 percent). At 115 percent, the first accelerating contactor closes, shorting out part of the resistance, and the inrush current immediately increases, only to decrease again as the motor speed picks up. This increases the counter emf, and the second accelerating contactor closes when the current decreases to the preset value, etc.

Current-limit acceleration is entirely dependent on load. Light loads will accelerate rapidly, while heavier loads will require a longer time. For this reason, current-limit starting is not so satisfactory for varying loads. It is entirely possible that, under heavy-load starting, the speed will not pick up to the point where the current will be reduced enough to actuate the succeeding accelerating contactor. Thus, the motor will operate at the point with starting resistance still in the armature circuit. Current-limit starting is most desirable for motors driving high inertia loads. Fig. 6-26 shows the diagram of a typical current-limit acceleration starter.

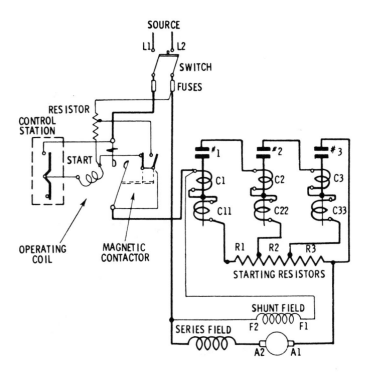

Fig. 6-26. Diagram of a typical DC current-limit acceleration starter.

Counter-emf Starters — Counter-emf acceleration uses a form of current-limit acceleration. The acceleration contactors close in succession as the voltage across the armature increases, until the motor is finally brought up to full speed with the closing of the accelerating contactor. This type of control (Fig. 6-27) is usually limited to motors rated up to 5 hp.

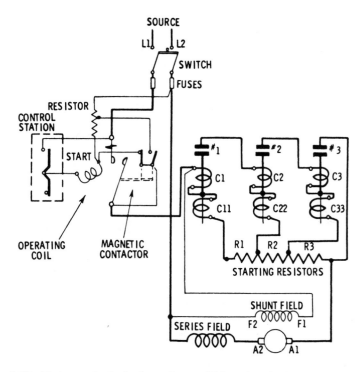

Fig. 6-27. Diagram of a typical counter-emf DC motor starter.

Advantages of Magnetic Controllers — Magnetic controllers are generally used because of their advantage over other forms in that the acceleration (time element) required for the motor starting is removed from the physical effort of the operator. With the face-plate type of controller, for example, the operator may accelerate the motor too rapidly, resulting in the motor taking excessive current. This usually results in burned-out fuses or resistances, or the opening of circuit breakers.

With magnetic controllers, the starting resistance may be cut out at the maximum safe rate by means of magnetically operated switches. The operator's duty is to push a button; the magnetic switch starts the motor and accelerates it by cutting out the starting resistance. In stopping or reversing the motor, pressing the button marked *stop* or *reverse* opens or closes the correct contactor or contactors to perform the desired operation.

Magnetic starting is especially adapted to frequent starting operations — when the motor is located at some distance from the operator, when space does not permit mounting a manual controller near the operator, or when large motors require commutation of heavy circuits.

DC Motor Troubleshooting Chart

Symptom and Possible Cause	*Possible Remedy*
Motor Will Not Start	
(a) Open circuit in controller	(a) Check controller for open starting resistor, open switch, or open fuse.
(b) Low terminal voltage	(b) Check voltage with nameplate rating.
(c) Bearing frozen	(c) Recondition shaft and replace bearing.
(d) Overload	(d) Reduce load or use larger motor.
(e) Excessive friction	(e) Check bearing lubrication to make sure that the oil has been replaced after installing motor. Disconnect motor from driven machine and turn motor by hand to see if trouble is in motor. Strip and reassemble motor; then check part by part for proper location and fit. Straighten or replace bent or sprung shaft (machines under 5 hp [3.73 kW]).

Symptom and Possible Cause	*Possible Remedy*

Motor Stops After Running Short Time

(a) Motor not getting power	(a) Check voltage at the motor terminals; also fuses, coils, and overload relay.
(b) Motor started with weak or no field	(b) If adjustable-speed motor, check rheostat for correct setting. If correct, check condition of rheostat. Check field coils for open winding. Check wiring for loose or broken connection.
(c) Motor torque insufficient to drive load.	(c) Check line voltage with nameplate rating. Use larger motor or one with suitable characteristic to match load.

Motor Runs Too Slow Under Load

(a) Line voltage too low	(a) Check and remove any excess resistance in supply line, connections, or controller.
(b) Brushes ahead of neutral	(b) Set brushes on neutral.
(c) Overload	(c) Check to see that load does not exceed allowable load on motor.

Motor Runs Too Fast Under Load

(a) Weak field	(a) Check for resistance in shunt-field circuits. Check for grounds.
(b) Line voltage too high	(b) Correct high-voltage condition.
(c) Brushes back of neutral	(c) Set brushes on neutral.

Symptom and Possible Cause	*Possible Remedy*

Sparking at Brushes

(a) Commutator in bad condition	(a) Clean and reset brushes.
(b) Commutator eccentric or rough	(b) Grind and true commutator. Undercut mica.
(c) Excessive vibration	(c) Balance armature. Check brushes to make sure they ride freely in the holders.
(d) Broken or sluggish brush-holder spring	(d) Replace spring and adjust pressure to manufacturer's recommendations.
(e) Brushes too short	(e) Replace brushes.
(f) Machine overloaded	(f) Reduce load or install larger motor.
(g) Short circuit in armature	(g) Check commutator and remove any metallic particles between segments. Check for short between adjacent commutator risers. Test for internal armature short and repair.

Brush Chatter or Hissing Noise

(a) Excessive clearance of brush holders	(a) Adjust holders.
(b) Incorrect angle of brushes	(b) Adjust to correct angle.
(c) Incorrect brushes for the service	(c) Get manufacturer's recommendation.
(d) High mica	(d) Undercut mica.
(e) Incorrect brush spring pressure	(e) Adjust to correct value.

Symptom and Possible Cause	*Possible Remedy*

Selective Commutation (one brush takes more load than it should)

(a) Insufficient brush spring pressure	(a) Adjust to correct pressure, making sure brushes ride free in holders.
(b) Unbalanced circuits in armature	(b) Eliminate high resistance in defective joints by inserting armature or equalizer-circuit or commutator risers. Check for poor contacts between bus and bus rings.

Excessive Sparking

(a) Poor brush fit on commutator	(a) Sand-in brushes and polish commutator surface.
(b) Brushes binding in the brush holder	(b) Remove and clean holders and brushes with carbon tetrachloride. Remove any irregularities on surfaces of brush holders or rough spots on the brushes.
(c) Insufficient or excessive pressure on brushes	(c) Adjust brush spring pressure.
(d) Brushes off neutral	(d) Set brushes on neutral.

Sparking at Light Loads

(a) Paint spray, chemical, oil or grease, or other foreign material on commutator	(a) Use motor designed for application. Clean commutator, and provide protection against foreign matter. Install an enclosed motor designed for the application.

Symptom and Possible Cause	*Possible Remedy*

Field Coils Overheat

(a) Short circuit between turns or layers

(a) Replace defective coil.

Commutator Overheats

(a) Brushes off neutral
(b) Excessive spring pressure on brushes

(a) Adjust brushes.
(b) Decrease brush spring pressure but not to the point where sparking is introduced.

Grooving of Commutator

(a) Brushes not properly staggered

(a) Stagger brushes.

Rapid Brush Wear

(a) Rough commutator

(b) Excessive sparking

(a) Resurface commutator and undercut mica.
(b) Make sure brushes are in line with commutating fields.

Armature Overheats

(a) Motor overloaded

(b) Motor installed where ventilation is restricted
(c) Armature winding shorted

(a) Reduce load to correspond to allowable load.
(b) Arrange for free circulation of air around motor.
(c) Check commutator and remove any metallic particles between segments. Test for internal shorts in armature and repair.

Summary

Direct-current motors are classified as: (1) series-wound; (2) shunt-wound; (3) compound-wound; and (4) permanent-magnet (PM). In series-wound motors, the field coils and armature are connected in series, and the entire current flows through the field coils. In the shunt-wound motor, the field coils and the armature are connected in shunt (parallel); therefore, the field current is only a small portion of the total or line current. The compound-wound motor incorporates both the series-wound and the shunt-wound windings. A PM motor substitutes permanent magnets for the field coils. The armature, brushes, and commutator are the same as for other DC motors.

In construction, all DC motors have a stationary field member (frame or yoke) and a rotating armature member. The frame, which is made of cast or fabricated steel, serves as a support for the motor and is a part of the magnetic circuit connecting the field poles and the commutating poles. The armature is made of wire-wound coils embedded in parallel slots on the surface of the armature core.

The series motor is suitable for applications where it is necessary to supply a large torque with a moderate increase in current, such as in traction work, crane operation, etc. The speed varies greatly with a change in load. Since there is a possibility of dangerously high speed at light loads, the series motor is not suitable for belt drive or where the torque may drop below 15 percent of full-load torque.

The shunt motor operates at nearly the same speed at any load within its capacity, and slows only slightly when greatly overloaded. There is only a slight drop in speed from no load to full load. Shunt motors without speed control are used to drive machinery that is designed to run continuously at a constant speed.

A compound motor is produced by adding a cumulative series field winding to the shunt field. The speed changes with the load, but the change is less than in the series motor and much more than in a shunt motor. Thus, heavy loads can be started, as with the series motor; yet, like the shunt motor, the compound motor will not run away with light loads.

Compound motors are commonly used for elevators, air compressors, ice machines, certain types of hoisting and conveying machinery, printing presses, paper cutters, pumps, and other machinery where the load fluctuates suddenly or periodically and where constant speed is not essential.

The permanent-magnet (PM) motor is another type of DC motor. It has high starting torque. The PM motor has no field windings to produce an armature reaction. Permanent-magnet motors produce relatively high torques at low speeds. Permanent-magnet motors cannot be continuously operated at the high torques they are capable of generating because they will overheat.

Permanent-magnet motors provide some self-braking when power is removed. They require only two leads from the power supply.

Disadvantages of the PM motor include some loss of magnetism at lower temperatures when they become increasingly susceptible to demagnetizing forces. Some of the PM motors require a keeper when they are disassembled.

PM motors work well with electronic speed controls.

DC motor controllers are divided into two types: (1) manual; and (2) magnetic. Three types of manual controllers are faceplate, multiswitch, and drum controllers. The magnetic controllers are semi-automatic and automatic.

**Table 6-1. Mechanical and Electrical
Characteristics of DC Motors**

	Machine		
Characteristic	**Series**	**Shunt**	**Compound**
Torque	Very high starting torque	Normal starting torque	High starting torque
Speed	Varying speed	Adjustable speed	Adjustable varying speed
Usual HP Range	Large HP sizes	Large HP sizes	Large HP sizes
Typical Applications	Hose take-up machines, Large factory cranes, Laundry Extractors	Looms, Textile finishing machines, Coke oven door machines	Paint mixers, Elevators, Turntables, Unloaders

(Courtesy Lincoln Electric)

Review Questions

1. What are the four types of DC motors?
2. Describe the types of windings found in the various DC motors.
3. Describe the operating characteristics of the various DC motors.
4. What factors determine the speed of a DC motor?
5. List the uses for the compound motor.

CHAPTER 7

Fractional-Horsepower Motors

Fractional-horsepower motors are manufactured in a large number of types to suit various applications. Because of its use in a great variety of household appliances, the fractional-horsepower motor is perhaps better known than any other type. It is nearly always designed to operate on single-phase AC standard frequencies, and is reliable, easy to repair, and comparatively low in cost.

Single-phase motors were one of the first types developed for use on alternating current. They have been perfected through the years from the original repulsion type into many improved types, such as:

1. Split-phase.
2. Capacitor-start.
3. Permanent-capacitor.
4. Repulsion.
5. Shaded-pole.

6. Universal.
7. Reluctance synchronous.
8. Hysteresis synchronous.

Split-Phase Induction Motors

The split-phase induction motor is one of the most popular of the fractional-horsepower types. It is most commonly used in sizes ranging from $1/30$ hp (24.9W) to $1/2$ hp (373 W) for applications such as fans, business machines, automatic musical instruments, buffing machines, etc. As shown in Fig. 7-1A, the motor consists

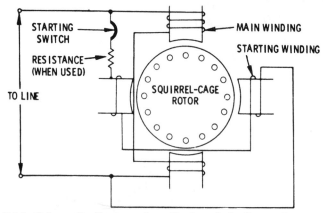

Fig. 7-1A. Schematic diagram of a split-phase induction motor.

Fig. 7-1B. Split-phase induction motor. *(Courtesy Leeson)*

essentially of a squirrel-cage rotor and two stator windings (a main winding and a starting winding). The main winding is connected across the supply line in the usual manner, and has a low resistance and a high inductance. The starting or auxiliary winding, which is physically displaced in the stator from the main winding, has a high resistance and a low inductance. This physical displacement, in addition to the electrical phase displacement produced by the relative electrical resistance values in the two windings, produces a weak rotating field that is sufficient to provide a low starting torque. The characteristics of a resistance-type split-phase motor are shown in Fig. 7-2.

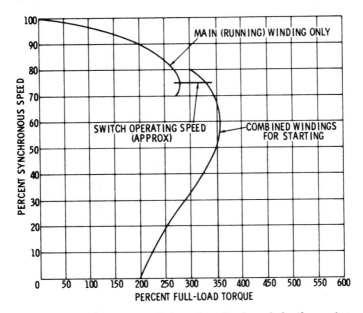

Fig. 7-2. Speed-torque characteristics of a split-phase induction motor.

After the motor has accelerated to 75 or 80 percent of its synchronous speed, a starting switch (usually centrifugally operated) opens its contacts to disconnect the starting winding. The function of the starting switch (after the motor has started) is to prevent the motor from drawing excessive current from the line and to protect the starting winding from damage due to heating. The motor may be started in either direction by reversing the connections to *either* the main or auxiliary winding, *but not to both.*

Starting Switches for AC Motors

Only the AC single-phase motors require centrifugally operated starting switches. Centrifugal force from the motor causes a cone or a slider on the shaft of the motor to be thrown outward from the windings. This movement is harnessed to cause a switch to open. Once the motor is turned off, the switch will close again when the cone or sliding device comes to rest near the windings or closer to the rotor.

Fig. 7-3 shows one way of representing the centrifugally operated switch. Note the four connections to the windings of the motor. Trace the connections from 2 to 3 to see how the switch is placed in the circuit with the starting coils.

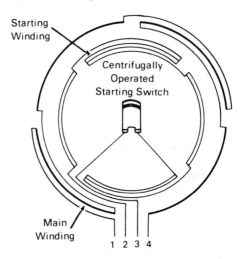

Fig. 7-3. Centrifugally operated starting switch.

Fig. 7-4A shows an older type of rotor with a centrifugal device mounted on the shaft. It throws out the arms when it comes up to speed. These arms operate the pressure-sensitive switch mounted on the frame of the motor. Fig. 7-4B shows the newer type rotor with sliding cone operating switch. Fig. 7-5 shows a different arrangement where springs are used to control the amount of movement in the switch mechanism.

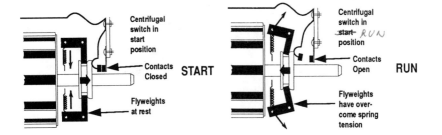

Fig. 7-4A. Older-type rotor with centrifugally-operated switch mechanism on shaft.

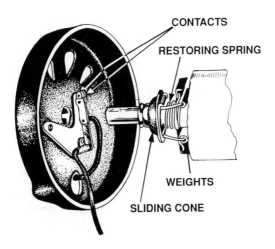

Fig. 7-4B. Newer-type rotor with sliding cone operating switch.

Fig. 7-5. Location of the centrifugal switch mechanism on a more recent motor model.

Fig. 7-6A is a switch operated by the start mechanism shown previously. Note how the spring action of the switch mechanism can cause the start cone on the rotor to snap back in place once the motor has been turned off and has to come to a resting position.

These switches take a variety of shapes. They are designed by the motor manufacturer to fit a particular motor, so every manu-

Fig. 7-6. (A) Another centrifugally operated switch. Contacts are located on the left. The rest of the metal part serves as a spring to force the start mechanism back toward the rotor when the motor stops.

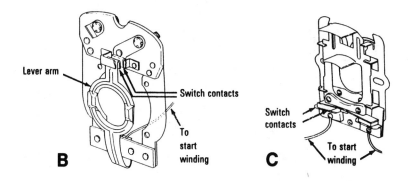

Fig. 7-6. (B) Centrifugal switch unit. (C) Another type of centrifugal switch unit.

facturer has a different design. Fig. 7-7 shows just five of the many shapes used as replacement switches.

Newer Switches — Modern motors used for clothes washers and dryers are designed with the start switch outside of the motor. This way it can be replaced when it shorts or opens with little or no effort on the part of the serviceperson. Fig. 7-8 shows a split-phase motor with the start switch mounted on the outside of the motor frame. The centrifugal force causes the cone to move outward on the shaft. As it moves outward, it puts pressure on a springlike arm that operates the switch located inside the enclosure sitting on top of the motor.

Resistance-Start Motors

A resistance-start motor is a form of split-phase motor having a resistance connected in series with the auxiliary winding. The auxiliary circuit is opened by a starting switch when the motor has attained a predetermined speed.

Reactor-Start Motors

A reactor-start motor is a form of split-phase motor designed for starting with a reactor in series with the main winding. The reactor is short-circuited, or otherwise made ineffective, and the auxiliary (starting) circuit is opened when the motor has attained a predetermined speed. A circuit arrangement for this type of motor is shown in Fig. 7-9. The function of the reactor is to reduce the

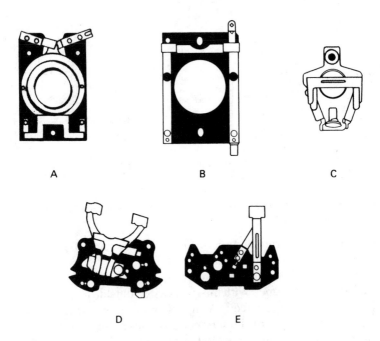

Fig. 7-7. Replacement switches for various makes of general-purpose split-phase and capacitor-start motors.

starting current and to increase the angle of lag of the main-winding current behind the voltage. This motor will develop approximately the same torque as the split-phase motors discussed previously. The centrifugally operated starting switch must be of the single-pole double-throw type for proper functioning.

Capacitor-Start Motors

The capacitor-start motor is another form of split-phase motor having a capacitor connected in series with the auxiliary winding. The auxiliary circuit is opened when the motor has attained a predetermined speed. The circuit in Fig. 7-10 shows the winding arrangement.

The rotor is of the squirrel-cage type, as in other split-phase motors. The main winding is connected directly across the line, while the auxiliary or starting winding is connected through a capacitor, which is connected into the circuit through a centrifugally operated starting switch. The two windings are approximately 90° apart electrically.

This type of motor has certain advantages over the previously described types in that it has a considerable higher starting torque accompanied by a high power factor.

Fig. 7-8. Centrifugally operated switch located on the outside of the motor frame.

Permanent-Capacitor Motors

A permanent-capacitor type of motor has its main winding connected directly to the power supply, and the auxiliary winding connected in series with a capacitor. Both the capacitor and auxiliary winding remain in the circuit while the motor is in operation. There are several types of permanent-capacitor motors, differing from one another mainly in the number and arrangement of capacitors employed. The running characteristics of this type of motor are extremely favorable, and the torque is fixed by the amount of additional capacitance, if any, added to the auxiliary winding during starting.

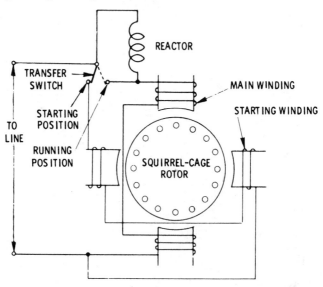

Fig. 7-9. Schematic diagram of a reactor-start motor.

The simplest of this type of motor is the low-torque, permanent-capacitor motor shown in Fig. 7-11. Here, a capacitor is permanently connected in series with the auxiliary winding. This type of motor can be arranged for an adjustable speed by the use of a tapped winding or an autotransformer regulator.

High-torque motors are usually provided with one running and one starting capacitor connected as shown in Fig. 7-12, or with an autotransformer connected to increase the voltage across the capacitor during the starting period, as indicated in Fig. 7-13.

Repulsion Motors

Repulsion motors have a stator winding arranged for connection to the source of power and a rotor winding connected to a commutator. As shown in Fig. 7-14, brushes on the commutator are short-circuited and are so placed that the magnetic axis of the rotor winding is inclined to the magnetic axis of the stator winding. This type of motor has a varying speed characteristic.

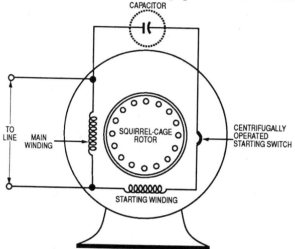

Fig. 7-10. Schematic diagram of a capacitor-start motor.

Principally, it has a stator like that of a single-phase motor, but has a rotor like the armature of a DC motor, with the opposite brushes on the armature short-circuited — that is, connected by a connector with a negligible resistance.

The brushes are placed so that a line connecting them makes a small angle with the neutral axis of the magnetic field of the stator. The stator induces a current in the armature; the current produces an armature field with poles in the neighborhood of the brushes. These armature fields have the same polarity as the adjacent field poles, and are repelled by them so that this repulsion causes the armature to revolve. The motor derives its name from this action.

A repulsion motor of this original type has characteristics (Fig. 7-15) similar to those of the series motor. It has a high starting torque and moderate starting current. It has a low power factor, except at high speeds. For this reason, it is often modified into the *compensated repulsion motor,* which has another set of brushes placed midway between the short-circuited set and connected in series with the stator winding.

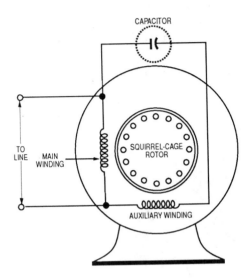

Fig. 7-11. Schematic diagram of a permanent-capacitor split-phase motor.

Repulsion-Start Induction Motors

A repulsion-start induction motor is a single-phase motor having the same windings (Fig. 7-16) as a repulsion motor. At a predetermined speed, however, the rotor winding is short-circuited, or otherwise connected, to give the equivalent of a squirrel-cage winding. This type of motor starts as a repulsion motor, but runs as an induction motor with constant-speed characteristics.

The repulsion-start induction motor has a single-phase distributed-field winding, with the axis of the brushes displaced from the axis of the field winding. The armature has an insulated winding. The current induced in the armature, or rotor, is carried by the brushes and commutator, resulting in a high starting torque. When nearly synchronous speed is attained, the commutator is short-circuited, so that the armature is then similar in function to a squirrel-cage armature.

The photographs in Figs. 7-17 and 7-18 show the working principles of the mechanism for simultaneously lifting the brushes and short-circuiting the commutator to change the operation from repulsion to induction. The object of lifting the brushes is to eliminate wear on the commutator during the running periods, since it makes no difference electrically whether the brushes are in contact or not after the motor comes up to speed.

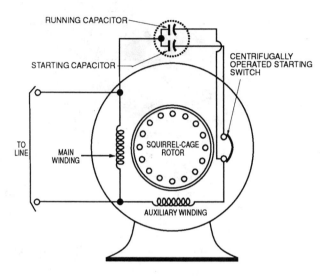

Fig. 7-12. Schematic diagram of a high-torque capacitor motor.

This type of motor has gone through many stages of improvement since its first appearance on the market, although its general principle has remained the same. The general reliability of this type of motor is largely governed by the reliability of the short-circuiting mechanism. For this reason, it has been the constant aim of engineers to improve on the principle and construction of the short-circuiting switch.

Since the motor starts on the repulsion principle, it has the same starting characteristics as the repulsion motor described previously — namely, high starting torque and low starting current. As the motor speeds up, the torque falls off rapidly. At some point on the speed-torque curve, usually at about 80 percent of synchronization, the commutator is automatically short-circuited, producing the effect of a cage winding in the armature, and the motor comes up to speed as an induction motor. After the commutator has been short-circuited, the brushes do not carry current and therefore

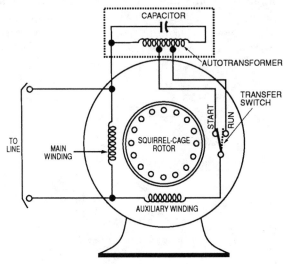

Fig. 7-13. Schematic diagram of a high-torque capacitor motor using an autotransformer.

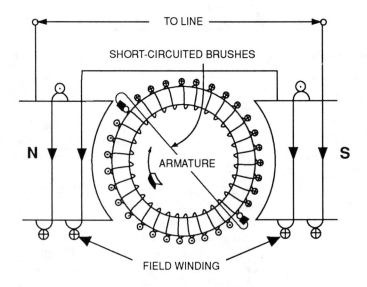

Fig. 7-14. Simplified schematic of a series repulsion motor.

may be lifted from the commutator, but lifting is not absolutely necessary.

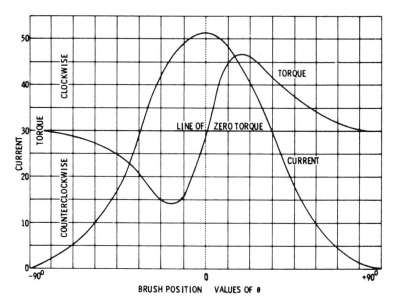

Fig. 7-15. Characteristic curves of a repulsion motor showing the effect of shifting brushes on current and torque.

The curve in Fig. 7-19 shows the speed-torque characteristics of a typical repulsion-start induction motor. The short-circuiting mechanism operates at point A, at which time the induction-motor torque is greater than the repulsion-motor torque. This means that if the repulsion winding has sufficient torque to bring the load up to this speed, there will be sufficient torque as an induction motor to bring the load up to full speed. The higher the speed at which the short-circuiting mechanism operates, the lower will be the induction-motor current at that point, and, consequently, the less disturbance to the line. After the commutator has been short-circuited, the motor has the same characteristics as the single-phase induction motor previously described.

If the short-circuiting mechanism operates before the repulsion curve crosses the induction-motor curve, and if the torque of the induction motor is less than that required to accelerate the load, the motor may slow down until the short circuit is removed from the commutator, in which case the motor will again operate on the repulsion principle. The armature will then speed up until the commutator is again short-circuited, after which the armature will slow down until it again becomes repulsion. This cycle will be repeated over and over again until some changes take place.

The efficiency and the maximum running torque of the repulsion-

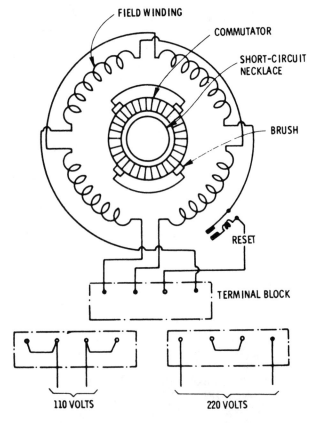

Fig. 7-16. Coil relationship in a single-phase repulsion-start induction motor.

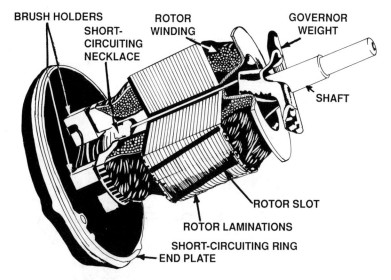

Fig. 7-17. The position of the short-circuiting necklace and brush mechanism while the motor is starting.

start induction motor are usually less than those of a cage-wound induction motor of comparative size. In other words, a repulsion-start induction motor must be larger than a cage-wound motor of the same rating to give the same performance.

Repulsion-Induction Motors

A repulsion-induction motor is another form of repulsion motor that has a squirrel-cage winding in the rotor in addition to the repulsion winding. A motor of this type may have either a constant-speed or a varying-speed characteristic. Specifically, this motor is a combination of the repulsion and induction types and operates on the combined principles of both repulsion and induction. It is sometimes termed a squirrel-cage repulsion motor. In this motor, the desirable starting characteristics of the repulsion motor and the constant-speed characteristic of the induction motor are obtained. It is, of course, impossible to combine the two types of motors and obtain only the desirable characteristics of each.

As shown in Fig. 7-20, the field has the same type of windings used in the repulsion-start induction motor, and the armature has two separate and independent windings. They are:

1. Squirrel-cage winding.
2. Commutated winding.

Both of these armature windings function during the entire period of operation of the motor. There are no automatic devices, such as the starting switch of the split-phase motor or the short-circuited device of the repulsion-start induction motor.

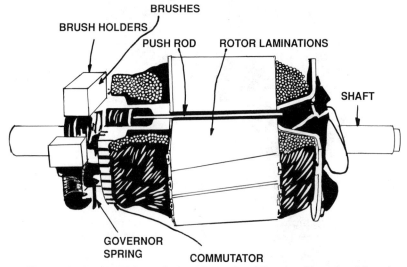

Fig. 7-18. The position of the short-circuiting necklace and brush mechanism after the governor weights have operated.

The cage winding is located in slots below those that contain the commutated winding. The slots that contain the two windings may or may not be connected by a narrow slot. Usually, there are the same number of slots for the two windings. It is not, however, absolutely essential that they be the same. Because of its construction, the squirrel-cage winding inherently has a high inductance. Its reactance, with the armature at rest, is therefore high.

The commutated winding has a low reactance, with the result that most of the current will flow in this winding. The ideal condition at starting would be for all of the flux to pass beneath the commutated winding and none of it beneath the cage winding. If this condition could be obtained, this motor would have the same starting characteristics as the repulsion-start induction motor. However, at full-load speed, which is slightly below synchronization, the reactance of the cage winding is low, and most of the mutual flux passes beneath the case winding. Both windings produce torque, and the output of the motor is the combined output of the cage winding and the commutated winding.

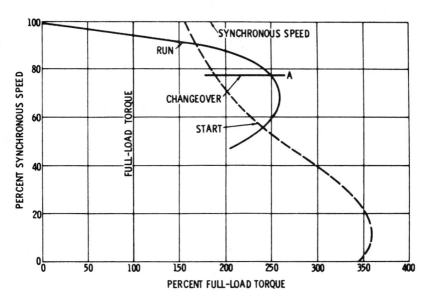

Fig. 7-19. Speed-torque characteristics of a typical repulsion-start induction motor.

The commutation of the motor is good at all speeds. The no-load speed is above synchronization and is limited by the combined effect of the field winding on the commutated winding and the cage winding, and by the action of the two armature windings on each other. At synchronous speed, a squirrel-cage motor has no torque. At synchronous speed, and for a short distance above synchronous speed, the torque of a repulsion-induction motor is greater than that of the commutated winding alone. This shows that because of the interaction between the two armature windings, the squirrel cage supplies torque instead of acting as a brake.

At full-load speed, and up to about the maximum running torque point, the torque of this motor is greater than the sum of the torques of the cage winding and the commutated winding. The inherent locked-torque curve of a repulsion-induction motor, shown in Fig. 7-21, is similar to that of the repulsion motor. At soft neutral, the primary winding carries the squirrel-cage current in addition to the exciting current of the motor.

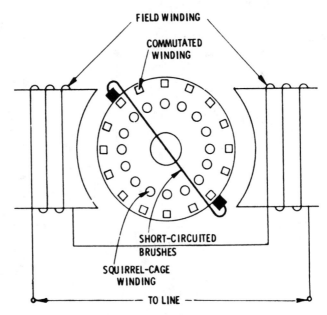

Fig. 7-20A. Schematic diagram of a repulsion-induction motor.

Since its starting current is low, the repulsion-induction motor may be operated from lighting circuits when driving frequently started devices. The repulsion-induction motor is especially suitable for such applications as household refrigerators, water systems, garage air pumps, gasoline pumps, compressors, etc.

Shaded-Pole Motors

A shaded-pole motor is a single-phase induction motor equipped with an auxiliary winding displaced magnetically from, and connected in parallel with, the main winding. This type of motor is manufactured in fractional-horsepower sizes, and is used in a variety of household appliances, such as fans, blowers, hair driers, and other applications requiring a low starting torque. It is oper-

Fig. 7-20B. (Top) Note location of shorting rings. (Bottom) Three-phase wound rotor. Note the three rings.

ated only on alternating current, is usually nonreversible, is low in cost, and is extremely rugged and reliable. The diagram of a typical shaded-pole motor is shown in Fig. 7-22.

The shading coil (from which the motor has derived its name) consists of low-resistance copper links embedded in one side of each stator pole, which are used to provide the necessary starting torque. When the current increases in the main coils, a current is

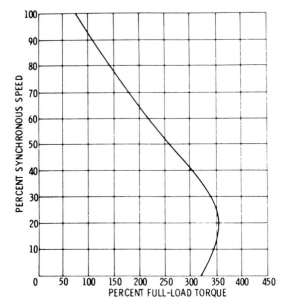

Fig. 7-21. Speed-torque characteristic of a typical repulsion-induction motor.

induced in the shading coil. This current opposes the magnetic field building up in the part of the pole pieces they surround. This produces the condition shown in Fig. 7-23, where the flux is crowded away from that portion of the pole piece surrounded by the shading coil.

When the main-coil current decreases, the current in the shading coils also decreases until the pole pieces are uniformly magnetized. As the main-coil current and the magnetic flux of the pole piece continue to decrease, the current in the shading coils reverses and tends to maintain the flux in part of the pole pieces. When the main-coil current drops to zero, current still flows in the shading coils to give the magnetic effect that causes the coils to produce a rotating or magnetic field that makes the motor self-starting.

Universal Motors

A universal motor is a series-wound or compensated series-wound motor that may be operated either on direct current or on single-phase alternating current at approximately the same speed

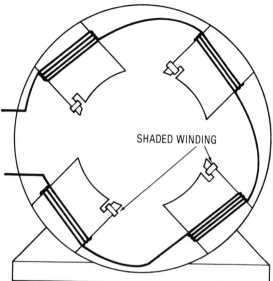

Fig. 7-22. A single-phase motor with shading coils for starting.

and output. These conditions must be met when the DC and AC voltages are approximately the same, and the AC frequency is not greater than 60 hertz.

Universal motors are commonly manufactured in fractional-horsepower sizes, and are preferred because of their use on either AC or DC, particularly in areas where power companies supply both types of current.

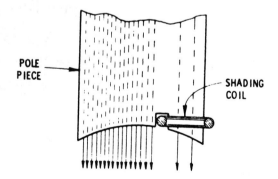

Fig. 7-23. Action of a shading coil in an AC motor.

As previously noted, all universal motors are series wound — therefore, their performance characteristics are very much like those of the usual DC series motor. The no-load speed is quite high, but seldom high enough to damage the motor, as in the case with larger DC series motors. When a load is placed on the motor, the speed decreases and continues to decrease as the load increases. Although universal motors of several types of construction are manufactured, they all have the varying-speed characteristics just mentioned.

Because of the difficulty in obtaining like performance on AC and DC at low speeds, most universal motors are designed for operation at speeds of 3,500 rpm and higher. Motors operating at a load speed of 8,000 to 10,000 rpm are common. Small stationary vacuum cleaners and the larger sizes of portable tools have motors operating at 3,500 to 8,000 rpm.

The speed of a universal motor can be adjusted by connecting a resistance of proper value in series with the motor. The advantage of this characteristic is obvious in an application such as a motor-driven sewing machine, where it is necessary to operate the motor over a wide range of speeds. In such applications, adjustable resistances are used by which the speed is varied at will.

When universal motors are to be used for driving any apparatus, the following characteristics of the motor must be considered:

1. Change in speed with change in load.
2. Change in speed with change in frequency of power supply.
3. Change in speed due to change in applied voltage.

Since most small motors are connected to lighting circuits, where the voltage conditions are not always the best, this last item is of the utmost importance. This condition should also be kept in mind when determining the proper motor to use for any application, regardless of type. In general, the speed of the universal motor varies with the voltage. The starting torque of universal motors is usually much more than required in most applications and not to be considered.

Universal motors are manufactured in two types. They are:

1. Concentrated-pole, noncompensated.
2. Distributed-field, compensated.

Most motors of low-horsepower rating are of the concentrated pole, noncompensated type, while those of higher ratings are of the distributed-field, compensated type. The dividing line is approximately ¼ hp (186.5 W), but the type of motor to be used is determined by the severity of the service and the performance required. All of the motors have wound armatures similar in construction to an ordinary DC motor.

The concentrated-pole, noncompensated motor is exactly the same in construction as a DC motor except that the magnetic path is made up of laminations. The laminated stator is made necessary because the magnetic field is alternating when the motor is operated on alternating current. The stator laminations are punched, with the poles and the yoke in one piece.

The compensated type of motor has stator laminations of the same shape as those in an induction motor. These motors have stator windings in one of two different types.

The noncompensated motor is simpler and less expensive than the compensated motor and would be used over the entire range of ratings if its performance were as good as that of the compensated motor. The noncompensated type is used for the higher speeds and lower horsepower ratings only. Figs. 7-24 and 7-25 show the speed-torque curves for a compensated and noncompensated motor, respectively. It will be noted in Fig. 7-24 that, although the rated speed is relatively low for a universal motor, the speed torque curves for the various frequencies lie very close together up to 50 percent above the rated torque load.

In Fig. 7-25, the performance of a much higher speed, noncompensated motor is shown. For most universal-motor applications, the variation in speed at rated loads, as shown on this curve, is satisfactory. However, the speed curves separate rapidly above full load. If this motor had been designed for a lower speed, the tendency of the speed-torque curves to separate would have been more pronounced. The chief cause of the difficulty in keeping the speeds the same is the reactance voltage that exists when the motors are operated on AC. Most of this reactance voltage is produced in the field windings by the main working field. However, in the noncompensated motor, some of it is produced in the armature winding by the field produced by the armature ampere-turns. The true working voltage is obtained by subtracting the reactance voltage vectorially from the line voltage. If the reactance is high, the performance at a given load will be the same as if there were no reactance voltage and as if the applied voltage had been reduced, with consequent reduction in speed.

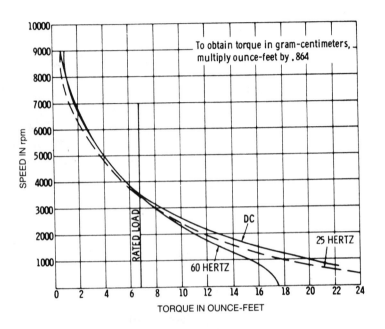

Fig. 7-24. Speed-torque characteristics of a typical ¼ hp (186.5 W), 3,400 rpm compensated universal motor.

Reluctance Synchronous Motors

The reluctance synchronous motor is really a variation on the classic squirrel-cage rotor. The rotor is modified to provide equally spaced areas of high reluctance (Fig. 7-26). Note how the rotor is designed (Fig. 7-27). Notches or flats are placed in the rotor periphery. The number of notches corresponds to the number of poles in the stator winding. Salient poles are the sections of the rotor periphery between the high-reluctance areas. These poles create a relatively low reluctance path for the stator flux. They are attracted to the poles of the stator field. The stator field rotates at the synchronous speed.

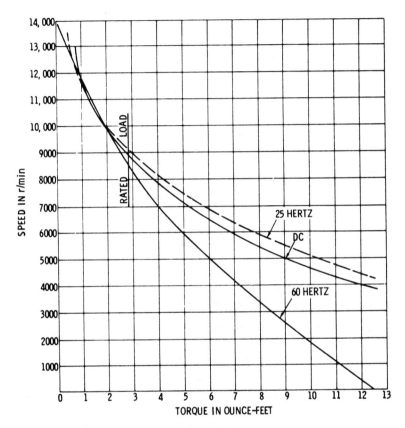

Fig. 7-25. Speed-torque characteristics of a typical ¼ hp (186.5 W), 8,700 rpm, noncompensated universal motor.

The reluctance synchronous rotor starts and accelerates like a regular squirrel-cage rotor. But as this rotor approaches the rotational speed of the field, a critical point is reached. This is where an increased acceleration takes place and the rotor "snaps" into synchronization with the stator field.

A load that is too great will prevent the rotor from pulling in and synchronizing with the rotating field in the stator. Too great a load will cause the motor to operate rough and produce a nonuniform operation.

Fig. 7-26. Reluctance synchronous motor rotor. *(Courtesy Bodine)*

Fig. 7-27. Cutaway view of a reluctance synchronous motor. *(Courtesy Bodine)*

A load on the reluctance synchronous motor produces an effect on the magnetic field. The magnetic field of lines coupling the rotor to the stator field is stretched by the application of a load. This increases the coupling angle, and continues until eventually the coupling between the rotor and stator breaks. This causes the rotor to pull-out of synchronization.

Motor Designs

Reluctance synchronous motors may be designed for operation on three-phase and two-phase power sources. They may also be designed for single-phase operation. In single-phase operation they may take the split-phase, capacitor-start, or permanent-split capacitor configurations. They have the same characteristics as these other nonsynchronous motors.

Note in Fig. 7-26 the skew of a rotor that aids in providing improved smoothness of operation when compared to capacitor-start and split-phase motors.

Hysteresis Synchronous Motors

The hysteresis synchronous motor has a rotor that is made of a heat-treated cast permanent magnet alloy cylinder. It has a non-magnetic support securely mounted to the shaft (Fig. 7-28).

The stator is wound much like that of the conventional squirrel-cage motor. The rotor design gives the hysteresis synchronous motor its characteristics.

This type of motor starts on the hysteresis principle and accelerates at a fairly constant rate until it reaches the synchronous speed of the rotating field. Instead of the permanently fixed poles found in the rotor of the reluctance synchronous design, the hysteresis rotor poles are *induced* by the rotating magnetic field. During the acceleration period, the stator field will rotate at a speed faster than the rotor. The poles that it induces in the rotor will shift around its periphery. When the rotor speed reaches that of the magnetic field, the rotor poles will take a fixed position.

The coupling angle in this motor is not rigid. As the load increases beyond the capacity of the motor, the poles on the periphery of the rotor core will shift as the rotor speed slips below that of the field. This means they take up new positions. If the load is then reduced to the *pull-in* capacity of the motor, the poles will take up fixed positions until the motor is overloaded once again, or stopped and restarted.

The hysteresis motor has the ability to *lock-in* at an infinite number of positions with respect to the stator field. The reluctance rotor locked in only at the same number of positions as it had poles.

One of the advantages of the hysteresis motor is its ability to pull into synchronization any load that is within its capacity to start and accelerate. It will do this with a more gradual characteristic than other motors.

Advantages

Synchronous motors are known for their constant speed characteristic. They will operate at a fixed speed that is determined by the number of stator poles and the frequency of the power supply. The hysteresis synchronous motor has a uniform acceleration characteristic. It can pull into synchronization any load that is within its ability to start and accelerate.

Disadvantages

The reluctance motor requires increased acceleration of the rotor at the critical point when it approaches the rotational speed of the field. That means it is possible that while the reluctance motor may start a high inertia load, it may not be able to accelerate the load enough to pull it into synchronization with the rotating field.

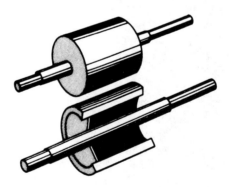

Fig. 7-28. Hysteresis synchronous motor rotor views. Solid and cutaway showing the supported cylinder. *(Courtesy Bodine)*

If the reluctance motor does not synchronize it will continue to operate, but will operate as an ordinary induction motor. This means it will operate at low efficiency and at a very irregular angular velocity. This irregular angular velocity is easily detected by the pounding noise the motor produces. As you can see, it is important that the motor is not overloaded so it can come up to speed and lock-in on the rotating field.

Synchronous motors are usually larger and more costly than nonsynchronous motors. Therefore, they should be chosen for jobs where the load needs to be driven at the exact speed of the rotating field and design speed of the motor.

Fig. 7-29 shows the comparison of typical speed curves for hysteresis and reluctance synchronous motors of identical frame size.

Speed Control

The control method used in single-phase motors depends on the type of motor selected for a certain application. Speed control is usually accomplished by means of a centrifugal switch on motors having speeds above 900 rpm. A comparison of typical speed-torque characteristics of fractional-horsepower, single-phase motors is given in Fig. 7-30.

Where speeds of 900 rpm and below are involved, capacitor motors utilizing voltage relays for changeover are preferable to motors using centrifugal switches, because of the sluggishness of a centrifugal switch of normal design at the lower speeds. A voltage relay for changeover is not subject to the same limitations.

Speed control, because of its more general use, has brought about the standardization of fan speeds, permitting the use of high-slip motors. A squirrel-cage motor with 8- to 10-percent slip and with low maximum torque operates satisfactorily, not only for constant-speed drives, but also for adjustable-speed drives through the use of voltage control on the primary winding. Within the working range of a high-slip squirrel-cage motor having adjustable voltage applied to the primary winding, the speed-torque characteristics resemble the characteristics of a wound-rotor motor having different amounts of external resistance in the secondary circuit.

A tapped autotransformer connected to the motor through a multiposition snap switch offers the simplest form of control. The transformer may have a number of taps, only two or three of which are brought out to the switch. This provides two or three speeds, any or all of which may be changed to suit the needs of the application by the proper selection of taps to connect to the switch. Additional contacts on the switch make it a complete starter and speed regulator. Two such transformers connected in open-delta for three-phase, or in each phase of a two-phase circuit, may be used to control the speed of polyphase motors. For polyphase use, the snap switch must have duplicate contacts for the transformers.

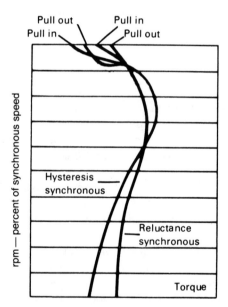

Fig. 7-29. Comparison of typical speed curves for hysteresis and reluctance synchronous motors of identical frame size. *(Courtesy Bodine)*

In applying speed control by means of line-voltage adjustment, it must be remembered that the speed of the drive will vary with the load. Therefore, this method of control is not suitable for centrifugal fans, especially those with damper control which affect the fan load. Two-speed pole-changing motors offer a solution

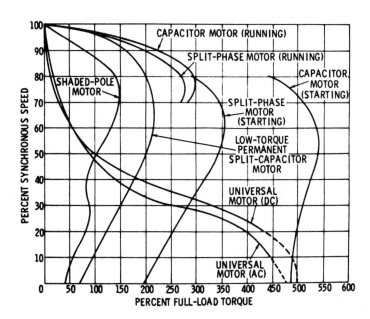

Fig. 7-30. Speed-torque characteristics of common types of fractional-horsepower motors.

for centrifugal-fan drives, inasmuch as the speed on either pole combination is affected only slightly by the change in load. Motor speeds in the ratio of 2 to 1 are obtainable as 3,600/1,800 and 1,800/900 rpm; 3 to 2 as 1,800/1,200 rpm; and 4 to 3 as 1,200/900 rpm. Other pole combinations are possible.

There is still another means of obtaining speed reduction on induction motors. The equivalent of building the transformer into the motor is obtained by a suitable tap on the primary winding, as shown in Fig. 7-31. By this means, normal speed and a single reduced speed are provided. A simple pole-changing controller is all that is required to complete the installation. Such motors are not generally available, but are provided through propeller-fan manufacturers. Each motor is designed for the characteristics of the fan it is to drive, and the tap on the winding is located to give the desired low speed.

When speed controllers are used with low-torque capacitor motors, it must be remembered that the available 50-percent starting torque, with full voltage applied, is reduced in proportion to the square of the voltage involved. The result is low breakaway and low accelerating torque on the low-speed setting. In manually operated control, one remedy for this limitation is a progressive starting switch that provides full voltage on the first (high-speed) position and reduced voltage on the intermediate and low-speed positions (Fig. 7-32).

Where temperature control is used to start and stop the motor, a preset full-voltage starting controller is usually provided. In this case, a relay is energized by the rising voltage of the motor auxiliary winding as it accelerates; the relay disconnects it to the transformer tap for which the controller is set. In high-torque motors, this relay serves to disconnect the starting capacitor and to change over from full to reduced voltage on the motor.

Under the general subject of adjustable-speed fan motors, the question of the amount of speed reduction is often debated, and although it is generally agreed that 30- to 35-percent speed reduction will meet almost any air-conditioning requirement, as much as 50-percent speed reduction is sometimes specified. In most cases, a lower percentage of speed reduction is found to be acceptable. It is fairly easy to obtain 50-percent reduction on a fan if the characteristic curve follows the law of horsepower load, increasing as the cube of the speed, and if the fan fully loads the motor at full speed. It is quite difficult to obtain 50-percent speed reduction with stability in cases where the drive is 20 to 25 percent overmotored.

In considering the wide range of air-conditioning equipment, from unit heaters and coolers to barn ventilators and incubator ventilating fans, with the many types of equipment between these extremes, the foremost impression is one of economy coupled with simplicity and safety. Simplicity promotes economy. Any additional expense imposed in the interest of safety is abundantly justified from the standpoint of good operating practice. It is not advisable to use a motor of the "long annual service" classification where the operation is infrequent, when a motor for short annual service will meet the requirements at a lower first cost with no appreciable increase in operating cost. Similarly, a fractional-horsepower split-phase motor is less expensive than a capacitor motor.

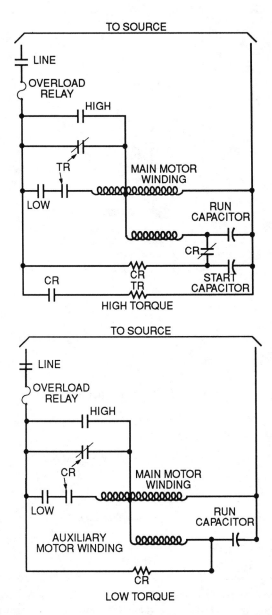

Fig. 7-31. Starting and running circuits of low-torque and high-torque adjustable-speed capacitor motors with tapped main windings.

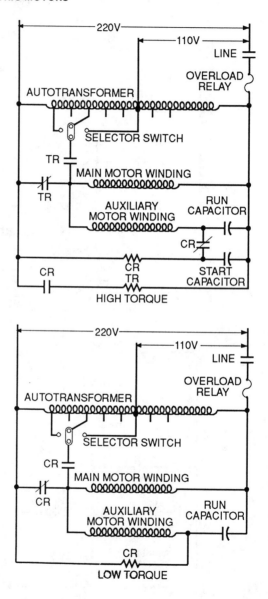

Fig. 7-32. Starting and running circuits of low-torque and high-torque adjustable-speed capacitor motors with manually operated transformer speed regulator.

It is essential to avoid misapplications such as low-torque motors on belted drives, or tapped-winding and transformer speed control in place of multispeed motors for centrifugal drives. The drive should always be closely motored with the correct size and type of motor to give the best results.

Application

Although single-phase motors are most commonly used in fractional-horsepower sizes, certain applications and power-supply conditions make use of single-phase motors in integral horsepower sizes. For extremely small capacities up to $1/40$ hp (18.7 W), the shaded-pole type of motor is most frequently used. It provides sufficient torque for fans, blowers, and other similar equipment, and the starting current is not objectionable on lighting lines.

The split-phase motor is the most commonly used in sizes ranging from $1/30$ hp (24.9 W) to $1/2$ hp (373 W), particularly for fans and similar drives where the starting torque is low. In this type of motor, a built-in centrifugal switch is usually provided to disconnect the starting winding as the motor comes up to speed.

In sizes above $1/4$ hp (186.5 W), a capacitor motor with 300 percent starting torque may be used to advantage, especially for pump and compressor drives. It may also be used in the overlapping ratings for fan drives in the lower capacities at higher speeds. However, it does not offer significant advantages over the split-phase motor to warrant its higher cost.

At low speeds (below 900 rpm, where centrifugal switches are less successful), and in ratings from $3/4$ hp (0.5595 kW) to 3 hp (2.238 kW) at all speeds, the capacitor motor finds its widest field of application in fan drives. The running characteristics of this motor are extremely good, and the starting torque is fixed by the amount of additional capacitance added to the auxiliary motor winding during the starting period.

A low-torque capacitor motor, in which no capacitance is added to the auxiliary winding during the starting period, provides approximately 50-percent starting torque. This is considered sufficient for directly connected fans, if the unit is one of constant speed or is always started on the high-speed position of the starting switch. Where the fan is coupled or belted to the motor, the high-torque type of motor with the additional starting capacitance is preferable. The changeover from start to run of the high-torque type may be accomplished either through the use of a centrifugal switch, or by electrical means responsive to current decay or voltage rise as the motor approaches normal running speed.

The repulsion-induction and the repulsion-start motors have ratings paralleling those of the capacitor types. These motors

Table 7-1. Typical Characteristics of AC Motors.

Motor Types	SPLIT-PHASE General-purpose	SPLIT-PHASE Special Service	CAPACITOR-START Special Service	CAPACITOR START General-purpose	POLYPHASE 1 HP & Below
Starting Torque (% Full Load Torque)	130%	175%	250%	350%	275%
Starting Current	Normal	High	Normal	Normal	Normal
Service Factor (% of Rated Load)	135%	100%	100%	135%	135%
Comparative Price Estimate (Based on 100% for Lowest Cost Motor)	110%	100%	135%	150%	150%
Remarks	Low starting torque. High service factor permits continuous loading—up to 35% over nameplate rating. Ideal for applications of medium starting duty.	Moderate starting torque, but has service factor of 1.0. Apply where load will not exceed nameplate rating for any extended duration of time. Because of higher starting current, use where starting is infrequent.	High starting torque but 1.0 service factor. Use only where load will not exceed nameplate rating for any extended duration of time. Starting current is normal.	Very high starting torque. High service factor permits continuous loading up to 35% over nameplate rating. Ideal for powering devices with heavy loads, such as conveyors.	Normal start current for polyphase is low compared to single-phase motors. High starting ability. High service factor permits continuous loading up to 35% over nameplate rating. Direct companion to general-purpose capacitor-start motor.

Permanent-Split-Capacitor Motors (KCP) are normally designed for direct-drive fans and blower applications. Because of the uniqueness of design and applications, these types of motors are excluded from the above table.

(Courtesy General Electric)

have commutators to provide the starting torque. Under normal running conditions, the commutator of the repulsion-start, induction-run type is short-circuited, and the motor operates as an induction motor. This differs from the repulsion-induction motor in which the commutator is not short-circuited. A squirrel-cage winding deep in the rotor is inactive at starting, but takes up the load as the rotor accelerates to full speed, where the normal load is about equally divided between the repulsion and the squirrel-cage winding. The starting efficiency of both types is high, and the 300 percent or more starting torque that is available makes them suitable for compressor drives.

Table 7-2. Individual Branch Circuit Wiring for Single-Phase Induction Motors

Motor Data		Copper Wire Size (Minimum AWG No.)				
		Branch Circuit Length				
Hp	Volts	0-25 ft.	50 ft.	100 ft.	150 ft.	200 ft.
1/6	115	14	14	14	12	10
	230	14	14	14	14	14
1/4	115	14	14	12	10	8
	230	14	14	14	14	14
1/3	115	14	12	10	8	6
	230	14	14	14	14	12
1/2	115	14	12	10	8	6
	230	14	14	14	14	12
3/4	115	12	10	8	6	4
	230	14	14	14	12	10
1	115	12	10	8	6	4
	230	14	14	14	12	10
1 1/2	115	10	10	6	4	4
	230	14	14	12	10	8
2	115	10	8	6	4	
	230	14	12	12	10	8
3	115	6	6	4		
	230	10	10	10	8	8
5	230	8	8	8	6	4

The universal-type motor, because of its ability to run on DC as well as on AC, is preferred on appliances that are required to operate on either an AC or a DC circuit. Popular applications for the universal-type motor are in vacuum cleaners, sewing machines, portable drills, saws, routers, home motion-picture projectors, and business machines of all kinds.

Whenever it becomes necessary to substitute a motor on any appliance, a good deal of grief and disappointment may be avoided if a motor of the identical size and type is reinstalled. Particular attention should be observed with respect to the nameplate data giving all the necessary information. A sound rule to follow is to copy the entire nameplate reading. This observation holds true whether it is a complete motor or only a spare part is required.

Table 7-1 shows the typical characteristics that relate to the application of motors.

Table 7-2 provides information needed to wire the motor properly so that overheating due to low voltage does not occur.

Fractional-Horsepower Motor Troubleshooting Chart
Split-Phase Induction Motors

Symptom and Possible Cause	Possible Remedy

Failure to Start

(a) No voltage

    (a) Check for voltage at motor terminals with test lamp or voltmeter. Check for blown fuses on meter. Check for blown fuses or open overload device in starter. If motor is equipped with a slow-blow fuse, see that the fuse plug is not open and that it is screwed down tight.

(b) Low voltage

    (b) Measure the voltage at the motor terminals with the switch closed. Voltage should read within 10 per-

Symptom and Possible Cause　　　*Possible Remedy*

cent of the voltage stamped on the motor nameplate. Overload transformers or circuits may cause low voltage. If the former, check with the power company. Overloaded circuits in the building can be found by comparing the voltage at the meter with the voltage at the motor terminals with the switch closed.

(c) Faulty cutout switch operation

(c) Cutout switch operation may be observed by removing the inspection plate in the front end bracket. The mechanism consists of a cutout switch mounted on the front end bracket, and a rotating part called the governor weight assembly, which consists of a Bakelite® disc so supported that it is moved back and forth along the shaft by the operation of the governor weights. At standstill, the disc holds the cutout switch closed. If the disc does not hold the switch closed, the motor cannot start. This may call for adjustment of the end-play washers. Dirty contact points may also keep the motor from starting. See that the contacts are clean. After the motor has accelerated to a predetermined speed, the disc is withdrawn from the switch, allowing it to open.

Symptom and Possible Cause   *Possible Remedy*

With the load disconnected from the motor, close the starting switch. If the motor does not start, start it by hand and observe the operation of the governor as the motor speeds up, and also when the switch has been opened and the motor slows down. If the governor fails to operate, the governor weights may have become clogged. If it operates too soon or too late, the spring is too weak or too strong. Remove motor to service shop for adjustment. Governor weights are set to operate at about 75 percent of synchronous speed. Place rotor in balancing machine and, with a tachometer, determine if the governor operates at the correct speed.

(d) Open overload device

(d) If the motor is equipped with a built-in micro switch, or similar overload device, remove the cover plate in the end bracket on which the switch is mounted and see if the switch contacts are closed. Do not attempt to adjust this switch or to test its operation with a match. Doing so may destroy it. If the switch is per-

Symptom and Possible Cause	*Possible Remedy*
	manently open, remove the motor to the service shop for repairs.
(e) Grounded field	(e) If the motor overheats, produces shock when touched, or if idle watts are excessive, test for a field ground with a test lamp across the field leads and frame. If grounded, remove the motor to the service shop for repairs.
(f) Open-circuited field	(f) These motors have a main and a phase (starting) winding. Apply current to each winding separately with a test lamp. Do not leave the windings connected too long while rotor is stationary. If either winding is open, remove the motor to the repair shop for repairs.
(g) Short-circuited field	(g) If the motor draws excessive watts, and, at the same time lacks torque, overheats, or hums, a shorted field is indicated. Remove to the service shop for repairs.
(h) Incorrect end play	(h) Certain types of motors have steel-enclosed cork washers at each end to cushion the end thrust. Too great an end thrust, hammering on the shaft, or ex-

Symptom and Possible Cause   *Possible Remedy*

cessive heat may destroy the cork washers and interfere with the operation of the cutout switch mechanism. If necessary, install new end-thrust cushion bumper assemblies. End play should not exceed 0.01 in. (0.254 mm); if it does, install additional steel end-play washers. End play should be adjusted so that the cutout switch is closed at standstill and open when the motor is operating.

(i) Excessive load

(i) This may be approximately determined by checking the ampere input with the nameplate marking. Excessive load may prevent the motor from accelerating to the speed at which the governor acts, and cause the phase winding to burn up.

(j) Tight bearings

(j) Test by turning armature by hand. If adding oil does not help, bearings must be replaced.

Motor Overheats

(a) Grounded field

(a) Test for a field ground with a test lamp between the field and motor frame. If grounded, remove the motor to the service shop for repair.

Symptom and Possible Cause	*Possible Remedy*
(b) Short-circuited field	(b) Test for excessive current draw, lack of torque, and presence of hum. Any of these symptoms indicates a shorted field. Remove the motor to the service shop for repair.
(c) Tight bearings	(c) Test by turning armature by hand. If oiling does not help, new bearings must be installed.
(d) Low voltage	(d) Measure voltage at motor terminals with switch closed. Voltage should be within 10 percent of nameplate voltage. Overloaded transformers or power circuits may cause low voltage. Check with power company. Overloaded building circuits can be found by comparing the voltage at the meter with the voltage at the motor terminals with the switch closed.
(e) Faulty cutout switch	(e) See Paragraph (c) under *Failure to Start.*
(f) Excessive load	(f) See Paragraph (i) under *Failure to Start.*

Motor Does Not Come Up to Speed

Some possible causes and possible remedies as under *Motor Overheats.*

Symptom and Possible Cause	*Possible Remedy*

Excessive Bearing Wear

(a) Belt too tight	(a) Adjust belt to tension recommended by manufacturer.
(b) Pulleys out of alignment	(b) Align pulleys correctly.
(c) Dirty, incorrect, or insufficient oil	(c) Use type of oil recommended by manufacturer.
(d) Dirty bearings	(d) Clean thoroughly. Replace worn bearings.

Excessive Noise

(a) Worn bearings	(a) See paragraphs (a), (b), (c), and (d) under *Excessive Bearing Wear.*
(b) Excessive end play	(b) If necessary, add additional end-play washers.
(c) Loose parts	(c) Check for loose hold-down bolts, loose pulleys, etc.
(d) Misalignment	(d) Align pulleys correctly.
(e) Worn belts	(e) Replace belts.
(f) Bent shaft	(f) Straighten shaft, or replace armature or motor.
(g) Unbalanced rotor	(g) Balance rotor.
(h) Burrs on shaft	(h) Remove burrs.

Motor Produces Shock

(a) Grounded field	(a) See paragraph (e) under *Failure to Start.*
(b) Broken ground strap	(b) Replace ground strap.
(c) Poor ground connection	(c) Inspect and repair ground connection.

Rotor Rubs Stator

(a) Dirt in motor	(a) Thoroughly clean motor.
(b) Burrs on rotor or stator	(b) Remove burrs.

Symptom and Possible Cause	*Possible Remedy*
(c) Worn bearings	(c) Replace bearings and inspect shaft for scoring.
(d) Bent shaft	(d) Repair and replace shaft or rotor.

Radio Interference

(a) Poor ground connection	(a) Check and repair any defective grounds.
(b) Loose contacts or connections	(b) Check and repair any loose contacts on switches or fuses, and loose connections on terminals.

Repulsion Start Induction Brush-Lifting Motors

Failure to Start

(a) Fuses blown	(a) Check capacity of fuses. They should not be greater in ampere capacity than recommended by the manufacturer, and in no case smaller than the full-load ampere rating of the motor, and with a voltage capacity equal to or greater than the voltage of the supply circuit.
(b) No voltage or low voltage	(b) Measure voltage at motor terminals with switch closed. See that it is within 10 percent of the voltage stamped on the nameplate of the motor.

Symptom and Possible Cause	*Possible Remedy*
(c) Open-circuited field or armature	(c) Indicated by excessive sparking in starting, refusal to start at certain positions of the rotor, or by a humming sound when the switch is closed. Examine for broken wires, loose connections, or burned segments on the commutator at the point of loose or broken connections. Inspect the commutator for a foreign metallic substance that might cause a short between the commutator segments.
(d) Incorrect voltage or frequency	(d) Requires new motor built for operation on local power supply. DC motors will not operate on AC circuit, or vice versa.
(e) Worn or sticking brushes	(e) When brushes are not making proper contact with the commutator, the motor will have a weak starting torque. This can be caused by the worn brushes, brushes sticking in holders, weak brush springs, or a dirty commutator. The commutator should be polished with fine sandpaper (never use emery). The commutator should never be oiled or greased.

Symptom and Possible Cause	*Possible Remedy*
(f) Improper brush setting	(f) Unless a new armature has been installed, the brush holder or rocker arm should be opposite the index and locked in position. If a new armature has been installed, the position may be slightly off the original marking.
(g) Improper line connection	(g) See that the connections are made according to the connection diagram sent with the motor. The motor may, through error, be wired for a higher voltage.
(h) Excessive load	(h) If the motor starts with no load, and if all the foregoing conditions are satisfactory, then failure to start is most likely due to an excessive load.
(i) Shorted field	(i) Take separate current readings on each of the two halves of the stator winding. Unequal readings indicate a short. Shorted coil may also feel much hotter than the normal coil. An increase in hum may also be caused by a shorted winding.
(j) Shorted rotor	(j) Remove the brushes from the commutator and impress full voltage on the stator. If there is one or more points at which the rotor

Symptom and Possible Cause	*Possible Remedy*
	"hangs" (fails to revolve easily when turned), the rotor is shorted. Forcing the rotor to the position where it is most difficult to hold will cause the shorted coil to become hot. Do not hold in position too long or the coil will burn out.

Motor Operates Without Lifting Brushes

(a) Dirty commutator	(a) Clean with fine sandpaper. Do not use emery.
(b) Governor mechanism or brushes sticking, or brushes worn too short for good contact	(b) See that brushes move freely in slots and that governor mechanism operates freely by hand. Replace worn brushes.
(c) Frequency of supply circuit incorrect	(c) Run motor idle. After brushes throw off, speed should be slightly in excess of full-load speed shown on nameplate. An idle speed varying more than 10 percent from nameplate speed indicates that motor is being used on an incorrect supply frequency. A different motor will be required.
(d) Low voltage	(d) See that voltage is within 10 percent of nameplate voltage with the switch closed.
(e) Line connection improperly or poorly made	(e) See that contacts are good and that connections correspond with diagram sent with motor.

Symptom and Possible Cause	*Possible Remedy*
(f) Incorrect brush setting	(f) Check to see that rocker-arm setting corresponds with index mark.
(g) Incorrect adjustment of governor	(g) The governor should operate and lift brushes at approximately 75 percent of speed stamped on nameplate. Below 65 percent or over 85 percent indicates incorrect spring tension.
(h) Excessive load	(h) An excessive load may be started but not carried to and held at full-load speed, which is beyond where the brushes lift. Tight motor bearings may contribute to overload. This is sometimes indicated by brushes lifting and returning to the commutator.
(i) Shorted field	(i) See Paragraph (i) under *Failure to Start*.

Excessive Bearing Wear

(a) Belt too tight, or an unbalanced line coupling	(a) Correct the mechanical condition.
(b) Improper, dirty, or insufficient oil	(b) The lubrication system of most small motors provides for supplying the right amount of filtered oil to the bearings. It is necessary only for the user to keep the wool yarn saturated with a good grade of machine oil.

Symptom and Possible Cause   *Possible Remedy*

Failure to Start

(c) Dirty bearings

(c) When bearings become clogged with dirt, the motor may need protection from excessive dust. The application may be such that a specially constructed motor should be used.

Motor Runs Hot

(a) Bearing trouble

(a) See Paragraphs (a), (b), and (c) under *Excessive Bearing Wear.*

(b) Short-circuited coils in stator

(b) Make separate wattmeter reading on each of the two halves of the stator winding. Sometimes the shorted coil may be located by the fact that one coil feels much hotter than the other. An increase over normal in the magnetic noise (hum) may also indicate a shorted stator.

(c) Rotor rubbing stator

(c) Extraneous matter may be between the rotor and the stator, or the bearings may be badly worn.

(d) Excessive load

(d) Be sure proper pulleys are on the motor and the machine. Driving the load at higher speeds requires more horsepower. Take an ammeter reading. If current draw exceeds the nameplate amperes for full load, the answer is evident.

Symptom and Possible Cause	*Possible Remedy*
(e) Low voltage	(e) Measure the voltage at the motor terminals with the switch closed. The reading should not vary more than 10 percent from the value stamped on the nameplate.
(f) High voltage	(f) See (e) above.
(g) Incorrect line connection to the motor	(g) Check the connection diagram sent with the motor.

Motor Burns Out

(a) Frozen bearing	(a) See Paragraphs (a), (b), and (c) under *Excessive Bearing Wear.*
(b) Some condition of prolonged excessive overload	(b) Before replacing the burned-out motor, locate and remove the cause of the overload. Certain jobs that present a heavy load will, under unusual conditions of operation, apply prolonged overloads that may destroy a motor and be difficult to locate unless examined carefully. On jobs where it is assumed somewhat intermittent service will normally prevail, and that consequently are closely monitored, the load cycle should be especially checked as a change in this feature will easily produce excessive overload on the motor.

Symptom and Possible Cause   *Possible Remedy*

Motor Is Noisy

(a) Unbalanced rotor

(a) When transportation handling has been so rough as to damage the heavy shipping case, it is well to test the motor for unbalanced conditions at once. It is even possible (though it rarely happens) that a shaft may be bent. In any event, the rotor should be rebalanced dynamically.

(b) Worn bearings

(b) See Paragraphs (a), (b), and (c) under *Excessive Bearing Wear.*

(c) Rough commutator, or brushes not seating properly

(c) Noise from this cause occurs only during the starting period, but conditions should be corrected to avoid consequent trouble.

(d) Excessive end play

(d) Proper end play is as follows: $\frac{1}{3}$ (248.7 W) and smaller — 0.127 mm to 0.762 mm; $\frac{1}{2}$ (373 W) to 1 hp (0.746 kW) — 0.254 mm to 1.905 mm. Washers supplied by the factory should be used. Be sure to tell factory all figures involved. Remember that too little end play is as bad as too much.

(e) Motor not properly aligned with the driven machine

(e) Correct the mechanical condition.

(f) Motor not firmly fastened to mounting base

(f) All small motors have steel bases so they can be firmly

Symptom and Possible Cause   *Possible Remedy*

	bolted to their mounting without fear of breaking. It is, of course, not to be expected that the base should be strained out of shape in order to make up for roughness in the mounting base.
(g) Loose accessories in motor	(g) Such parts as oil covers, guards (if any), end plates, etc., should be checked, especially if they have been removed for inspection of any sort. The conduit box should be tightened when the top is fitted after the connections are made.
(h) Air gap not uniform	(h) This results from a bent shaft or an unbalanced rotor. See Paragraph (a).
(i) Amplified motor noises	(i) When this condition is suspected, set the motor on a firm floor. If the motor is now quiet, then the mounting is acting as an amplifier to bring about certain noises in the motor. Frequently the correction of slight details in the mounting will eliminate this, but rubber mounts are the surest cure.

Excessive Brush Wear

(a) Dirty commutator   (a) Clean with fine sandpaper (never use emery).

Symptom and Possible Cause	Possible Remedy
(b) Poor contact with commutator	(b) See that the brushes are long enough to reach the commutator, that they move freely in the slots, and that the spring tension gives firm but not excessive pressure.
(c) Excessive load	(c) If the brush wear is due to overload, it can usually be checked by noting the time required for the brushes to lift from the commutator. The proper time is less than 10 seconds.
(d) Failure to lift promptly and stay off during the running period	(d) Examine for conditions listed under *Motor Operates Without Lifting Brushes.*
(e) High mica	(e) Examination will show this condition. Take a very light cut off the commutator face and polish with fine sandpaper. Undercut the mica.
(f) Rough commutator	(f) True up on lathe.

Brush-Holder or Rocker-Arm Wear

(a) Failure to lift properly and stay off during the running period	(a) No noticeable wear of this part should occur during the life of the motor. Troublesome wear indicates faulty operation. See under *Motor Operates Without Lifting Brushes.*

Symptom and Possible Cause   *Possible Remedy*

Radio Interference

(a) Faulty ground

(a) Check for poor ground connections, and repair. Static electricity generated by the belts may cause radio noises if the motor frame is not thoroughly grounded. Check for loose connections or contacts in the switch, fuses, or starter.

Capacitor-Start Induction Motors

Failure to Start

(a) Blown fuses or overload device tripped

(a) Examine motor bearings. Be sure that they are in good condition and properly lubricated. Be sure the motor and driven machine both turn freely. Check the circuit voltage at the motor terminals against the voltage stamped on the motor nameplate. Examine the overload protection of the motor. Overload relays operating on either magnetic or thermal principles (or a combination of the two) offer adequate protection to the motor. Ordinary fuses of sufficient size to permit the motor to start do not protect against burnout.

Symptom and Possible Cause	*Possible Remedy*
	A combination fuse and thermal relay, such as *Buss Fusetron,* protects the motor and is inexpensive. If the motor does not have overload protection, the fuses should be replaced with overload relays or *Buss Fusetrons.* After installing suitable fuses and resetting the overload relays, allow the machine to go through its operating cycle. If the protective devices again operate, check the load. If the motor is excessively overloaded, take the matter up with the manufacturer.
(b) No voltage or low voltage	(b) Measure the voltage at the motor terminals with the switch closed. See that it is within 10 percent of the voltage stamped on the motor nameplate.
(c) Open-circuited field	(c) Indicated by a humming sound when the switch is closed. Examine for broken wires and connections.
(d) Incorrect voltage or frequency	(d) Requires motor built for operation on power supply available. AC motors will not operate on DC circuit, or vice versa.

Symptom and Possible Cause	*Possible Remedy*
(e) Cutout switch faulty	(e) The operation of the cutout switch may be observed by removing the inspection plate in the end bracket. If the governor disc does not hold the switch closed, the motor cannot start. This may call for additional end-play washers between the shaft shoulder and the bearing. Dirty or corroded contact points may also keep the motor from starting. See that the contacts are clean. With the load disconnected from the motor, close the starting switch. If the motor does not start, start it by hand and listen for the characteristic click of the governor as the motor speeds up and also when the switch has been opened and the motor slows down. Absence of this click may indicate that the governor weights have become clogged, or that the spring is too strong. Continued operation under this condition may cause the phase winding to burn up. Remove the motor to the service shop for adjustment.

Symptom and Possible Cause	*Possible Remedy*
(f) Open field	(f) These motors have a main and phase winding in the stator. With the leads disconnected from the capacitor, apply current to the motor. If the main winding is all right, the motor will hum. If the main winding tests satisfactorily, connect a test lamp between the phase lead (the black lead) from the capacitor and the other capacitor lead. Close the starting switch. If the phase winding is all right, the lamp will glow and the motor may attempt to start. If either winding is open, remove the motor to the service shop for repairs.
(g) Faulty capacitor	(g) If the starting capacitor (electrolytic) is faulty, the motor starting torque will be weak and the motor may not start at all, but may run if started by hand. A capacitor can be tested for open circuit or short circuit as follows: Charge it with DC (if available), preferably through a resistance or test lamp. If no discharge is evident on immediate short circuit, an open or a short is indicated. If no DC is available,

Symptom and Possible Cause   *Possible Remedy*

charge with AC. Try charging on AC several times to
make certain that the ca-
pacitor has had a chance to
become charged. If the capacitor is open, short-
circuited, or weak, replace
it. Replacement capacitors
should not be of a lower
capacity or voltage than the
original. In soldering the
connections, *do not use
acid flux.*
Note 1 — Electrolytic capacitors, if exposed to
temperatures of 20°F (-6.7°)
and lower, may temporarily
lose enough capacity so that
the motor will not start, and
may cause the windings to
burn up. The temperature of
the capacitor should be
raised by running the motor
idle, or by other means.
Capacitors should not be
operated in temperatures
exceeding 165°F (74°C).
Note 2 — The frequency of
operation of electrolytic
capacitors should not exceed
two starts per minute of
three seconds' acceleration
each, or three to four starts
per minute at less than two
seconds' acceleration, provided the total accelerating

Symptom and Possible Cause	*Possible Remedy*
	time (i.e., the time before the switch opens) does not exceed one to two minutes per hour. This may be approximately determined by checking the ampere input with the nameplate marking. Excessive load may prevent the motor from accelerating to the speed at which the governor acts, and thus cause the phase winding to burn up.

Radio Interference

(a) Faulty ground	(a) Check for poor ground connections. Static electricity generated by the belts may cause radio noises if the motor frame is not thoroughly grounded.
(b) Loose connections	(b) Check for loose connections or contacts in the switch, fuses, or starter. Capacitor motors ordinarily will not cause radio interference. Sometimes vibration may cause the capacitor to move so that it touches the metal container. This may cause radio interference. Open the container, move the capacitor, and replace the paper packing so that the capacitor cannot shift.

Summary

The fractional-horsepower motor is usually designed to operate on single-phase AC at standard frequencies, and is reliable, easy to repair, and comparatively low in cost. The various types of single-phase motors developed for use on alternating current are: (1) split-phase; (2) capacitor-start; (3) permanent-capacitor; (4) repulsion; (5) shaded-pole; (6) universal; (7) reluctance synchronous; and (8) hysteresis synchronous.

The *split-phase* induction motor is one of the most popular of the fractional-horsepower motors. It is used in sizes ranging from $1/30$ hp (24.9 W) to $1/2$ hp (373 W) on fans, business machines, automatic musical instruments, buffing machines, grinders, etc. The *resistance-start* motor and the *reactor-start* motor are forms of the split-phase induction motor.

The *capacitor-start* motor is another form of split-phase motor having a capacitor connected in series with the auxiliary winding. This circuit is opened when the motor has attained a predetermined speed.

The *permanent-capacitor* motor has its main winding connected directly to the power supply, and the auxiliary winding connected in series with a capacitor. Both the capacitor and the auxiliary winding remain in the circuit while the motor is in operation.

The *repulsion* motor has a stator winding arranged for connection to the source of power, and a rotor winding connected to a commutator. This type of motor has a varying speed characteristic. The *repulsion-start-induction* motor and the *repulsion-induction* motor are forms of the repulsion motor.

A *shaded-pole* motor is a single-phase induction motor equipped with an auxiliary winding displaced magnetically from, and connected in parallel with, the main winding. This type of motor is manufactured in fractional-horsepower sizes, and is used in a variety of household appliances, such as fans, blowers, hair driers, and other applications requiring a low starting torque. It operates only on alternating current, is usually nonreversible, is low in cost, and is extremely rugged and reliable.

A *universal* motor is a series-wound or compensated series-wound motor that may be operated either on direct current or on single-phase alternating current at approximately the same speed and output. Universal motors are commonly manufactured in fractional-horsepower sizes, and are preferred because they can be used either on AC or on DC, especially in areas that supply both types of current. Common applications for the universal-type motor are vacuum cleaners, sewing machines, portable drills, saws, routers, home motion-picture projectors, and business machines of all types.

The *reluctance synchronous motor* is really a variation of the class squirrel-cage rotor. The rotor is modified to provide equally spaced areas of high reluctance. The reluctance synchronous rotor starts and accelerates like a regular squirrel-cage rotor. Reluctance synchronous motors may be designed for operation on single- or three-phase power. In single-phase operation, they may take the split-phase, capacitor-start, or permanent-split capacitor configuration. The skew of the rotor aids in providing smoothness of operation when compared to the capacitor-start and split-phase motors.

The *hysteresis synchronous motor* has a rotor that is made of a heat-treated cast permanent-magnet alloy cylinder. It has a non-magnetic support securely mounted to the shaft. This type of motor starts on the hysteresis principle and accelerates at a fairly constant rate until it reaches the synchronous speed of the rotating field. This type is strictly dependent on the frequency of the power source for its speed.

Synchronous motors are known for their constant speed characteristic. They operate at a speed determined by the frequency of the power source. They are more expensive and heavier than their single-phase counterparts, so they are used only when speed is the absolute factor in the design of the machine being driven by the motor.

Review Questions

1. Describe the basic construction of the split-phase induction motor.
2. List the six types of split-phase motors.
3. What are the two types of repulsion motors?
4. What is the chief advantage of the universal-type motor?

5. What are the applications of the universal-type motor?
6. What factor determines the speed of a reluctance synchronous motor?
7. What factor determines the speed of a hysteresis synchronous motor?
8. What happens when a reluctance motor is overloaded?
9. What happens when a hysteresis motor is overloaded?
10. What is meant by "pull-in"?

CHAPTER 8

Magnetic Contactors

A contactor may be defined as a device, operated other than by hand, for repeatedly establishing and interrupting an electric power circuit. Contactors are employed in control, as well as in power circuits, and as single units or in groups, with load currents on the main contacts ranging from fractions of an ampere to thousands of amperes, depending on the particular application.

Although the construction of a contactor is similar to that of a circuit breaker, its action is the exact opposite. A circuit breaker usually employs an electromagnet for its opening when a fault occurs, whereas a contactor employs an electromagnet for opening or closing a circuit under normal operating conditions. Thus, a contactor is never designed to operate on short-circuit currents. This function is delegated to circuit breakers or fuses, as the case may be.

Interlocks

There are two main types of interlocks usually incorporated in magnetic contactors, and they are named according to their function:

1. Electrical interlocks.
2. Mechanical interlocks.

Electrical Interlocks

Electrical interlocking generally consists of a system of wiring whereby, when one of the interlocked devices is operating, the control circuit to a second device cannot be completed; or the wiring may be such that the second device or starter cannot be energized until the first device or starter has been put into operation. This is accomplished by auxiliary circuit devices known as *electrical interlocks,* which are opened or closed as the device upon which they are used is operated. By definition, an electrical interlock is a device that is actuated by the operation of some other device (usually the main contactor) with which it is directly associated to govern succeeding operations of the same or some allied device.

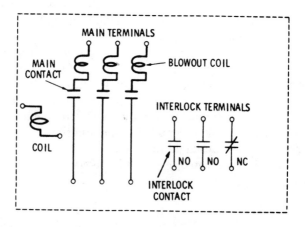

Fig. 8-1. Schematic diagram of a three-pole magnetic contactor with two normally open (NO) and one normally closed (NC) electrical interlocks. Contactor shown in its de-energized position.

Electrical interlocks (sometimes called *auxiliary contactors*) function as a result of the movement of the main contacts of the contactor to which they are mechanically connected. Thus, when the main contacts are closed, one or more of the electrical interlocks may be in the open position, and one or more in the closed position. On the other hand, when the main contacts are opened, the interlock contacts that were formerly in the open positions will now be closed, and vice versa (Figs. 8-1 and 8-2).

Electrical interlocks are usually termed normally closed (NC) and normally open (NO), depending on their function in the circuit they operate. This terminology refers to the position of the contacts of the interlock when the main contacts of the contactor are in a de-energized or nonoperated position.

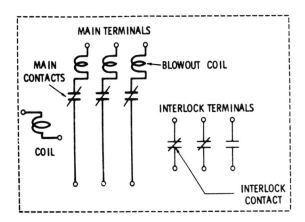

Fig. 8-2. Schematic diagram showing the contactor in Fig. 8-1 in its energized position.

Mechanical Interlocks

This type of interlocking may be defined as a mechanical arrangement whereby one of the interlocked devices is mechanically prevented from functioning when the other is energized. In certain applications it is necessary to prevent one contactor from closing when a second contactor is closed. In order to prevent this, a mechanical interlock consisting of a piece of metal is hinged at the center and arranged in such a way that, when both contactors are in the open position, the center piece is in the neutral position.

Closing one contactor causes it to strike one end of the interlock, moving it toward the panel, thus causing the other end of the interlock to take such a position as to prevent the closing of the second contactor.

A common type of mechanical interlocking is that associated with doors of enclosures housing high-voltage control apparatus, whereby the door cannot be opened while the contactor or circuit breaker is in the energized position. Electrical and mechanical interlocking are both frequently used in the same device, one of the best examples being an AC reversing motor starter.

Shading Coils

The magnetic structure in an AC contactor must be laminated to reduce core losses, whereas the magnetic structure in a DC contactor is of solid material. Another difference between the two types of contactors is that the AC type produces a distinct hum or chattering, pitched at twice the frequency of the circuit, whereas the DC contactor is entirely noiseless.

The chattering effect is usually reduced to a minimum in an AC contactor by the use of a *pole shader* or *shading coil*. A pole shader may be defined as a short-circuited coil embedded in one or both pole faces of an AC magnet. Its function is to divide the pole faces (Fig. 8-3) so that only a part of the flux links the short-circuited turn, with the result that the flux from each part of the pole face will exert an independent holding force on the contactor armature. The resistance and reactance of the shading coil are in such proportion that the induced current in the coil is out of phase with the main flux. Thus, the pull on the armature can be maintained without the pulsating effect caused by the normal AC frequency.

Contactor-Coil Power Supply

DC magnets operated from solid-state rectifiers are commonly used on the larger sizes of AC contactors. This makes it possible to use a solid magnetic structure rather than one that is laminated. When this method is used, all troubles incidental to pole shaders, as well as the disturbing AC hum, are eliminated. This method also results in reduced power consumption in the magnet coil, and a longer contactor life.

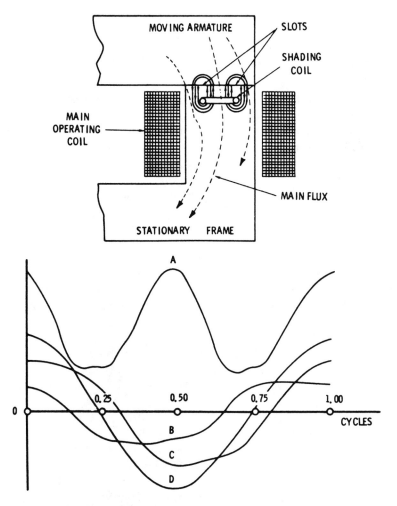

Fig. 8-3. AC contactor with shading coil and the graph of the flux during one cycle of the alternating current. Curve A represents the magnetic pull on the magnet armature; Curve B, the flux through the unshaded portion of the pole face; Curve C, the flux through the shaded portion of the pole face; and Curve D, the total flux.

Arc Suppression

An electric arc is the discharge of electricity between electrodes. In order for an arc to be established between the electrodes, it is essential that there be a surrounding medium of gas or vapor.

Magnetic Blowout

The various manufacturers of electrical contactors have conducted considerable research that has resulted in great improvements, especially in the field of arc extinguishing, resulting in more effective contactors and a substantial saving in material and space requirements.

Since an arc occurs between two contactor tips as they open on normal operation, heating and burning of the tips occur. It is therefore highly desirable to extinguish the arc as quickly as possible to keep it from hanging on any part of the tip surface where the heating would be concentrated. When relatively small currents are being controlled, the tip can be opened sufficiently to stretch the arc, so that it is quickly extinguished. Double-break tips that give the same length of gap with only one-half the travel of the movable tip on a single-break device are also employed for fast extinction of the arc. Double-break tips have an added advantage in that no flexible shunts are required as on a single-break contactor.

When it is designed to employ a contactor for interruption of heavier current, *blowout coils* and arc chutes are used to produce a long arc in a relatively small space. The blowout coil is mounted so that its magnetic field is in the path of the arc. The coil is connected in series with the contactor itself so that the motor current flows through the coil as long as the contactor is closed or as long as there is an arc between the contacts. The field set up by the blowout coil magnetically repels the arc upward, as in Fig. 8-4, and quickly stretches it within the arc chute to such a length that it is extinguished. The extinction is assisted by the cooling action of the restricted section of the arc chutes. The blowout coil is used effectively on most contactors carrying a current of 25 amperes or more, and where it is desirable to interrupt the circuit as rapidly as possible.

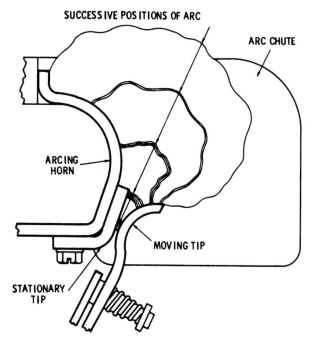

SUCCESSIVE POSITIONS OF ARC

ARC CHUTE

ARCING HORN

MOVING TIP

STATIONARY TIP

Fig. 8-4. The behavior of an electric arc in a correctly designed switch equipped with a blowout coil.

Deion Arc Quenching

Until quite recently, the principal method employed in contactor arc extinction in both AC and DC circuits was the use of magnetic blowouts. Here, the arc is extinguished by a powerful blowout field and by stretching the arc across arc splitters, thereby lengthening it until instability and extinction occur. In AC circuits where it is required to interrupt large amounts of current, however, the breaker structure must be relatively large, and special precautions must be observed to keep the arc within the limits of safety.

The deion type of contactor has been designed as a result of the continued demand for a contactor of increased interrupting and current capacity for use in small enclosures. The word *deion* has been coined to designate any type of arc-extinguishing apparatus that functions by rapidly restoring the insulating properties of air or gases in which it operates, by means other than simply stretching or cooling the arc. By the use of a properly designed deionizing chamber, the arc may be extinguished in a relatively small space, and the gases associated with the arc allowed to escape from the deionizing chamber only after being rendered nonconductive and harmless.

The elimination of the magnetic blowout coil also decreases the heat losses produced in the blowout structure, permitting the current-carrying parts to operate cooler, which permits a reduction in the size of the enclosure for a given temperature rise. The deion type of contactor thus makes it possible to have a more compact line starter, more tightly enclosed, with a higher rating, and capable of performing more difficult tasks than those built conventionally.

Shunt Contactors

A shunt contactor is one that is operated by a shunt coil energized from a constant-voltage power source. For DC operation, such contactors are usually of the single- or double-pole type.

A single-pole shunt contactor is suitable for the remote control of circuits carrying large currents and for the full-voltage control of small motors. These contactors are also used extensively for the control of miscellaneous DC circuits in such applications as industrial heating, battery charging, control interlocking, and others. When used with such accessories as float-switches, pressure governors, pushbutton stations, etc., they provide undervoltage protection or release, depending on the contact arrangement of the accessory used.

The single-pole contactor consists of a front-connected contactor mounted on a compound base that is usually equipped with feet for wall mounting. The magnet frame consists of a steel casing, with the coil mounted on a steel core.

The circuit for a pushbutton-operated single-pole contactor is shown in Fig. 8-5, and a double-pole type in Fig. 8-6. With the coil de-energized, the contactor is in its normal, or open, position. When the pushbutton is pressed, the coil attracts the movable armature to the stationary core, thus closing the contactor. The

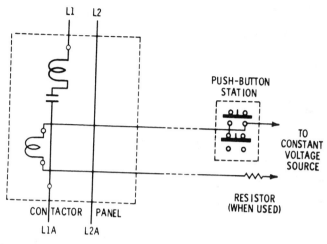

Fig. 8-5. Wiring diagram of a single-pole shunt-type circuit breaker controlled from a pushbutton station.

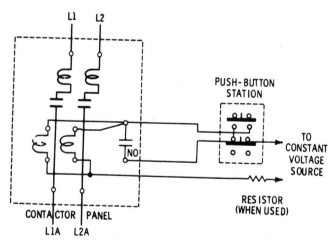

Fig. 8-6. Wiring diagram of a double-pole shunt-type circuit breaker controlled from a pushbutton station.

stationary contact is provided with a sizable arc shield or horn that serves to limit the arc and direct it upward and away from the contacts. The movable contact has a heavy, flexible cable connected to an independent stud through the insulating base, thus avoiding the necessity of carrying the current through the steel frame of the contactor.

Lockout Contactors

Lockout contactors are used principally as a means of obtaining current-limit acceleration on DC motors. Such contactors for motor control are constructed so that the contacts will not close until the circuit current has dropped below a definite value. They are made in two general forms:

1. The single coil.
2. The double coil.

The single-coil type is usually referred to as a *series contactor*, while the two-coil type is called a *series lockout contactor*. Both of these types, however, possess the lockout feature of the contactor; that is, they are held open until the current has dropped to some definite value.

The general principle of operation of the single-coil type can be better understood by referring to Fig. 8-7. The current passing through the operating coil produces a magnetic field, which takes two paths. One path is through the core of the coil, through a section of the restricted area surrounded by the copper sleeve, through the armature and the main air gap, and back to the core of the coil. The other path is through the core of the coil, down through the adjustable air gap, through the tail of the armature, and through the armature and main air gap back to the core of the coil.

When no current is passing through the coil, the armature is in the open position. When a large current is passing through, the magnet field becomes so strong that the path surrounded by the copper sleeve becomes saturated so that the flux through this part cannot exceed a certain amount. The magnetic flux through the adjustable air gap will vary with the current through the operating coil, and this path will not become saturated.

Any flux passing across the adjustable air gap produces an attracting force on the tail of the armature. This force tends to hold

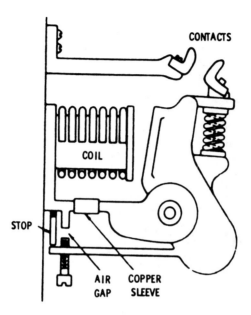

Fig. 8-7. Principal parts of a single-pole, single-coil, lockout contactor.

the armature open against the closing force exerted between the core of the coil and main part of the armature. As the current decreases in value, the flux across the adjustable air gap also decreases, and thus the holding-out force on the tail of the armature is reduced.

After the current has decreased to a certain value (determined by the setting of the adjustable air gap), the holding-out force on the tail of the armature is not sufficient to overcome the closing force exerted by the core of the coil on the main part of the armature, and the contactor closes.

In the two-coil type of series contactor (Fig. 8-8), two coils are connected in series with each other and with the circuit to be controlled. The armature is pivoted so that the larger coil acts upon one side of the armature with a force that tends to close the contactor. The smaller coil acts on the other side of the armature with a force that tends to hold the contactor open. The areas of the magnetic circuits are so designed that the force tending to close the contactor remains practically constant, regardless of the amount

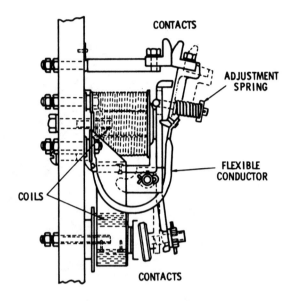

Fig. 8-8. A two-coil magnetic lockout contactor.

of current, while the force tending to hold the contactor open
varies with the current. Therefore, after the current has decreased
to a certain value, the holding-out force is not sufficient to overcome the closing force, and the contactor closes. The air gap of
the locking-out circuit can be adjusted so that the current above
which the contactor holds open can be adjusted over a wide range.

Inductive-Type Contactors

Inductive-type contactors are similar in appearance to the previously described double-coil contactor. They differ only with
respect to the arrangement of the lockout magnet, which is constructed of heavy iron.

In operation, the relative strengths of the main closing and holdout coils are so adjusted that the contactor will remain open, with
full voltage on the main closing coil and approximately 1 percent
of the operating voltage on the holdout coil. When the magnetic
contactor is operated by pressing the *start* button, both coils are
energized simultaneously and the holdout coil is short-circuited.

The magnetic circuit comprising the holdout coil is constructed so that a certain time interval is required for the residual flux to decrease to a value permitting the contactor to close its circuit. This time interval is adjusted by changing the length of an air gap in the holdout magnetic circuit.

Contactor Ratings

The rating of a pair of current-carrying contacts is generally expressed in terms of *make*, *break*, and *carry*.

Make is defined as the value of transient or steady current that the contacts can establish without welding. The current that a set of tips can make is not constant, but depends to a considerable degree on the design of the device on which the tips are used. Some sparking may be visible at the moment of make because of certain irregularities in the tip surface, and because of the relative motion of the tips as they travel from the first contact point to their final position.

The statement that a device will *break* a certain current means that it is capable of interrupting the flow of that amount of current without excessive arcing or burning:

The *Underwriters' Laboratories* specify:

1. That a contactor rated in amperes shall be tested for durability by interrupting it at its rated current 6,000 times at intervals of 1 second.

2. That it shall be able to interrupt 150-percent rated current 50 times at intervals of 10 seconds.

It is further specified that DC devices shall be tested with a noninductive resistance load, while AC devices shall be tested with an inductive load.

The term *carry* designates the load current that can be continuously conducted without excessive temperature rise. The *Underwriters' Laboratories* specify that, with a motor load, the contacts shall carry 115 percent of their ampere rating continuously until their temperatures are constant without exceeding a 167°F (75°C) rise, and that the contacts shall carry 100 percent of their rating and meet these conditions on all other loads.

The eight-hour rating given devices was agreed upon by the industry in recognition of the fact that most standard devices cannot be energized continuously for days without exceeding a safe temperature rise. It requires the periodic cleaning action of both the arcing and the physical opening and closing of the contacts to remove the high-resistance oxides that would otherwise form.

The NEMA standard eight-hour open rating for AC contactors shall be 15, 25, 50, 100, 150, 300, and 600 amperes.

The open eight-hour rating for DC contactors shall be 25, 50, 100, 150, 300, and 600 amperes.

The eight-hour open rating for AC and DC contactors rated more than 600 amperes has been suggested as 1,200 and 2,400 amperes.

The ratings for various AC and DC contactors are given in Tables 8-1 through 8-13.

Maintenance

Proper maintenance of contactors plays a very important role in their satisfactory functioning and will increase their life span and the economy of their operation. One important factor to remember is that the use of lubricants on the contacts or bearings should be avoided. Oil quickly collects dust and, unless the parts are frequently cleaned, will interfere with the operation of the contactor. Oily surfaces that collect dust may also result in an arc between live parts of the contactor.

Most contactors are furnished with flexible braided-copper shunts that are designed to give complete freedom to the moving armature with sufficient current-carrying capacity. Burned or badly corroded shunts should be promptly replaced.

Range of Operating Voltage

Contactors generally are designed to operate properly if the line voltage is within 85 to 110 percent of the nameplate rating for AC circuits, and within 80 to 100 percent for DC circuits. Wider ranges usually require a special device.

Where there is a continuous 10-percent increase in voltage,

Table 8-1. Contactor Ratings for Line, Reversing, and Final Accelerating for General-Purpose and Machine-Tool DC Magnetic Starters

Open 8-Hour Ampere Rating	Horsepower Rating (Kilowatts)		
	115 Volts	230 Volts	550 Volts
25	3 (2.24 kW)	5 (3.73 kW)	
50	5 (3.73 kW)	10 (7.46 kW)	20 (14.92 kW)
100	10 (7.46 kW)	25 (18.65 kW)	50 (37.3 kW)
150	20 (14.92 kW)	40 (29.84 kW)	75 (55.95 kW)
300	40 (29.84 kW)	75 (55.95 kW)	150 (111.9 kW)
600	75 (55.95 kW)	150 (111.9 kW)	300 (223.8 kW)

Intermediate accelerating contactors shall be selected so that the open 8-hour ampere rating shall not be less than one-quarter the maximum current for that step.

Table 8-2. Contactor Ratings for Continuous-Duty Steel Mill Auxiliary Standard DC Accelerating Contactors

Open 8-Hour Contactor Ampere Rating	Horsepower Rating (Continuous) (Kilowatts)	Minimum Number of Accelerating Contactors
100	25 (18.65 kW)	2
150	40 (29.84 kW)	2
300	75 (55.95 kW)	2
600	150 (111.9 kW)	3

Table 8-3. Contactor Ratings for Mill Duty Accelerating Contactors

Open 8-Hour Contactor Ampere Rating	Rating Mill Contactor	Horsepower Mill Rating (Kilowatts)	Minimum Number of Accelerating Contactors
100	133	35 (26.11 kW)	2
150	200	53 (39.54 kW)	2
300	400	110 (82.06 kW)	2
600	800	225 (167.85 kW)	2° or 3°

*As specified by the user.

The minimum contactor rating to be offered as a standard with control for dc steel mill auxiliaries shall be the 100-ampere 8-hour rating.

Table 8-4. Contactor Ratings for
Crane Standard DC Contactors

Open 8-Hour Ampere Rating	CRANE RATING		Minimum Number of Accelerating Contactors
	Amperes	Horsepower at 230 Volts (Kilowatts)	
100	133	35 (26.11 kW)	3
150	200	55 (39.54 kW)	3
300	400	110 (82.06 kW)	3
600	800	225 (167.85 kW)	4

Note 1—General-purpose control will be considered as suitable for steel-mill accessory machines.
Note 2—The minimum contactor rating to be offered as standard with crane control shall be the 100-ampere, 8-hour rating.
Note 3—Accelerating contactors shall be the same rating as the line contactors and shall be equipped with blowouts.
Note 4—The number of accelerating contactors is exclusive of the plugging contactor.

suitable coils should be ordered because, with this increase in voltage, there is approximately a 50-percent increase in wattage. The latter greatly increases the heating of the coil, causing a rapid deterioration of the insulation, and shortens the ultimate life of the coil. There is also an approximate increase of 20 percent in the pounding effect, resulting in a more rapid deformation of the armature, crystallization of the magnet parts, breaking of the contact tips, and an increase in noise.

Table 8-5. Contactor Ratings for DC Crane Protective Panels

Open 8-Hour Ampere Rating	Total Horsepower All Motors at 230 Volts Kilowatts	Largest Individual Motor at 230 Volts (Kilowatts)
100	55 (39.54 kW)	35 (26.11 kW)
150	80 (59.68 kW)	55 (39.54 kW)
300	160 (119.36 kW)	110 (82.06 kW)
600	320 (238.72 kW)	225 (167.85 kW)

Note 1—For motors of other voltage ratings, the 8-hour rating of the main line contactor shall not be less than 50 percent of the combined one-half or one-hour rating of the motors, nor less than 75 percent of the one-half or one-hour rating of the largest individual motor.
Note 2—In no case shall the main line contactor of the protective panel be of smaller rating than the largest contactor used on any of the controllers protected.

Table 8-6. Contactor Ratings for AC Crane and Hoist Duty

Open 8-Hour Ampere Rating	CRANE RATING		
	Amperes	(Contactors in Motor Primary)	
		Horsepower at 220 Volts (Kilowatts)	Horsepower at 440 and 550 Volts (Kilowatts)
100	133	40 (29.84 kW)	75 (55.95 kW)
150	200	60 (44.76 kW)	125 (93.25 kW)
300	400	150 (111.9 kW)	300 (223.8 kW)
600	800	300 (223.8 kW)	

Note 1—The minimum contactor rating to be offered as standard with ac crane control shall be the 100 ampere, 8-hour rating

Note 2—Accelerating contactors shall be equipped with blowouts, and shall have a crane rating of not less than the full-load secondary current of the motor

Note 3—When used for motor secondary control, the ampere rating of a three-pole ac contactor with its poles connected in delta shall be 1.5 times its standard crane rating

Note 4—The number of accelerating contactors is exclusive of the plugging contactor for reversing controllers and the low-torque contactor for hoist controllers.

Table 8-7. Number of Accelerating Contactors for AC Crane and Hoist Duty

Motor Horsepower Rating	Minimum Number of Accelerating Contactors
75 hp (55.95 kW) and less	3
76 to 200 hp (56.696 kW to 149.2 kW)	4
200 hp (149.2 kW and Above)	5

Contact Movement

The armatures should seal when the proper voltage is applied to the coils, and should open by gravity when the power is cut off. All contacts should, when closed, make line contact near the bottom of the face. On opening, the final break should be at the top. The rolling and wiping motions when closing and opening keep the contacts in good condition by removing any oxidation that might tend to accumulate on their faces.

Interlock Adjustment

Electrical interlocks are usually factory adjusted to make contact at approximately the same time that the main-contactor tips touch, or even a trifle later. However, for some special applications, the interlocks may make contact before the main tips touch. To change this adjustment, the usual procedure is to loosen the nuts on the front and back of the base, and to screw the stud in and out to suit the conditions.

Mechanical interlocks are usually so adjusted that, with one contactor in the sealed closed position, there is a little play on the other contactor. This play must not be great enough to permit the moving contacts of the second contactor to touch the corresponding stationary tips when the tips of the first contactor are just touching.

Table 8-8. Contactor Ratings for AC Crane Protective Panels

Open 8-Hour Ampere Rating	Total Horsepower All Motors (Kilowatts)		Largest Individual Motor (Kilowatts)	
	220 Volts	440 and 550 Volts	220 Volts	440 and 550 Volts
100	65 (48.49)	125 (93.25)	40 (29.84)	75 (55.95)
150	110 (82.06)	225 (167.85)	60 (44.76)	125 (93.25)
300	225 (167.85)	450 (335.7)	150 (111.9)	300 (223.8)
600	450 (335.7)		300 (223.8)	

Note 1—For motors of other voltage ratings, the 8-hour rating of the main line contactor shall not be less than 50 percent of the combined one-half or one-hour rating of the motors, nor less than 75 percent of the one-half or one-hour rating of the largest individual motor.

Note 2—In no case shall the main contactor of the protective panel be of smaller rating than the largest primary contactor used on any of the controllers protected thereby.

Contact Pressure

Contactors for various classes of service are designed with a definite contact-spring pressure. To avoid undue heating of the contacts, it is important that this spring pressure be maintained at a constant rate. The pressure can be checked with a properly calibrated spring balance, as shown in Fig. 8-9.

Contact Heating

A contactor has several bolted and spring-closed contacts. Excessively high resistance at these contacts can be the cause of

Table 8-9. Contactor Ratings for Polyphase Multispeed Motors

Size Number	Open 8-Hour Ampere Rating	Horsepower Rating (Kilowatts)		
		110 Volts	220 Volts	440 and 550 Volts
00		¾ (.5595)	¾ (.5595)	
0	15	1 (.746)	1½ (1.119)	2 (1.492)
1	25	1½ (1.119)	3 (2.238)	5 (3.73)
2	50	5 (3.73)	10 (7.46)	25 (14.92)
3	100	10 (7.46)	20 (14.92)	40 (29.84)
4	150	15 (11.19)	30 (22.38)	60 (44.76)
5	300		75 (55.95)	150 (111.9)

Note 1—For constant hp (2, 3, and 4 speeds).
Note 2—For constant and variable torque motors (2, 3, and 4 speeds), the horsepower ratings shall be the ratings standard for across-the-line starters shown in Table 12.

very high temperatures (100°C to 200°C) [212°F to 392°F] when the contactor is carrying less than the rated current. The most likely point of high resistance is at the contacts where the movable tips meet the stationary tips.

High resistance, however, may occur at any of the several bolted joints of the contactor. Therefore, if any of these devices begin to develop an excessive temperature, a millivoltmeter should be employed to ascertain which joints have a high voltage drop across them.

When copper contacts with a high resistance have been located,

Table 8-10. Ratings of AC Low-Voltage Contactors Used with Synchronous Motors, Starters, and Controllers

8-Hour Contactor Ampere Rating	Maximum Horsepower (Kilowatts)			
	220 Volts		440 and 550 Volts	
	1.0 PF	0.8 PF	1.0 PF	0.8 PF
50	20 (14.92)	15 (11.19)	30 (22.38)	25 (18.65)
100	40 (29.84)	30 (22.38)	60 (44.76)	50 (37.3)
150	60 (44.76)	50 (37.3)	125 (93.25)	100 (74.6)
300	125 (93.25)	100 (74.6)	250 (186.5)	200 (149.2)
600	250 (186.5)	200 (149.2)	500 (373)	400 (298.4)
1200	500 (373)	400 (298.4)	1000 (746)	800 (596.8)

Note 1—The accelerating contactor of single-step primary-resistor, autotransformer, and reactor starters shall have the same rating as the line contactor when the line contactor has an 8-hour rating of 150 amperes or less. When the line contactor has an 8-hour rating of 300 amperes or larger, an accelerating contactor having the next lower rating may be used.
Note 2—The accelerating contactors of multiple-step starters shall have the same rating as multiple-step ac starters.

Table 8-11. Ratings of AC Reduced-Voltage General-Purpose Magnetic Starters

NEMA Size Number	Open 8-Hour Ampere Rating of Con-troller	Horsepower Ratings (Kilowatts)					
		110 Volts		220 Volts		440 and 550 Volts	
		Three-Phase	Single-Phase	Three-Phase	Single-Phase	Three-Phase	Single-Phase
1	25	3 (2.238)	1½(1.119)	5 (3.37)	3 (2.238)	7½(5.595)	5 (3.73)
2	50	7½(5.595)	3 (2.238)	15 (11.19)	7½(5.595)	25 (18.650)	10 (7.46)
3	100	15 (11.190)	7½(5.595)	30 (22.38)	15 (11.190)	50 (11.190)	25 (18.65)
4	150	25 (18.650)		50 (37.30)		100 (74.600)	
5	300			100 (74.60)		200 (149.200)	
6	600			200 (149.20)		400 (298.400)	

Note 1—The single-phase ratings apply to either two- or three-pole contactors.

Note 2—The standard horsepower rating adopted for three-phase starters shall apply to two-phase, four-wire and two-phase, three-wire starters. A current rating of 90 percent of the 8-hour rating of the contactor employed shall apply, based on the average of the currents in the three legs of the two-phase circuit.

Note 3—The accelerating contactor of single-step, primary-resistor, autotransformer, and reactor starters, shall have the same rating as the line contactor when the line contactor has an 8-hour rating of 150 amperes or less. When the line contactor has an 8-hour rating of 300 amperes or larger, an accelerating contactor having the next lower rating may be used. The accelerating contactors of multiple step starters shall have the same rating as for network starters.

Note 4—The intermediate accelerating contactors of secondary-resistor starters shall be selected so that the 8-hour open rating will not be less than one-sixth of the accelerating peak current. The 8-hour open rating of three-pole, delta-connected secondary contactors is 150 percent of the normal star-connected rating.

correction can be made by opening the contacts and removing the oxide film with a file (not with sandpaper or carborundum paper). It is unusual to find a high resistance in a bolted joint unless the contactor has previously reached excessive temperatures. When excessive resistances are found in joints, however, the cause should be removed.

Since high resistances will most commonly be found in the active contact, it is a very simple matter to inspect these tips weekly or monthly. If the temperature is unduly high, the tips should be lightly filed. The foregoing remarks apply particularly to copper contacts, because they oxidize readily and because the copper oxide thus formed has a very high resistance. A file will remove the oxide and again reduce the resistance to a low value.

Electrical interlocks may also fail because of the oxidation of copper contacts. Sometimes such failure occurs because of dirt between the contacts. Using one hemispherical and one flat tip made of silver will overcome both of these troubles. The face of the hemispherical unit will make contact without trapping foreign particles between it and the flat tip.

Table 8-12. Ratings of AC Across-the-Line Magnetically Operated Starters

NEMA Size Number	Open 8-Hour Ampere Rating of Contactor	Horsepower at 100 Volts (Kilowatts)		Horsepower at 220 Volts (Kilowatts)		Horsepower at 440 and 550 Volts (Kilowatts)	
		Three-Phase	Single-Phase	Three-Phase	Single-Phase	Three-Phase	Single-Phase
00		¾(0.5595)	¼(0.373)	1 (0.746)	¾(0.5595)	1 (0.746)	
0	15	1¼(1.1190)	1 (0.746)	2 (1.492)	1¼(1.1190)	2 (1.492)	1¼(1.119)
1	25	3 (2.2380)	1¼(1.119)	5 (3.730)	3 (2.2380)	7½(5.595)	5 (3.730)
2	50	7½(5.5950)	3 (2.238)	15 (11.190)	7½(5.5950)	25 (18.650)	10 (7.460)
3	100	15 (11.1900)	7½(5.595)	30 (22.380)	15 (11.1900)	50 (37.300)	25 (18.650)
4	150	25 (18.6500)		50 (37.300)		100 (74.600)	
5	300			100 (74.600)		200 (149.200)	
6	600			200 (149.200)		400 (298.400)	

Note 1—This table applies to all starting, reversing, and throw-over switches furnished with any type of enclosure, either with or without disconnecting means or other accessories. The enclosures may have operating shafts, push buttons in the case, reset rods passing through the case, slots for test jacks (with covers thereover) or similar openings that are filled by the shaft rod or other member passing through them.

Note 2—The single-phase ratings apply to either two- or three-pole starters. Standard two-pole magnetic starters, reversing and nonreversing, for single-phase motors will have overload protection in one leg only.

Excessive wear on contact tips of a contactor indicates that it is operated frequently. Silver contacts should not be used for conditions of this kind, or where the current exceeds 25 amperes, because they do not stand up as well as copper under severe service. If the service is usually heavy, and if the current is equivalent to no more than three-fourths of the contactor's rating, the tip can be faced with an alloy such as copper-tungsten that will last several times longer than copper.

Rough Contacts

Numerous maintenance people have the erroneous impression that contact tips that have been roughened by service should be kept smooth. A roughened tip will carry current just as well as a smooth tip. Of course a large projection on a tip, caused by unusual arcing, should be removed. However, a tip that has been roughened by ordinary arcing need not be serviced. If a copper tip becomes overheated, oxide will form and should be removed.

Contactor Coils

A large percentage of contactor-coil troubles are caused by overheating. Therefore, if the temperature can be reduced, coil

Table 8-13. Ratings of AC Across-the-Line Manually Operated Starters

Size Number	Open 8-Hour Ampere Rating	Horsepower at 110 Volts (Kilowatts)		Horsepower at 220 Volts (Kilowatts)		Horsepower at 440 and 550 Volts (Kilowatts)	
		Three-Phase	Single-Phase	Three-Phase	Single-Phase	Three-Phase	Single-Phase
00		¾(0.5595)	¾(0.5595)	¾(0.5595)	¾(0.5595)		
0	15	1½(1.1190)	1 (0.7460)	2 (1.4920)	1½(1.1190)	2 (1.492)	1½(1.119)
1	20	3 (2.2380)	1½(1.1190)	5 (3.7300)	2 (2.2380)	7½(5.595)	5 (3.730)
2	50	7½(5.5950)		15 (11.90)		25 (18.650)	
3	100	15 (11.1900)		30 (22.380)		50 (37.300)	
4	150	25 (18.6500)		50 (37.300)		100 (74.600)	

troubles can be greatly minimized. Since the heating of a DC coil will vary as the square of the voltage, and the heating of an AC coil will vary approximately as the cube, it follows that coils should be wound for the voltage that exists on the line. If the ambient temperature is high, this precaution is all the more important.

When an AC magnet, such as a solenoid, is supplied with constant-voltage excitation, a large inrush of current is required to close the armature. When the armature closes, the coil current drops to a normal value. Sometimes the armature may not close (because of excess friction or some other reasons) and the resulting large inrush current may burn out the coil within a few seconds. Such mishaps can be prevented by the use of a thermal cutout. When the thermal cutout opens because the armature fails to close, it is merely necessary to replace a small link made of two pieces of metal held together by a low-melting-point solder.

Some of the contributing causes of overheating of AC contactor coils are:

1. Overvoltage.
2. Low frequency.
3. Excessive magnetic gap caused by faulty assembly of the magnet parts, or an accumulation of dirt, rust, or paint on the pole faces.
4. Failure of the armature to close completely because of low voltage or mechanical interference.
5. Too frequent operation.

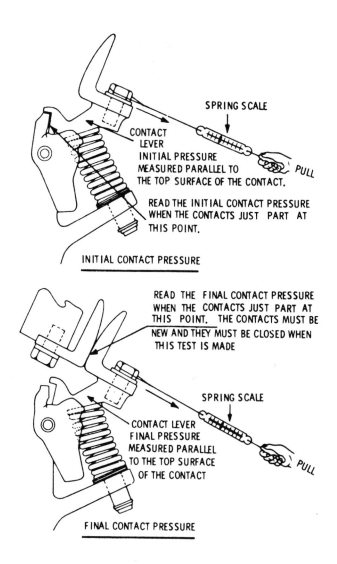

Fig. 8-9. Method of testing contactors for initial and final contact pressure.

The current in AC shunt coils is at a minimum value when the armature seals properly with the stationary frame. Any gap in the magnetic circuit lowers the impedence of the coil, allowing the coil to draw more current. If the magnet is accidentally blocked open, or if the voltage is so low that the magnet cannot close, the current is likely to be several times greater than the minimum-gap current; if this condition occurs, the coil may soon be damaged.

Unfortunately, the leads on coil are often used as carrying handles. Although the leads are strong enough to support the weight of the coil, this practice should be discouraged. It is quite easy to completely or partially fracture the lead wire or joint, resulting in either an open-circuited coil or one that may open shortly after being installed.

If the coils become wet, it is advisable to bake them as soon as possible in an oven at a temperature of 100°C to 125°C (230°F to 257°F). This should also be done if the coils have been soaked in any degreaser to remove grease or oil. If necessary to paint the coils, only an approved insulating paint or varnish should be used. Some paints or varnishes contain thinners that will rapidly attack the insulation. Apply the paint while the coils are still warm from baking.

Circuit Wiring

Table 8-14 shows the wire sizes needed for individual branch circuit wiring for three-phase motors.

Table 8-14. Individual Branch Circuit Wiring for Three-Phase Squirrel-Cage Induction Motors

Motor Data		Copper Wire Size (Minimum AWG No.)				
	Volts	Branch Circuit Length				
		0–25 ft.	50 ft.	100 ft.	150 ft.	200 ft.
½	230	14	14	14	14	14
	460	14	14	14	14	14
¾	230	14	14	14	14	14
	460	14	14	14	14	14
1	230	14	14	14	14	14
	460	14	14	14	14	14
1½	230	14	14	14	14	12
	460	14	14	14	14	14
2	230	14	14	14	12	12
	460	14	14	14	14	14
3	230	14	14	12	10	10
	460	14	14	14	14	14
5	230	12	12	10	10	8
	460	14	14	14	14	14
7½	230	10	10	10	8	6
	460	14	14	14	14	12
10	230	8	8	8	6	4
	460	12	12	12	12	12
15	230	6	6	6	4	4
	460	10	10	10	10	10
20	230	4	4	4	4	3
	460	8	8	8	8	8
25	230	4	4	4	3	2
	460	8	8	8	8	8
30	230	3	3	3	2	1
	460	6	6	6	6	6
40	230	1	1	1	1	0
	460	6	6	6	6	6
50	230	00	00	00	00	00
	460	4	4	4	4	4

Summary

A contactor is a device, operated other than by hand, for repeatedly establishing and interrupting an electric power circuit. Contactors are employed in control circuits and in power circuits, and as single units or in groups with load currents on the main contacts ranging from a fraction of an ampere to thousands of amperes, depending on the application.

Electrical interlocking generally consists of a system of wiring whereby, when one of the interlocked devices is operating, the control circuit to a second device cannot be completed; or the wiring may be such that the second device cannot be completed; or the wiring may be such that the second device or starter cannot be energized until the first device or starter has been placed in operation. *Mechanical interlocking* consists of a mechanical arrangement whereby one of the interlocked devices is prevented mechanically from functioning when the other is energized; or, in some applications, where it is necessary to prevent closing of one contactor when a second contactor is closed.

The AC type of contactor produces a distinct hum or chattering pitched at twice the frequency of the circuit, whereas the DC contactor is entirely noiseless. The chatter effect is usually reduced to a minimum in an AC contactor by means of a *pole shader* or *shading coil*.

Formerly, the principal method employed in contactor arc extinction in both AC and DC circuits was the use of magnetic blowout coils. The arc is extinguished by a powerful blowout field and by stretching the arc across the splitters, thereby lengthening it until instability and extinction occur. The *deion* type of contactor has been designed to satisfy the demand for a contactor of increased interrupting and current capacity for use in small enclosures. The term is used to designate any type of arc-extinguishing apparatus that functions by rapidly restoring the insulating properties of air or gases in which it operates, by some other means other than simply stretching or cooling the arc.

A *shunt contactor* is operated by a shunt coil energized from a constant-voltage power source. For DC operation, these contactors are usually of the single- or double-pole type.

Lockout contactors are used principally as a means of obtaining current-limit acceleration on DC motors. These contactors are constructed so that the contacts do not close until the current in the circuit has dropped below a definite value. The single-coil type is called a *series contactor* and the two-coil type is called a *series lockout contactor*.

The rating of current-carrying contacts is generally expressed in terms of *make, break,* and *carry*. The value of transient or steady current that the contacts can establish without welding is known as *make*. The statement that a device can *break* a certain current means that it can interrupt the flow of that amount of current without excessive arcing or burning. The term *carry* indicates the load current that can be conducted continuously without excessive temperature rise.

The distance from the power source as well as the horsepower rating of the motor determines the correct wire size for 3-phase motors.

Review Questions

1. Give the definition for a contactor.
2. What are the most common types of relays?
3. What is the purpose of the shading coil in an AC contactor?
4. What two methods are used to suppress the arc in electrical contactors?
5. What is the purpose of a lockout contactor?
6. How are magnetic contactors rated?

CHAPTER 9

Methods of Motor Control

Since over 90 percent of all motors are used on AC, DC motors and their controls will not be discussed in this chapter.

Wound-rotor motors and AC commutator motors have a limited application and are also excluded. Since the squirrel-cage induction motor is the most widely used, its control is the subject of this chapter. The use of higher voltages (2,400, 4,800, and higher) introduces requirements that are additional to those for 600-volt equipment, and although the basic principles are unchanged, these additional requirements are not covered here.

Selection of Motor Control

The motor, machine, and motor controller are interrelated. They need to be considered as a package when choosing a specific device for a particular application. In general, five basic factors influence the selection of a controller:

1. Electrical service.
2. Motor.
3. Operating characteristics of the controller.
4. Environment.
5. National codes and standards.

Electrical Service — Establish whether the service is DC or AC. If AC, determine the number of phases and the frequency — in addition to the voltage.

Motor — The motor should be matched to the electrical service and correctly sized for the machine load (horsepower rating). Other considerations include the motor speed and torque. To select the proper protection for the motor, its full load current rating, service factor, and time rating must be known.

Operating Characteristics of the Controller — The fundamental job of a motor controller is to start and stop the motor. It should also protect the motor, machine, and operator.

The controller might also be called upon to provide supplementary functions, which could include reversing, jogging or inching, plugging, operating at several speeds or at reduced levels of current and motor torque.

Environment — Controller enclosures serve to provide protection for operating personnel by preventing accidental contact with live parts. In certain applications, the controller itself must be protected from a variety of environmental conditions, which might include:

1. Water, rain, snow, or sleet.
2. Dirt or noncombustible dust.
3. Cutting oils, coolants, or lubricants.

Both personnel and property require protection in environments made hazardous by the presence of explosive gases or combustible dusts.

National Codes and Standards — Motor control equipment is designed to meet the provisions of the *National Electrical Code (NEC)*. Code sections applying to industrial control devices are *Article 430* on motors and motor controllers and *Article 500* on hazardous locations.

The 1970 *Occupational Safety and Health Act (OSHA)*, as amended in 1972, requires that each employer furnish employment free from recognized hazards likely to cause serious harm. Provisions of the act are strictly enforced by inspection. Standards established by the *National Electrical Manufacturers Association (NEMA)* assist users in the proper selection of control equipment. NEMA standards provide practical information concerning construction, test, performance, and manufacture of motor control devices such as starters, relays, and contactors.

One of the organizations that actually tests for conformity to national codes and standards is *Underwriter's Laboratories (UL)*. Equipment tested and approved by UL is listed in an annual publication, which is kept current by means of bimonthly supplements that reflect the latest additions and deletions.

Motor Controller

A motor controller will include some or all of the following functions: starting, stopping, overload protection, overcurrent protection, reversing, changing speed, jogging, plugging, sequence control, and pilot light indication. The controller can also provide the control for auxiliary equipment such as brakes, clutches, solenoids, heaters, and signals. A motor controller may be used to control a single motor or a group of motors (Fig. 9-1).

Starter

The terms *starter* and *controller* mean practically the same thing. Strictly speaking, a *starter* is the simplest form of controller and is capable of starting and stopping the motor and providing it with overload protection (Fig. 9-2).

Squirrel-Cage Motor

The workhorse of industry is the AC squirrel-cage motor. Of the thousands of motors used today in general applications, the vast majority are of the squirrel-cage type. Squirrel-cage motors are simple in construction and operation — merely connect power lines to the motor and it will run.

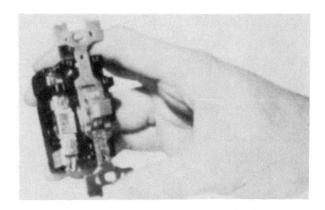

Fig. 9-1. Motor controllers can be simple or complex. Both the small fractional horsepower manual starter above and the special control panel in Fig. 9-2 qualify as motor controllers. *(Courtesy Square D Co.)*

The squirrel-cage motor gets its name because of its rotor construction, which resembles a squirrel cage and has no wire winding.

Full-Load Current (FLC) — The current required to produce full-load torque at rated speed is called FLC.

Locked-Rotor Control (LRC) — During the acceleration period at the moment a motor is started, it draws a high current called the *inrush* current. The inrush current when the motor is connected directly to the line (so that full line voltage is applied to the motor) is called the *locked-rotor* or *stalled-rotor* current. The locked-rotor current can be from 4 to 10 times the motor full-load current. The vast majority of motors have an LRC of about 6 times FLC, and therefore this figure is generally used. The "6 times" value is often expressed as 600 percent of FLC.

Motor Speed — The speed of a squirrel-cage motor depends on the number of poles on the motor's winding. On 60 hertz, a two-pole motor runs at about 3,450 rpm, a four-pole at 1,725 rpm, a six-pole at 1,150 rpm. Motor nameplates are usually marked with actual full-load speeds, but frequently motors are referred to by their synchronous speeds — 3,600, 1,800, and 1,200 rpm respectively.

Torque — Torque is the turning or twisting force of the motor and is usually measured in pound-feet. Except when the motor is accelerating up to speed, the torque is related to the motor horsepower. If a motor is not able to furnish the proper amount of torque for a given load, it will draw an excess of current and overheat.

Ambient Temperature — The temperature of the air where a piece of equipment is situated is called the ambient temperature. Most controllers are of the enclosed type and the ambient temperature is the temperature of the air outside the enclosure, not inside. Similarly, if a motor is said to be in an ambient temperature of 30°C (86°F), it is the temperature of the air outside the

Fig. 9-2. A more complex motor controller. *(Courtesy Square D Co.)*

motor, not inside. Per NEMA standards, both controllers and motors are subject to 40°C (104°F) ambient temperature limit.

Temperature Rise — Current passing through the windings of a motor results in an increase in the motor temperature. The difference between the winding temperature of the motor when running and the ambient temperature is called the temperature rise.

The temperature rise produced at full load is not harmful provided the motor ambient temperature does not exceed 40°C (104°).

Higher temperature caused by increased current or higher ambient temperatures produces a deteriorating effect on motor insulation and lubrication. An old rule of thumb states that for each increase of 10°F above the rated temperature, motor life is cut in half.

Time (Duty) Rating — Most motors have a continuous duty rating permitting indefinite operation at rated load. Intermittent duty ratings are based on a fixed operating time (5, 15, 30, 60 minutes) after which the motor must be allowed to cool.

Motor Service Factor — If the motor manufacturer has given a motor a service factor, it means that the motor can be allowed to develop more than its rated or nameplate horsepower, without causing undue deterioration of the insulation. The service factor is a margin of safety. If, for example, a 10-hp motor has a service factor of 1.15, the motor can be allowed to develop 11.5 hp. The service factor depends on the motor design.

Jogging — Jogging describes the repeated starting and stopping of a motor at frequent intervals for short periods of time. A motor would be jogged when a piece of driven equipment had to be positioned fairly closely — for example, when positioning the table of a horizontal boring mill during set-up. If jogging is to occur more frequently than 5 times per minute, NEMA standards require that the starter be derated.

For example, NEMA Size 1 starter has a normal duty rating of 7.5 hp at 230 volts, polyphase. On jogging applications, this same starter has a maximum rating of 3 hp.

Plugging — When a motor running in one direction is momentarily reconnected to reverse the direction, it will be brought to rest very rapidly. This is referred to as plugging. If a motor is plugged more than 5 times per minute, derating of the controller is necessary, due to the heating of the contacts. Plugging can be used only if the driven machine and its load will not be damaged by the reversal of the motor torque.

Sequence (Interlocked) Control — Many processes require a number of separate motors that must be started and stopped in a definite sequence. This happens, for example, in a system of conveyors. When starting up, the delivery conveyor must start first, with the other conveyors starting the sequence. This is to avoid a pileup of material. When shutting down, the reverse sequence must be followed with time delays between the shut-downs (except for emergency stops) so that no matieral is left on the conveyors. This is an example of a simple sequence control. Separate starters could be used, but it is common to build a special controller that incorporates starters for each drive, timers, control relays, etc.

Enclosures

NEMA and other organizations have established standards of enclosure construction for control equipment. In general, equipment would be closed for one or more of the following reasons:

1. To prevent accidental contact with live parts.
2. To protect the control from harmful environmental conditions.
3. To prevent explosion or fires that might result from the electrical arc caused by the control.

The following identifies some common types of NEMA enclosures according to their numbers.

NEMA 1 General Purpose

The general-purpose enclosure is intended primarily to prevent accidental contact with the enclosed apparatus. It is suitable for general-purpose applications indoors where it is not exposed to unusual service conditions. A NEMA 1 enclosure serves as protection against dust and light indirect splashing, but is not dusttight (Fig. 9-3).

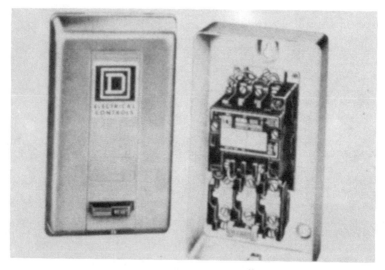

Fig. 9-3. NEMA 1, General-purpose enclosure. *(Courtesy Square D Co.)*

NEMA 3 Dusttight, Raintight

This enclosure is intended to provide suitable protection against specified weather hazards. A NEMA 3 enclosure is suitable for application outdoors, on ship docks, canal locks, and construction work, and for application in subways and tunnels. It is also sleet-resistant (Fig. 9-4).

NEMA 3R Rainproof, Sleet Resistant

This enclosure protects against interference in operation of the contained equipment due to rain, and resists damage from exposure to sleet. It is designed with conduit hubs and external mounting, as well as drainage provisions.

NEMA 4 Watertight

A watertight enclosure is designed to meet this hose test:

Enclosures shall be tested by subjection to a stream of water. A hose with one-inch nozzle shall deliver at least 65 gallons per minute. The water shall be directed on the enclosure from a

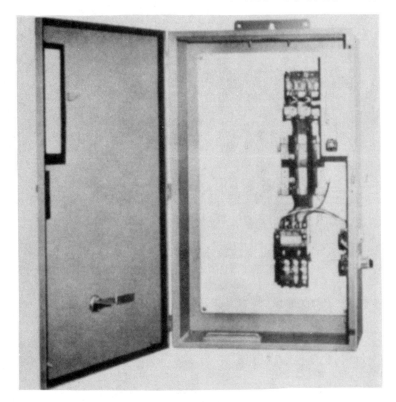

Fig. 9-4. NEMA 3, Dusttight, raintight enclosure. *(Courtesy Square D Co.)*

distance of not less than 10 feet and for a period of five minutes. During this period it may be directed in any one or more directions as desired. There shall be no leakage of water into the enclosure under these conditions.

A NEMA 4 enclosure is suitable for applications outdoors on ship docks and in dairies, breweries, etc. (Fig. 9-5).

NEMA 4X Watertight, Corrosion-Resistant

These enclosures are generally constructed along the lines of NEMA 4 enclosures except they are made of a material that is highly resistant to corrosion. For this reason, they are ideal in applications such as paper mills, meat packing, fertilizer, and

Fig. 9-5. NEMA 4, Watertight enclosure. *(Courtesy Square D Co.)*

plants, where contaminants would ordinarily destroy a steel enclosure over a period of time (Fig. 9-6).

NEMA 7 Hazardous Locations, Class I

Here the design is to meet the application requirements of the NEC for Class I hazardous locations. In this type of equipment, the circuit interruption occurs in air (Fig. 9-7).

NEMA 9 Hazardous Locations, Class II

Class II locations are those that are hazardous because of the presence of combustible dust. This enclosure has been designed to handle the requirements of Class II (Fig. 9-8).

Fig. 9-6. NEMA 4X, Watertight, corrosion-resistant enclosure. *(Courtesy Square D Co.)*

NEMA 12 Industrial Use

When it is necessary to exclude such materials as dust, lint, fibers and flyings, oil seepage, or coolant seepage, the NEMA 12 is required. There are no conduit openings or knockouts in the enclosure, and mounting is by means of flanges or mounting feet (Fig. 9-9).

NEMA 13 Oiltight, Dusttight

This enclosure is usually made by casting. It has a gasket and can be used in the same atmospheres and locations as NEMA 12. The essential difference is that, due to its cast housing, a conduit entry is provided as an integral part of the enclosure. Mounting is by means of blind holes, rather than mounting brackets (Fig. 9-10).

Fig. 9-7. NEMA 7, Class I, Groups C and D, Hazardous locations enclosure. Class I locations are those in which flammable gases or vapors are or may be present in the air in quantities sufficient to produce explosive or ignitable mixtures. *(Courtesy Square D Co.)*

Overcurrent Protection

The function of the overcurrent protective device is to protect the motor branch-circuit conductors, control apparatus, and motor from short circuits or grounds. The protective devices commonly used to sense and clear overcurrents are thermal magentic circuit breakers and fuses. The short-circuit device shall be capable of carrying the starting current of the motor but the device setting shall not exceed 250 percent of full-load current with no code letter on the motor, or from 150 to 250 percent of full-load current depending on the code letter of the motor. Where the value is not sufficient to carry the starting current, it may be increased, but shall in no case exceed 400 percent of the motor full-load current. The *National Electrical Code* requires (with few exceptions) a means to disconnect the motor and

Fig. 9-8. NEMA 9, Class II, Groups E, F, and G, Hazardous locations enclosure. The letter or letters following the type number indicates the particular group or groups of hazardous locations for which the enclosure is designed. The designation is incomplete without a suffix letter or letters. *(Courtesy Square D Co.)*

controller from the line, in addition to an overcurrent protective device to clear short circuit faults. The circuit breaker in Fig. 9-11 incorporates fault protection and disconnects in one basic device. When the overcurrent protection is provided by fuses, a disconnect switch is required. The switch and fuses are generally combined as shown in Fig. 9-13.

Overload Protection

The effect of an overload is a rise in temperature in the motor windings. The larger the overload, the more quickly the temperature will increase to a point damaging to the insulation and lubrication of the motor. An inverse relationship, therefore, exists between current and time — the higher the current, the shorter the time before motor damage or burnout can occur.

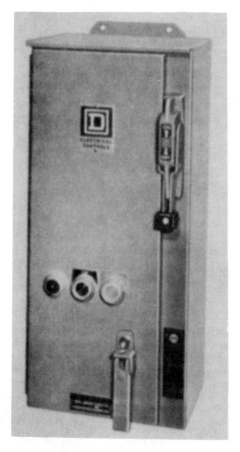

Fig. 9-9. NEMA 12 enclosure
(Courtesy Square D Co.)

All overloads shorten motor life by deteriorating the insulation. Relatively small overloads of short duration cause little damage but, if sustained, could be just as harmful as overloads of greater magnitude. The relationship between the magnitude (percent of full load) and duration (time in minutes) of an overload is illustrated by the Motor Heating Curve shown in Fig. 9-12.

The ideal overload overload protection for a motor is an element with current-sensing properties very similar to the heating curve of the motor, which could act to open the motor circuit when *full-load current* is exceeded. The operation of the protective device should be such that the motor is allowed to carry harmless overloads, but is quickly removed from the line when an overload has persisted too long.

Fuses are not designed to provide overload protection. Their basic function is to protect against short circuits. A fuse chosen on the basis of motor FLC would blow every time the motor started. On the other hand, if a fuse were chosen large enough to pass the starting or inrush current, it would not protect the motor against small, harmful overloads that might occur later.

Dual-element or time-delay fuses can provide motor overload protection. But they suffer the disadvantage of being nonrenewable and must be replaced.

Overload relays consist of a current-sensing unit connected in the line to the motor, plus a mechanism actuated by the sensing unit. This unit serves to break the circuit directly or indirectly. In a manual starter, an overload trips a mechanical latch, causing the starter contacts to open and disconnect the motor from the line. In magnetic starters, an overload opens a set of contacts within the overload relay itself. These contacts are wired in series with the

Fig. 9-10. NEMA 13, Oiltight, dusttight. *(Courtesy Square D Co.)*

starter coil in the control circuit of the magnetic starter. Breaking the coil circuit causes the starter contacts to open, disconnecting the motor from the line.

There are two classifications of overload relays — *magnetic* and *thermal*. Magnetic relays react only to current excesses and are not affected by temperature. Thermal relays can be further divided into two types — *melting alloy* and *bimetallic*.

Melting alloy thermal overload relays are also referred to as *solder pot relays*. The motor current passes through a small heater winding (Fig. 9-14). Under overload conditions the heat causes a special solder to melt. This allows a ratchet wheel to spin free. The ratchet wheel then opens the contacts. When this occurs the relay is said to trip. To obtain approximate tripping current for motors of different sizes, or different full-load currents, a range of thermal units (heaters) is available. A reset button is usually

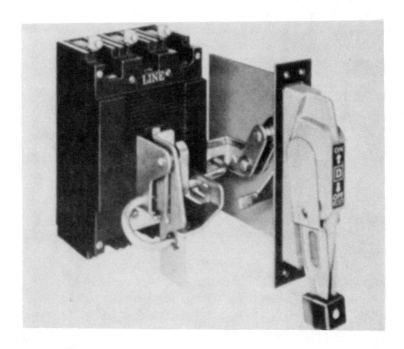

Fig. 9-11. Circuit breaker, overcurrent protective device. (*Courtesy Square D Co.*)

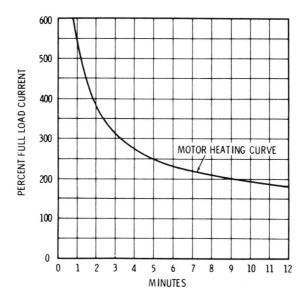

Fig. 9-12. Application of motor heating curve data.
Note: on 300% overload, the particular motor for which this curve is characteristic would reach its permissible temperature limit in 3 minutes. Overheating or motor damage would occur if the overload persisted beyond this time limit.

mounted on the cover of enclosed starters. Thermal units are rated in amperes and are selected on the basis of motor full-load current, not horsepower (Figs. 9-15, 9-16, and 9-17).

Bimetallic thermal overload relays employ a U-shaped bimetal strip that is associated with a current-carrying heater element (Fig. 9-18). When an overload occurs, the heat will cause the bimetal to deflect and open a contact. Different heaters give different trip points. In addition, most relays are adjustable over a range of 85 to 115 percent of the nominal heating rating.

A *magnetic overload relay* has a movable magnetic core inside a coil that carries the motor current. The flux set up inside the coil pulls the core upward. When the core rises far enough (determined by the current and the position of the core), it trips a set of contacts on the top of the relay. The movement of the core is

Fig. 9-13. Fusible disconnect, overcurrent protective device. *(Courtesy Square D Co.)*

slowed by a piston working in an oil-filled dashpot (similar to a shock absorber) mounted below the coil. This produces an inverse-time characteristic (Fig. 9-19). The effective tripping current is adjusted by moving the core on a threaded rod. The tripping time is varied by uncovering oil bypass holes in the piston. Because of the time and current adjustments, the magnetic overload relay is sometimes used to protect motors having long accelerating times or unusual duty cycles. The instantaneous trip magnetic overload relay is similar but has no oil-filled dashpot (Figs. 9-20 and 9-21).

ONE PIECE THERMAL UNIT

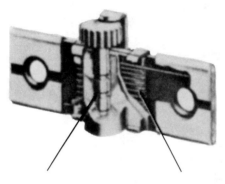

Solder pot (heat sensitive element) is an integral part of the thermal unit. It provides accurate response to overload current, yet prevents nuisance tripping.

Heater winding (heat producing element) is permanently joined to the solder pot, so proper heat transfer is always insured. No chance of misalignment in the field.

Fig. 9-14. One-piece thermal unit. Cutaway view of how it works. *(Courtesy Square D Co.)*

Manual Starter

A manual starter is a motor controller whose contact mechanism is operated by a mechanical linkage from a toggle handle or pushbutton, which is in turn operated by hand. A thermal unit and direct acting overload mechanism provides motor running overload protection. Basically, a manual starter is an "on-off" switch with overload relays.

Manual starters are generally used on small machine tools, fans, and blowers, pumps, compressors, and conveyors. They are the least expensive of all motor starters, have a simple mechanism, and

Fig. 9-15. Single-pole, melting alloy, thermal overload relay. *(Courtesy Square D Co.)*

Fig. 9-16. Three-pole, melting alloy, thermal overload relay. *(Courtesy Square D Co.)*

provide quiet operation with no AC magnet hum. Moving a handle or pushing the start button closes the contacts, which remain closed until the handle is moved to OFF, the stop button is pushed, or the overload-relay thermal unit trips.

Fractional Horsepower (FHP) Manual Starter — FHP manual starters are designed to control and provide overload protection for motors of 1 hp or less on 115 or 230 volts single phase. They are available in single- and two-pole versions and are operated by a toggle handle on the front.

Manual Motor-Starting Switches — Manual motor-starting switches provide on-off control of single-phase or three-phase AC

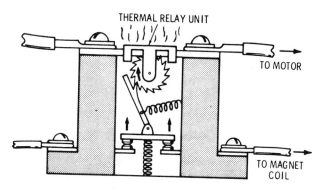

THERMAL RELAY UNIT

TO MOTOR

TO MAGNET COIL

Fig. 9-17. Operation of melting alloy overload relay. As heat melts alloy, ratchet wheel is free to turn — spring then pushes contacts open. *(Courtesy Square D Co.)*

motors where overload protection is not required or is separately provided. Two- or three-pole switches are available with ratings up to 5 hp, 600 volts, three-phase (Fig. 9-22). The continuous current rating is 30 amperes at 250 volts maximum and 20 amperes at 600 volts maximum.

The toggle operation of the manual switch is similar to the FHP

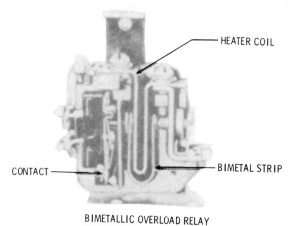

HEATER COIL

CONTACT

BIMETAL STRIP

BIMETALLIC OVERLOAD RELAY
WITH SIDE COVER REMOVED

Fig. 9-18. Cutaway view of the bimetallic thermal overlay relay. *(Courtesy Square D Co.)*

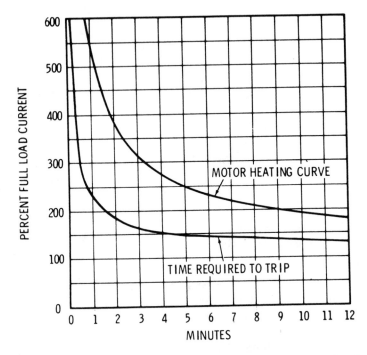

Fig. 9-19. Motor heating curve and overload relay trip curve. Overload relay will always trip out at a safe value. *(Courtesy Square D Co.)*

starter, and typical applications of the switch include small machine tools, pumps, fans, conveyors, and other electrical machinery that have separate motor protection. They are particularly suited to switch nonmotor loads, such as resistance heaters.

Integral Horsepower Manual Starter — The integral horsepower manual starter is available in two- or three-pole versions. It is used to control single-phase motors up to 5 hp and polyphase motors up to 10 hp respectively.

The two-pole starters have one overload relay. Three-pole starters usually have two overload relays but are available with three overload relays. When an overload relay trips, the starter mechanism unlatches, opening the contacts to stop the motor. The contacts cannot be reclosed until the starter mechanism has been reset by pressing the stop button or moving the handle to the reset position, after allowing time for the thermal unit to cool (Fig. 9-23).

Fig. 9-20. Magnetic overload relay. Note the size of the wire that makes up the coil.
(Courtesy Square D Co.)

Fig. 9-21. Magnetic overload relay.
(Courtesy Square D Co.)

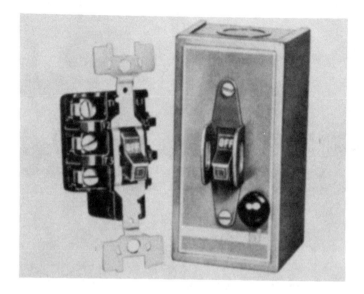

Fig. 9-22. Enclosed switch with handle guard and pilot light. *(Courtesy Square D Co.)*

Magnetic Control

A high percentage of applications require the controller to be capable of operation from remote locations, or to provide automatic operation in response to signals from pilot devices such as thermostats, pressure or float switches, limit switches, etc. Low-voltage release or protection might also be desired. Manual starters cannot provide this type of control, and therefore magnetic starters are used.

The operating principle that distinguishes a magnetic from a manual starter is the use of an electromagnet (Fig. 9-24). The electromagnet consists of a coil of wire placed on an iron core. When current flows through the coil, the iron bar, called the armature, is attracted by the magnetic field created by the current in the coil. To this extent both will attract the iron bar (the arms of the core). The electromagnet can be compared to the permanent magnet shown in Fig. 9-24.

Fig. 9-23. Integral HP manual starter in general-purpose enclosure with pilot light. *(Courtesy Square D Co.)*

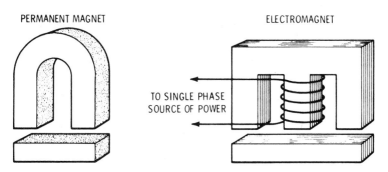

Fig. 9-24. An electromagnet and a permanent magnet. *(Courtesy Square D Co.)*

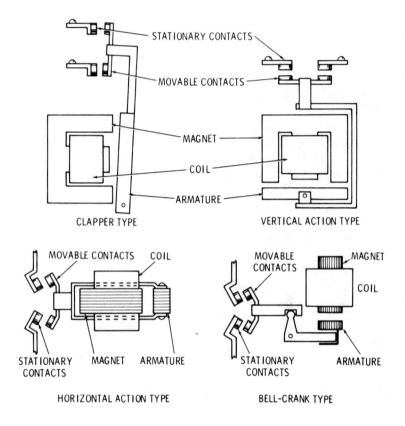

Fig. 9-25. Magnet frame and armature assemblies. *(Courtesy Square D Co.)*

The field of the permanent magnet, however, will hold the armature against the pole faces of the magnet indefinitely. The armature could not be dropped out except by physically pulling it away. In the electromagnet, interrupting the current flow through the coil of wire causes the armature to drop out due to the presence of an air gap in the magnetic circuit.

With manual control, the starter must be mounted so that it is easily accessible to the operator. With magnetic control, the push-button stations or other pilot devices can be mounted anywhere on the machine, and connected by control wiring into the coil circuit of the remotely mounted starter.

Magnet Frame and Armature Assemblies — In the construction of a magnetic controller, the armature is mechanically connected to a set of contacts, so that when the armature moves to its closed position, the contacts also close. The drawings in Fig. 9-25 show several magnet and armature assemblies in elementary form. When the coil has been energized, the armature has moved to the closed position, the controller is said to be "picked up," and the armature "seated" or sealed-in.

Magnetic Circuit — The magnetic circuit of a controller consists of the magnet assembly, the coil, and the armature. It is so named from a comparison with an electrical circuit. The coil and the current flowing in it cause magnetic flux to be set up through the iron in a similar manner to a voltage causing current to flow through a system of conductors. The changing magnetic flux produced by alternating currents results in a temperature rise in the magnetic circuit. The heating affect is reduced by laminating the magnet assembly and armature.

Magnet Assembly — The magnet assembly is the stationary part of the magnetic circuit. The coil is supported by and surrounds part of the magnet assembly in order to induce magnetic flux into the magnetic circuit.

Armature — The armature is the moving part of the magnetic circuit. When it has been attracted into its sealed-in position, it completes the magnetic circuit.

Air Gap — When a controller's armature has sealed-in, it is held closely against the magnet assembly. However, a small gap is always deliberately left in the iron circuit. When the coil is de-energized, some magnetic flux (residual magnetism) always remains — and if it were not for the gap in the iron circuit, the residual magnetism might be sufficient to hold the armature in the sealed-in position (Fig. 9-26).

Shading Coil — A shading coil is a single turn of conducting material (generally copper or aluminum) mounted in the face of the magnet assembly or armature (Fig. 9-27). The alternating main magnetic flux induces currents in the shading coil; these currents set up auxiliary magnetic flux, which is out of phase from the main flux. The auxiliary flux produces a magnetic pull that is out of phase from the pull due to the main flux, and this keeps the armature sealed-in when the main flux falls to zero (which occurs

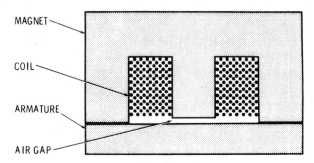

MAGNET

COIL

ARMATURE

AIR GAP

Fig. 9-26. Air gap shown in center section of the magnet. *(Courtesy Square D Co.)*

120 times per second with 60-hertz AC). Without the shading coil, the armature would tend to open each time the main flux went through zero. Excessive noise, wear on the magnet faces, and heat would occur (Fig. 9-28).

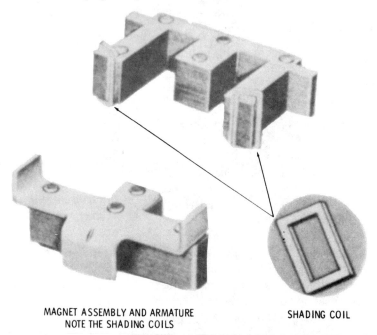

MAGNET ASSEMBLY AND ARMATURE
NOTE THE SHADING COILS

SHADING COIL

Fig. 9-27. Magnet assembly and armature. Note the shading coils.
(Courtesy Square D Co.)

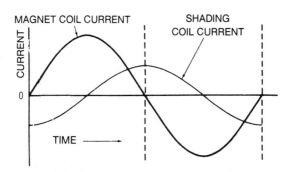

Fig. 9-28. Shading coil current vs. magnet coil current. *(Courtesy Square D Co.)*

Effects of Voltage Variation (Voltage Too Low) — Low control voltage produces low coil currents and reduced magnetic pull. On devices with vertical action assemblies, if the voltage is greater than pick-up voltage but less than seal-in voltage, the controller may pick up but will not seal. With this condition, the coil current will not fall to the sealed value. As the coil is not designed to carry continuously a current greater than its sealed current, it will quickly get very hot and burn out. The armature will also chatter. In addition to the noise, wear on the magnet faces results (Fig. 9-29).

AC Hum — All AC devices that incorporate a magnetic effect produce a characteristic hum. This hum or noise is due mainly to the changing magnetic pull (as the flux changes) inducing mechanical vibrations. Contactors, starters, and relays could become excessively noisy as a result of some of the following operating conditions:

1. Broken shading coil.
2. Operating voltage too low.
3. Wrong coil.
4. Misalignment between the armature and magnet assembly —the armature is then unable to seat properly.
5. Dirt, rust, filings, etc., on the magnet faces — the armature is unable to seal-in completely.
6. Jamming or binding of moving parts (contacts, springs, guides, yoke bars) so that full travel of the armature is prevented.

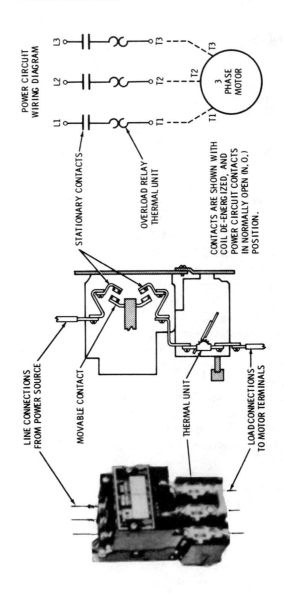

Fig. 9-29. Magnetic starter power circuit. *(Courtesy Square D Co.)*

7. Incorrect mounting of the controller, as on a thin piece of plywood fastened to a wall, for example, so that a "sounding board" effect is produced.

Holding Circuit Interlock

The holding circuit interlock is a normally open (NO) auxiliary contact provided on standard magnetic starters and contactors. It closes when the coil is energized to form a holding circuit for the starter after the *start* button has been released. As a matter of economics, vertical action contactors and starters in the smaller NEMA sizes (size 0 and size 1) have a holding interlock that is physically the same size as the power contacts (Fig. 9-30).

Electrical Interlocks — In addition to the main or power contacts that carry the motor current, and the holding circuit interlock, a starter can be provided with externally attached auxiliary contacts, commonly known as electrical interlocks. Interlocks are rated to carry only control circuit currents, not motor currents. NO and NC versions are available.

Among a wide variety of applications, interlocks can be used to control other magnetic devices where sequence operation is desired; to electrically prevent another controller from being energized at the same time; and to make and break circuits to indicating or alarm devices such as pilot lights, bells, or other signals.

Electrical interlocks are packaged in kit form, and can be easily added to the field (Fig. 9-31).

Control Device (Pilot Device) — A device that is operated by some nonelectrical means (such as the movement of a lever), and that has contacts in the control circuit of a starter, is called a control device. Operation of the control device will control the starter and hence the motor. Typical control devices are control stations, limit switches, foot switches, pressure switches, and float switches. The control device may be of the maintained contact or momentary contact type. Some control devices have a horsepower rating, and are used to directly control small motors through the operation of their contacts. When used in this way, separate overload protection (such as a manual starter) normally

Fig. 9-30. Electrical interlock, normally closed contact. *(Courtesy Square D Co.)*

Fig. 9-31. Magnetic contactor with externally attached electrical interlocks. *(Courtesy Square D Co.)*

should be provided, as the control device does not usually incorporate overload protection.

Maintained Contact — A maintained-contact control device is one that, when operated, will cause a set of contacts to open (or close) and stay open (or closed) until a deliberate reverse operation occurs. A conventional thermostat is a typical maintained-contact device.

Momentary Contact — A standard pushbutton is a typical momentary contact control device. Pushing the button will cause NO contacts to close and NC contacts to open. When the button is released, the contacts revert to their original states. Momentary-contact devices are used with three-wire control or jogging service.

Reversing Starter

Reversing the direction of motor shaft rotation is often required. Three-phase squirrel-cage motors can be reversed by reconnecting any two of the three line connections to the motor. By inter-wiring two contactors, an electromagnetic method of making the reconnection can be obtained.

As seen in the power circuit (Fig. 9-32A), the contacts *F* of the *forward* contactor, when closed, connect lines *1, 2*, and *3* to motor terminals T_1, T_2, and T_3 respectively. As long as the *forward* contacts are closed, mechanical and electrical interlocks prevent the *reverse* contactor from being energized.

When the *forward* contactor is de-energized, the second contactor can be picked up, closing its contacts *R*, which reconnect the lines to the motor. Note that by running through the *reverse* contacts, Line *1* is connected to motor terminal T_3 and line 3 is connected to terminal T_1. The motor will now run in the opposite direction.

Whether operating through either the forward or reverse contactor, the power connections are run through an overload relay assembly, which provides motor overload protection. A magnetic reversing starter, therefore, consists of a starter and contactor, suitably interwired, with electrical and mechanical interlocking to prevent the coil of both units from being energized at the same

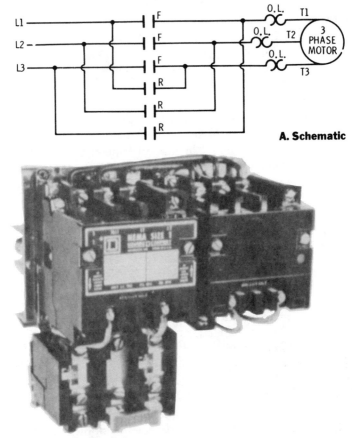

A. Schematic

B. Size 1, three-pole reversing starter.

Fig. 9-32. Reversing starter circuit. *(Courtesy Square D Co.)*

time. Fig. 9-32B shows a Size 1, three-pole reversing starter as it looks in operation.

Motor Reversing Characteristics

Various reversing characteristics of the different motor types should be checked before attempted to reverse any motor. Table 9-1 shows the reversing characteristics of a number of different motors.

Table 9-1. Motor Reversing Characteristics

	⇥	Duty	Typical Reversibility	Speed Character	Typical Start Torque*
POLYPHASE	A-C	Continuous	Rest/Rot.	Relatively Constant	175% & up
SPLIT PHASE SYNCH.	A-C	Continuous	Rest Only	Relatively Constant	125-200%
SPLIT PHASE Nonsynchronous	A-C	Continuous	Rest Only	Relatively Constant	175% & up
PSC Nonsynchronous High Slip	A-C	Continuous	Rest/Rot.†	Varying	175% & up
PSC Nonsynchronous Norm. Slip	A-C	Continuous	Rest/Rot.†	Relatively Constant	75-150%
PSC Reluctance Synch.	A-C	Continuous	Rest/Rot.†	Constant	125-200%
PSC Hysteresis Synch.	A-C	Continuous	Rest/Rot.†	Constant	125-200%
SHADED POLE	A-C	Continuous	Uni-Directional	Constant	75-150%
SERIES	A-C/ D-C	Int./Cont.	Uni-Directional●	Varying‡	175% & up
PERMANENT MAGNET	D-C	Continuous	Rest/Rot.§	Adjustable	175% & up
SHUNT	D-C	Continuous	Rest/Rot.	Adjustable	125-200%
COMPOUND	D-C	Continuous	Rest/Rot.	Adjustable	175% & up
SHELL ARM	D-C	Continuous	Rest/Rot.	Adjustable	175% & up
PRINTED CIRCUIT	D-C	Continuous	Rest/Rot.	Adjustable	175% & up
BRUSHLESS D-C	D-C	Continuous	Rest/Rot.	Adjustable	75-150%
D-C STEPPER	D-C	Continuous	Rest/Rot.	Adjustable	■

(Courtesy Bodine)

* Percentages are relative to full-load rated torque. Categorizations are general and apply to small motors.
■ Dependent upon load inertia and electronic driving circuitry.
● Usually unidirectional — can be manufactured bidirectional.
† Reversible while rotating under favorable conditions: generally when inertia of the driven load is not excessive.
‡ Can be adjusted, but varies with load.
§ Reversible down to 0°C after passing through rest.

Fig. 9-33. Timing relay.
(Courtesy Square D Co.)

Timers and Timing Relays

A pneumatic timer or timing relay is similar to a control relay, except that certain portions of its contacts are designed to operate at a preset time interval after the coil is energized or de-energized. A delay on energization is also referred to as "on delay." A time delay on de-energization is called "off delay."

A timed function is useful in applications such as the lubricating system of a large machine, in which a small oil pump must deliver lubricant to the bearings of the main motor for a set period of time before the main motor starts (Fig. 9-33).

In pneumatic timers, the timing is accomplished by the transfer of air through a restricted orifice. The amount of restriction is controlled by an adjustable needle valve, permitting changes to be made in the timing period.

Fig. 9-34. Drum switch. *(Courtesy Square D Co.)*

Drum Switch

A drum switch is a manually operated three-position, three-pole switch that carries a horsepower rating and is used for manual reversing of single- or three-phase motors. Drum switches are available in several sizes, and can be spring-return-to-off (momentary-contact) or maintained-contact. Separate overload protection, by manual or magnetic starters, must usually be provided, as drum switches do not include this feature (Fig. 9-34).

Control Station (Pushbutton Station)

A control station may contain pushbuttons, selector switches, and pilot lights. Pushbuttons may be momentary- or maintained-contact. Selector switches are usually maintained-contact, or can be spring-return to give momentary contact operation. Standard duty stations will handle the coil currents of contactors up to Size

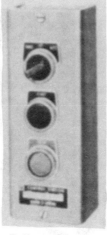

A. Standard push-
button station.

B. Heavy-duty
pushbutton station.

C. Heavy-duty oiltight
pushbutton station.

Fig. 9-35. Control pushbutton station. *(Courtesy Square D Co.)*

4. Heavy-duty stations have higher contact ratings; they provide greater flexibility through a wider variety of operators and interchangeability of units (Figs. 9-35A, B, and C).

Limit Switch

A limit switch is a control device that converts mechanical motion into an electrical control signal. Its main function is to limit movement, usually by opening a control circuit when the limit of travel is reached. Limit switches may be momentary-contact (spring-return) or maintained-contact types. Among other applications, limit switches can be used to start, stop, reverse, slow down, speed up, or recycle machine operations (Figs. 9-36A and B).

Snap Switch

Snap switches for motor control purposes are enclosed precision switches that require low operating forces and have a high repeat accuracy. They are used as interlocks and as the switch mechanisms for control devices such as precision limit switches

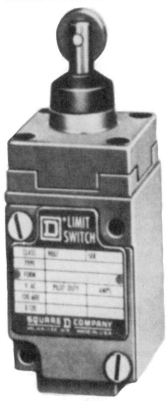

A. Heavy duty limit switch with lever arm operator.

B. Turret head limit switch with roller push rod operator.

Fig. 9-36. Limit switch. *(Courtesy Square D Co.)*

A. Double-pole, double throw snap switch.

B. Single-pole, double throw.

Fig. 9-37. Snap switches. *(Courtesy Square D Co.)*

and pressure switches. They are also available with integral operators for use as compact limit switches, door-operated interlocks, etc. Single-pole double-throw and two-pole double-throw switches are available (Figs. 9-37A and B).

Pressure Switch

The control of pumps, air compressors, welding machines, lube systems, and machine tools require control devices that respond to the pressure of a medium such as water, air, or oil. The control device that does this is a pressure switch. It has a set of contacts that are operated by the movement of a piston, bellows, or diaphragm against a set of springs. The spring pressure determines the pressures at which the switch closes and opens its contacts.

Float Switch

When a pump motor must be started and stopped according to changes in water (or other liquid) level in a tank or sump, a float switch is used. This is a control device whose contacts are controlled by movement of a rod or chain and counterweight, fitted with a float. For closed-tank operation, the movement of a float arm is transmitted through a bellows seal to the contact mechanism (Fig. 9-38).

Summary

The squirrel-cage induction motor is the most widely used. It is the workhorse of industry. An electric motor should be matched to the electrical service.

The fundamental job of a controller is to start and stop the motor. It should also protect the motor, machine, and operator. Controllers are enclosed to provide protection for operating personnel by preventing accidental contact with live parts.

Motor controls are designed to meet the provisions of the *National Electrical Code*. Standards established by the *National Electrical Manufacturers Association* assist in the proper selection of control equipment.

Fig. 9-38. Typical two-pole float switch. Float and float rod are not shown. *(Courtesy Square D Co.)*

Current required to produce full-load torque at rated speed is called FLC. Locked-rotor current is nothing more than the inrush current that occurs when the motor is first energized by line voltage. Locked-rotor current can be from 4 to 10 times the motor full-load current.

Motor speed of a squirrel-cage motor depends on the number of poles on the motor winding. The torque is the turning or twisting force of the motor and is usually measured in pound-feet or in metric kilogrammeters or gram-centimeters.

Motor controllers are used for jogging, plugging, and sequence control steps. Derating of the controls is sometimes called for when the frequency of jogging, plugging, or sequencing operation is abnormal.

Overcurrent protection is provided for motors to protect the motor branch circuit conductors, control apparatus, and motor from short circuits or grounds.

Overloads shorten motor life by deteriorating the insulation. The ideal overload protection for a motor is an element with current-sensing properties very similar to the heating curve of the motor. Fuses are designed to provide overload protection. Overload relays are actuated by a sensing unit. They may be taken out of the circuit by breaking the coil circuit. This opens the starter contacts, disconnecting the motor from the line.

The holding circuit interlock is a normally open auxiliary contact provided on standard magnetic starters and contactors. It closes when the coil is energized to form a holding circuit for the starter after the *start* button has been released.

Reversing the direction of motor rotation is often required. Three-phase squirrel-cage motors can be reversed by reconnecting any two of the three line connections to the motor. Other motors react differently to being reversed.

A drum switch is a manually operated three-position, three-pole switch that carries a horsepower rating and is used for manual reversing of a single- or three-phase motor. Limit switches, snap switches, and float switches are examples of other types of control devices.

Review Questions

1. What is a motor controller's main purpose?
2. What standards must the controller be designed to meet?
3. What is the Code that governs the use of controllers?
4. What is meant by derating a controller?
5. Can a shaded pole motor be reversed?
6. Can polyphase motors be reversed?
7. What's the typical start torque of a split phase motor?

Note: The information and pictures herein have been furnished by *Square D Company.* Some of the material has been copyrighted by *Square D* and this publication does not claim copyright for such information. *Square D* assumes no obligations or liabilities arising from reproduction.

CHAPTER 10

Motor Testing

The purpose of testing motors is primarily to make sure that they have been properly constructed, that they meet the manufacturer's guarantee, and that they will perform their assigned duty with safety under normal conditions of operation. Although the tests described herein deal principally with DC machines, they may, with certain modifications, be applied to AC machines as well.

The nature of the various tests and their functions are substantially as follows:

1. To ascertain that the machine complies with the specifications under which it was constructed.
2. To determine whether the machine is suitable for a given duty.
3. To determine the maximum capacity of the machine as limited by considerations of safety and liability to undue deterioration.

4. To determine the characteristics of the machine so that it may be known in advance how it will perform under given conditions.

5. To facilitate the location of trouble and to familiarize the operator with the characteristics of the machine.

Such tests as applied to commutating machines may be grouped and classified according to the nature of the test in the following manner.

1. Preliminary inspection and adjustment.
2. Resistance measurements.
3. Characteristics.
4. Efficiency.
5. Heating.
6. Miscellaneous.

Instruments

Prior to making a test, it is important that the necessary and correct instruments and accessories be available, in addition to the proper source of power. The instrument used depends on the nature of the test to be made. Instruments and accessories most commonly used are tachometers, frequency indicators, meggers, megohmmeters, ohmmeters, voltmeters, ammeters, shunts, graphic recording instruments, wattmeters, temperature meters, potential transformers, current transformers, etc.

Some of the precautions necessary when using instruments are as follows:

1. Electrical instruments should not be used if the date on the calibration card is over two weeks old.
2. The zero position of an instrument in use should be checked at least twice a day.
3. No instruments that are not suitable for the work should be used.
4. No readings should be taken at scale points below which there is a black or colored band.
5. Instruments should always be tested for stray fields.

6. Any portable instruments having switches should not be left continually in the circuit.
7. Voltmeters must not be left connected to inductive circuits.
8. A special switch for the purpose of disconnecting the voltmeter and table from the field of a synchronous motor when starting should always be used.
9. The voltmeter should be disconnected from the field of a synchronous converter before starting.
10. No field circuit should be opened while a voltmeter is connected across it.
11. Frequency meters should never be placed in a circuit until the voltage has been measured by a voltmeter, and no frequency meter or any other meter should be used on a circuit of higher voltage than the rated voltage of the instrument.
12. The residual voltage of a machine should never be used to operate a frequency meter.
13. Instruments should not be left lying around where they are unprotected or subject to injury.

Wiring

The power for testing is usually delivered through properly protected switchboards, and all precautions should be observed with regard to the safety of attendants, instruments, and machinery associated with the work. This is particularly important when dealing with high-voltage sources of power. A wire misplaced during a test, a wrong switch thrown, or any other careless action may result in very serious consequences to attendants or to the apparatus. Therefore, it is necessary to know the nature and wiring of a circuit before any switch or instrument manipulation takes place. It is particularly important to see that all cables and terminals are properly insulated, especially those carrying injurious voltages.

Machine Inspection

A careful inspection of the machine should be made prior to any test, and every point connected to the field, armature, commutator, brushes, brush rigging, terminal block, etc., should be noted to see that it conforms with the manufacturer's specifications. In addition, the general form of the field frame, the number of field coils and poles (main and interpoles), and the method used to connect the terminals of the windings of one field coil to another should be noted. The number of terminals on each coil will indicate the form of field winding — that is, whether *series, shunt,* or *compound.*

For a compound machine, it should be noted whether the two windings are separate or whether one is over the other, noting which one is on the outside. The method of ventilation, whether it involves open or enclosed frames or the use of fans, should be observed. Armature winding construction should be noted, whether it is *lap* or *wave,* and it should be noted whether the conductors are wound directly on the armature core or embedded in slots.

The commutator segments should be counted and the method of securing the armature winding to them noted — that is, whether they are clamped, screwed, or soldered to the commutator segments. The type and number of brushes should be noted, together with the mechanical features and method of securing the rocker arm.

When preparing commutating machines for test, the brushes must be equally spaced around the commutator, 180 electrical degrees apart. The brushes on the stud must align properly with each other and with the commutator bars from the front to the back end of the commutator. To space the brushes, place a strip of power tape around the commutator and mark the paper where the ends overlap. Remove the paper tape and divide it with a scale or dividers into as many equal divisions as there are poles on the machine. Then replace the paper around the commutator and place the overlapping ends together. Space the brushes by the marks on the paper, taking care that the holders are clamped to the stud in the proper position. In some cases, brushes are run trailing and in other cases leading, with reference to the direction of rotation. Radial brushes are also sometimes used (Fig. 10-1).

When the brushes have been set, fit them to the commutator surface, using a strip of sandpaper between the commutator and the brush surface. Coarse sandpaper should be used first to obtain

Fig. 10-1. Brush running positions used on various electric motors.

an approximate fit, followed with very fine sandpaper for a finish. A close and accurate fit on the commutator is essential to obtain good commutation results. When sandpapering, the sandpaper must be held close to the commutator to prevent rounding the tip of the brush when drawing the paper away. Sometimes the best results can be obtained by drawing the sandpaper in the direction of rotation, while at other times the best results can be obtained by drawing the sandpaper in the direction opposite to rotation.

When the sandpapering of the brushes is finished, the resulting carbon dust must be blown from the armature, or rotating part. The air blast should be directed away from the rotating part so that the carbon dust is carried completely away and cannot sift into the winding.

The following are some of the more common defects that may appear in motors with commutators:

1. Copper bridges formed between the bars over the side mica of the commutator due to improper turning.
2. Bent end conductors of commutator leads.
3. Improper brush staggering.
4. Improper alignment of brushes with commutator bars.
5. Damaged insulation of armature and field coils.
6. Broken insulation boards on fields.
7. Insufficient clearance between bare electrical terminals or conductors and ground.
8. Poor joints between electrical conductors.
9. Loose terminals, bus rings, or other connections improperly supported.
10. Brush pigtails too long or touching the armature risers.
11. Insufficient clearance between a brush stud or various parts of fittings and ground.

12. Incorrect brush spring pressure and defective spacing of collector ring taps.

The manufacturer's nameplate furnishes additional information as to the rating of the machine, terminal voltage, current at full load, type of winding, revolutions per minute of the armature, etc.

Lubrication

Manufacturers specify how often their motors should be lubricated. Shorter service intervals are recommended when the ambient temperature and the load conditions are other than normal as rated by the manufacturer. Keep in mind that excessive oiling can cause more damage than it prevents. Excess oil can contaminate windings, commutator, and internal switches. The oil attracts dirt and dust, which can build up to cause serious overheating problems.

Ball-Bearing Lubrication

The ball bearing and the roller bearing take very little lubricant. It has been estimated that only $^1/_{1000}$ of a drop will completely cover all surfaces of a 10-mm bearing. However, the lubricant must be constantly replenished. Lubricant must always be present to make the bearing operate properly. Manufacturers pass on the lubrication information for their motors, and indicate how long the bearing will last with proper lubrication and maintenance.

The lubricant for a ball bearing is usually grease or oil, often fortified with additives to improve its qualities. Keep in mind that greases in ball bearings may tend to harden or even separate when stored for long periods without use.

Temperature is the greatest enemy of ball-bearing lubricants. An increase in temperature can cause thinning of the lubricant. Grease is usually preferred over oil in the lubrication of electric motors with ball bearings.

Most ball bearings are prelubricated and will last for years without maintenance problems or attention. Larger motors have grease fittings that can increase the life of the bearings when properly utilized.

The major function of ball-bearing lubricants is as a heat dissipator. They dissipate heat caused by the friction of the bearing's being under load. The lubricant also protects the bearing members from rust and corrosion.

Ball-bearing failures are caused by:

1. Ineffective seals allowing dirt to contaminate the lubricant.
2. Excessive temperature causing a breakdown of the lubricant.
3. Contamination causing bearing failure.
4. An overload creating too much heat.
5. Too much grease creating excessive heat.

Heat is the most damaging of factors when it comes to the operation of a ball or roller bearing. Controlling the temperature of the motor and its load will increase motor life.

The best rule is to check the manufacturer's recommendations and follow them. Many factors influence the operation of bearings in motors; however, most manufacturers have tested their products and have come up with the best information on lubrication and lubricants available.

Sleeve-Bearing Lubrication

Sleeve bearings make a different demand on lubricants. The primary purpose of lubricants for sleeve bearings is to make sure there is no metal-to-metal contact between the bearing and shaft. The lubricant is expected to completely cover the shaft and prevent metal-to-metal contact.

This purpose is best served by using oil. Oil is spread all over the rotating shaft as it turns in the bearing. The oil forms a film automatically when the shaft starts to turn. A wedge-shaped gap is formed between the shaft and the bearing. Higher pressures are developed where the gap narrows, which in turn lifts the shaft and its load.

An oil film is made up of many layers (laminations) that slide on top of one another as the shaft rotates. Some friction is encountered when the oil films slide past one another. Film friction is measured in terms of viscosity. Viscosity of the oil indicates how easily it slides, or how much relative friction is present in the layers of film. The lighter-weight or lighter-viscosity oils are used in fractional horsepower motors, while larger horsepower motors may require a higher viscosity oil for lubrication purposes. Here again, temperature must be considered. Temperature affects the quality of lubrication obtained from an oil. High ambient temperatures and high operating motor temperatures have destructive effects on sleeve bearings lubricated with standard-temperature-range oils.

Special oils can be obtained for operation of motors in high temperatures. Special oils are also available to lubricate sleeve bearings where the motor operates at very low temperatures. Keep in mind that the temperature of the bearing will have a direct influence on the life of the motor.

Sleeve-bearing motors have a tendency to lose their oil film when stored for periods of one year or more.

How frequently you change the oil or clean the bearings depends on how severe the operating conditions are. The average recommended maintenance calls for periodic inspection of the oil level and cleaning and refilling with new oil every six months.

Older Motors

Older motors that are still in operating condition must also be given attention. The following information is made available for those who still have some old machines.

All machine bearings should be provided with oil wells of sufficient capacity. When oil-ring lubrication is used, the lubricating oil must not be allowed to fall to a level so low in the oil well that the ring does not dip into it. For high bearing pressures, or high speed, some form of forced lubrication is used. The oil is forced into the bearing, either on the bottom or the lower quarter, and enters the bearing in such a way that the revolving shaft draws the oil under the shaft.

Oil from forced lubricated bearings is usually returned to an external cooling tank, where its temperature is reduced before it is pumped back into the bearing. Oil rings and forced lubrication are occasionally used on the same bearings, so that if the oil pressure fails, the rings will supply enough oil to prevent danger until the oil pressure can be restored.

A properly designed bearing may run hot from the following causes: oil ring sticking, scarcity or poor quality of lubricating oil, excessive local pressure in the bearings, insufficient relief on the sides of the bearing, improper alignment and excessive belt pull, or current flowing from the shaft to the frame. The remedy for the greater part of these troubles is obvious. In the case of excessive local pressure in the bearing, or insufficient relief on the sides, the remedy is to remove the high spots of the babbitt or bearing metal with a scraper, and increase the side clearance.

Before starting a machine, all bearings must be filled with the proper amount of oil. Bearings that contain gage glasses should be filled until the glass is about two-thirds full. Bearings containing an overflow cup gage should be filled until the oil is about $1/2$ inch (12.7 mm) from the top of the gage. Bearings should be carefully filled, and no oil should be spilled on the housing or shell, or on other parts associated with the bearing, because a false impression may be obtained as to oil leakage or throwing when under test.

To give a bearing a critical test for oil leaking and throwing, the dividing line between the cap and bearing pedestal and between bearing brackets should be painted with whiting. The end of the commutator or field spider adjacent to the bearing should also be given a white coating, so that it is possible, after a comparatively short run, to detect the slightest leakage or throwing of oil.

Bearings with the end of the shell visible should be filled with oil until it touches the lower part of the shell at the end of the bearing housing. Overflow gages (Fig. 10-2) used on small machines should be filled to within $1/16$ inch (1.59 mm) of the overflow. In filling the bearings, a funnel must be used and the oil inserted into the sight holes for the oil rings or through the opening above the shaft at the end of the bearing housing.

During testing, no oil should be allowed to leak or be thrown from the bearings upon the rotating parts or windings. This is especially true with reference to commutating machines, where it is important that lubricating oil be kept away from the commutator, brushes, and fittings. Should oil leaking or throwing on these parts be detected during the test, the test should be immediately discontinued and the cause of leakage removed. If bearings under test rise 40°C (104°F) above room temperature, the fact should be referred to the engineer in charge of the test, for no properly designed bearing should rise more than 40°C (104°F) under normal conditions.

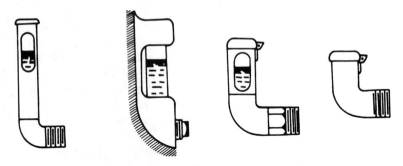

Fig. 10-2. Various types of oil gages.

Resistance Measurements

The various resistance tests made on commutating machines may be classified as follows:

1. Armature resistance.
2. Tests for locating defects.
3. Series-winding resistance.
4. Shunt-winding resistance.
5. Insulation resistance.
6. Dielectric strength.

Armature Resistance

Armature resistance includes not only the resistance of the armature itself, but also the resistance of the brushes and brush contacts — that is, the total resistance of the armature circuit from terminal to terminal. This resistance is nearly always measured by passing a known current throught the stationary armature and measuring the voltage drop between the brushes and also across the terminals of the machine. Noting the current flow and the voltage drop, it is a simple matter to calculate the resistance of the various parts of the circuit.

It should be noted that the brush-contact resistance is not constant, but varies approximately in inverse proportion to the current, from which it follows that the voltage drop of any brush is approximately constant regardless of the load. The brush-contact drop is approximately 1 volt for carbon, 0.75 volt for graphite and 0.10 volt or less for copper brushes.

Since any resistance will increase with temperature, the value obtained when the armature is cold will be considerably lower than that obtained when the armature is hot. Thus, the resistance value obtained during the test will be only approximately correct.

A simplified method of obtaining armature resistance is to pass a current through the armature and also through a standard resistance of approximately the same value and current-carrying capacity, or to compare the drop of potential across each, in which case the current need not be measured.

Tests for Locating Defects

DC windings may be tested for grounds, short circuits, high-resistance joints, wrong connections, wrong polarity, etc. If a low-resistance ground in armatures with commutators has developed, it may quickly and accurately be located by the following method: A low-voltage current sufficient to give a readable deflection on a galvanometer or millivoltmeter is passed through the armature windings from one commutator bar to the next, as shown in Fig. 10-3. A line is connected from the galvanometer to the ground, and the other galvanometer connection is placed on one of the commutator bars. The supply and galvanometer leads are then passed from segment to segment. If a full deflection is obtained, the coil is grounded.

Measuring the ohmic resistance of the winding will sometimes reveal a wrong connection, which, on a bar-to-bar measurement, would give a uniform deflection all around the commutator. Series or wave windings may sometimes have all conductors joined in series but in the wrong order, making the armature inoperative. In case of multiple or lap windings, double-, triple-, or even quadruple-spiral re-entrant windings are possible, whereas a single spiral is required. In taking a resistance measurement of the armature, see that the measurement is made from the proper commutator segments.

Special clips are attached to the commutator bars, and are spaced the number of bars apart equal to the total number of bars

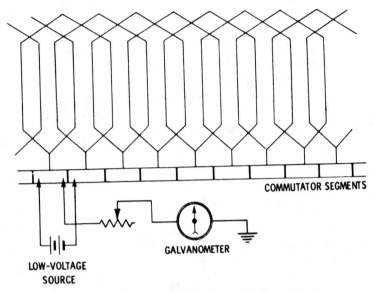

Fig. 10-3. Method of testing an armature winding for defects.

divided by the number of poles. Each alternate clip is connected in parallel to one terminal of a double bridge, and the remaining clips are connected in parallel and to the other terminal of the double bridge. The resistance reading then obtained is the true running resistance.

AC windings are similarly tested for grounds, short circuits, open circuits, wrong connections, polarity, etc. In testing for grounds, the methods and apparatus used are similar to those used to test DC machines, except that with AC, the voltage generated is usually higher. Consequently, the testing voltages are correspondingly higher, and greater care must be taken in testing. Place the testing equipment as near the apparatus as possible, as the additional capacitance of the testing lines may raise the voltage at the receiving end much higher than that at the operating end. Unless this precaution is taken, excessive voltage may be applied, which may damage the insulation.

If a ground develops, a resistance measurement will generally locate the point at which it occurs, unless each phase has two or more multiple circuits. In this case, it may be more readily located by opening one or more of the end-connection joints and separating the circuits.

A measurement may be taken as shown in Fig. 10-4, which represents a single circuit or phase of an AC machine with the ground as indicated. If the resistance between A and B is 1 ohm, betwen A and G is 0.35 ohm, and between B and G is 0.65 ohm, the location of the ground is 0.35 of the distance between A and B, from A. Since ten coils are in the circuit, the measurements show that the fourth coil, counting from A, is grounded.

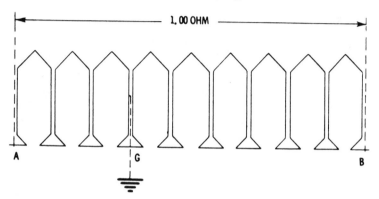

Fig. 10-4. Method of testing an AC winding for ground.

In the case of an AC winding, the ohmic resistance measurement will not always detect a wrong connection, such as a reversed coil, pole section, or phase. Although the copper resistance would be measured correctly, the total winding might be partly reversed, and therefore inoperative. Such faults can be discovered by a polarity or impedance test with alternating current. For this purpose, a single-phase current can be used, since a reading may be taken on the different circuits or between pairs of terminals successively, by shifting the testing lines until all windings have been tested.

Series-Winding Resistance

Since this resistance is of a very low value, special galvanometer measuring sets are generally used. A considerable percentage of the resistance of a series field may consist of contact resistance between the spools.

Shunt-Winding Resistance

This resistance is generally determined by passing a known current through the coils and measuring the drop of potential across the field terminals. With a given current flowing through the field, the voltage drop across any one coil of a DC machine should never be more than 5 percent higher or lower than the average drop.

Insulation Resistance

This test gives an indication of the condition of the insulation of the machine. The actual value of the resistance varies greatly in different apparatus, depending on the type, size, voltage rating, etc. The principal worth of such measurements, therefore, is in the relative values of insulation resistance that are even slightly higher than for the good material. Surface dirt or moisture that causes loss of insulation resistance may also cause low dielectric strength.

Insulation resistances will vary inversely with the temperature — that is, the insulation resistance will decrease with an increase in temperature. As an approximation, the insulation resistance will be halved by an 8°C to 10°C rise in temperature of the apparatus. As an example, when new insulation or insulation that is very damp is being dried, the resistance will generally fall rapidly as the temperature is raised to a drying value. After falling to the minimum for a given temperature, the resistance will rise as moisture is expelled from the insulation.

Measurements of insulation resistance are most readily obtained with a megger or an ohmmeter. For general maintenance work the megger should be of the 500-volt rating, although for special tests a 1,000-volt or even 2,500-volt megger may be required. When a 500-volt or 600-volt DC circuit is available, satisfactory results may be secured with a DC voltmeter having high internal resistance. The method of measurement is first to read the voltage of the line, and then to connect the resistance to be measured in series with the voltmeter and take a second reading. The measured resistance is then calculated by using the following formula:

$$R = \frac{r(V - v)}{v(1{,}000{,}000)}$$

where

V is the voltage of the line,

v is the voltage reading with the insulation in series with voltmeter,

r is the resistance of the voltmeter in ohms (generally marked on the label inside the instrument cover),

R is the resistance of the insulation in megohms (1,000,000 ohms).

The method of connecting the apparatus is shown in Fig. 10-5.

If a grounded circuit is used in making this measurement, care must be taken to connect the grounded side of the line to the frame of the apparatus to be measured, and the voltmeter between the windings and the other side of the circuit.

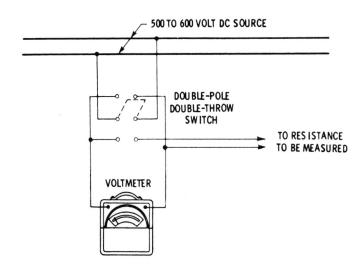

Fig. 10-5. Connections for measuring insulation resistance.

Voltmeters having a resistance of 1 megohm are now made for taking insulation resistance measurements. If one of these instruments is used, the calculation is somewhat simplified since $r = 1$ and the foregoing formula becomes:

$$R = \frac{V}{v} - 1$$

Example — In testing a motor insulated for 13,800-volt service with a line voltage, V, of 600, using a voltmeter having an internal resistance, *r*, of 1,000,000 ohms (1 megohm), and assuming the voltage, *v*, indication with the meter in series with the insulation to be 40, then:

$$R = \left(\frac{1,000,000 \, [600 - 40]}{40 \times 1,000,000} \right) - 1$$

or, since this is a megohm meter:

$$R = \frac{1,000,000 \, (560)}{40 \times 1,000,000} - 1 = 14$$

$$R = \frac{600}{40} - 1, \text{ or } R = 14 \text{ megohms}$$

(14 megohms minus 1 is still 14,000,000 for all practical purposes.)

A safe general rule is that the insulation resistance should be approximately 1 megohm for each 1,000 operating volts, with a minimum value of 1 megohm. No new apparatus should have an insulation resistance of less than 1 megohm.

Insulation resistance of apparatus in service should be checked periodically at *approximately the same temperature* and under similar conditions of humidity to determine possible deterioration of the insulation. When such measurements, made at regular intervals as a part of the general-maintenance routine, show wide variations in insulation resistance, the cause should be determined and abnormal conditions should be corrected before an insulation failure occurs.

Dielectric Strength

The purpose of the dielectric tests is to determine whether the insulation will withstand the voltage stresses occurring during normal or assumed abnormal conditions of operation. The assurance provided by a dielectric test may warrant the risk involved in applying it as preventive maintenance prior to a critical period of operation. The damage resulting from an insulation failure at the time of a properly applied dielectric or high-voltage test is likely to be slight compared to the loss caused by a breakdown while the apparatus is carrying an important load.

For example, in the chemical industry, such a breakdown might ruin the product of many days' operations, with more days being required to bring the process to the state at the time of the failure, even if other electrical apparatus were immediately available. Furthermore, the additional cost of repairs resulting from the failure under load must be given consideration.

The insulation of electrical apparatus that has been repaired or reconstructed should be given a dielectric test. Samples of insulating oil taken from transformers, induction regulators, and circuit breakers should be tested for breakdown value at regular intervals. Dielectric tests are a part of the regular test procedures of the manufacturers of electrical apparatus. *The Standards of the Institute of Electrical & Electronic Engineers (IEEE)* is the accepted authority for these tests.

Dielectric tests are generally made with alternating current of commercial frequency (25-60 hertz) for a period of 60 seconds. DC high-voltage tests are being used to a limited extent in testing insulation. Their principal use is for determining the dielectric characteristics of cables and for laboratory work. Research is being carried on to develop these test for the commercial testing of electrical-apparatus insulation. A decided advantage of high-voltage DC for maintenance insulation testing is the greatly reduced weight and size of the testing equipment required for checking large, high-voltage apparatus and machines.

Dielectric test voltage values have not been established for apparatus that has been operated in commercial service, nor is there authorization for dielectric tests for windings that have been repaired. It is rather general practice, however, to apply from 65 to 75 percent of the authorized test voltage for new apparatus. Because many of the conditions affecting such tests are extremely variable, the decision concerning procedure is generally made by the maintenance department engineer.

Motor Efficiency

The purpose of obtaining characteristic curves of a machine is to get a permanent record of its behavior by which to judge its performance and stability for different applications. The availability of a motor for any particular duty is determined almost entirely by two factors — the variation of its *torque with load* and the variation of its *speed with load.*

The more common characteristics of motors obtained by tests are:

1. Efficiency, torque, and speed as a function of current.
2. Saturation curves.
3. Speed-torque curves (series, shunt, cumulative, and differential compound).

In addition, friction, core loss, and miscellaneous load and heat-run tests are frequently made.

Efficiency — The efficiency of a motor is the ratio of its useful output to its total output, and it is written:

$$\text{Efficiency} = \frac{\text{output}}{\text{input}} \tag{1}$$

This may also be written:

$$\text{Efficiency} = \frac{\text{input} - \text{losses}}{\text{input}} \tag{2}$$

If the losses in a motor are summarized, equation (2) may be rewritten and the efficiency found by:

$$V_m = \frac{V_t I_L - (V_f I_f + I_s^2 R_s + I_a^2 R_a + P_{sp})}{V_t I_L} \tag{3}$$

where

V_m is the efficiency
V_t is the terminal voltage,
I_L is the line current,
$V_f I_f$ are the copper losses in the shunt field,
$I_s^2 R_s$ are the copper losses in the series field,
$I_a^2 R_a$ are the copper losses in the armature,
P_{sp} are the stray power losses.

Terms within parentheses are the combined total motor losses. The *copper losses* in a machine may be readily calculated in each case. The stray power losses (P_{sp}) consist of frictional losses in the bearings and brushes, windage resistance, and hysteresis and eddy currents in the armature and pole faces, but cannot be calculated directly, as they are all some function of speed or flux, or both.

Dynamometers

There are generally two classes of dynamometer, namely:

1. Absorption.
2. Transmission.

They differ mainly in that, while an absorption dynamometer absorbs the total power delivered by the motor being tested, the transmission dynamometer absorbs only that part represented by friction in the dynamomoter itself. A typical absorption dynamometer is represented by the Prony brake.

The Prony Brake

If the output of a motor is measured by a Prony brake (Fig. 10-6), then the output measured in horsepower is:

$$Bhp = \frac{2\pi(FR)N}{33,000}$$

where

Bhp is brake horsepower
F is force in pounds
R is radius of brake arm,
N is revolutions per minute.

Since FR is the torque (T), of the motor in foot-pounds (metric uses *Newton-meters*), it follows that:

$$Bhp = \frac{2\pi TN}{33,000}$$

To convert to kilogram-meters per minute, multiply hp by 4,564.

The principal forms of transmission dynomometers are the torsion and cradle types. In the torsion type, the deflection of a shaft or spiral spring which mechanically connects the driving and

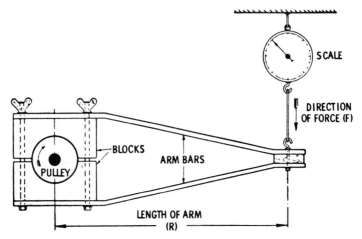

Fig. 10-6. A Prony brake. It is generally necessary to provide cooling water to the wood blocks and pulley surfaces to dissipate the heat caused by friction.

driven machines is used to measure the torque. The spring is calibrated statically by noting the angular twist corresponding to that of a known weight at the end of a measured lever arm perpendicular to the axis. The angle when testing may be measured by electrical or optical methods.

In the cradle type of dynamometer, the power is absorbed electrically by a generator whose shaft is connected to the motor under test. The term "cradle" is derived from the mounting of the generator in a trunnion-supported cradle. The pull exerted between the armature and the field tends to rotate the latter. This torque is counterbalanced and measured by weights moved along an arm in the conventional manner.

Speed-Torque Characteristics

Series Motor — The speed-torque characteristic curve of a series motor is obtained by running the motor at its rated voltage and absorbing the power by a brake. Readings of torque and speed are recorded (the speed should not be allowed to exceed four times the rated spped). The motor, under a moderate load to prevent its running away, is started by cutting out its series resistance.

The load should be gradually reduced by adjusting the brake until the maximum safe speed is recorded. At this point, the speed and torque should be recorded. Then the load should be increased step by step, taking simultaneous readings of speed and torque. The load should be increased until either the motor stops or the current exceeds the heating limit of the motor, which for a few minutes may be taken as two or three times the normal full-load current. An ammeter of sufficient rating should be connected to record the current. A voltmeter is required to make certain that the impressed voltage remains constant and equal to the rated voltage of the motor. The readings of the speed in rpm are plotted against the torque in pound-feet. Fig. 10-7 shows the characteristic curves of a series motor.

Shunt Motor — The motor is started by an ordinary shunt-motor starting rheostat, the brake being adjusted to absorb no power, and the speed of the motor measured. Then the load is increased, step by step, until the maximum safe overload or breakdown torque is reached. A line ammeter and voltmeter to record the current and voltage, respectively, are required. The readings of the speed in rpm are plotted against the torque in foot-pounds. Fig. 10-8 shows the characteristic curves of a shunt motor.

Compound Motor — If the series field in a compound-wound motor is connected so that it aids the shunt field, the motor is said to be cumulatively compounded; if the series field is connected to oppose the shunt field, the motor is said to be differentially compounded.

In a differential-compound motor, the series field opposes the shunt field so that the magnetic flux is decreased when the load is applied. This results in the speed remaining substantially constant or even increasing with an increase in load. This speed characteristic is obtained with a corresponding decrease in the rate at which the torque increases with the load.

The cumulative-compound motor has characteristics that are a combination of shunt and series characteristics. As the load is applied, the series ampere-turns increase the magnetic flux, causing the torque for any given current to be greater than it would be for a shunt motor. On the other hand, this increase in flux causes the speed to decrease more rapidly than it does in the shunt motor. This type of motor develops a high torque with a sudden increase of load, has a definite no-load speed, and has no tendency to run away when the load is removed. The speed-torque curves are

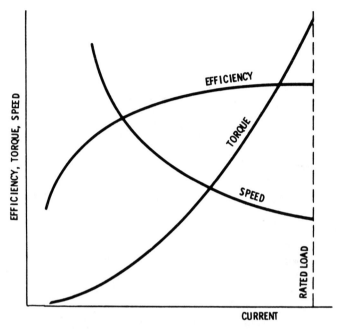

Fig. 10-7. Characteristic curves of a typical series motor.

obtained in a manner similar to that previously described for the series and shunt motor.

Loading Back Tests

When a machine is loaded on a brake or a rheostat, the total power input is converted into heat, and therefore wasted. This waste of power, particularly in the case of large machines, represents a considerable expense, and may in some cases exceed the available amount of power in the testing room or factory. To overcome this loss of energy, various testing methods have been devised. These consist commonly in loading one machine with another, so that only the losses in the two machines need be supplied by the power source.

The connection arrangement for Kapp's loading back test is shown in Fig. 10-9. With this method, the two machines are electrically connected in parallel and mechanically connected by means of a belt. One machine is run as a motor and the other as a

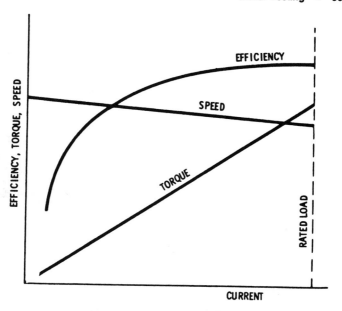

Fig. 10-8. Characteristic curves of a typical shunt motor.

generator. The output of the generator is fed back to the motor, so that the supply source has to supply only the difference between the input to the motor and the output of the generator.

In starting the test, the switch connecting the two machines is opened, and the motor is brought up to speed in the usual manner. The other machine is now being run as a generator through the belt connection. The generator field rheostat is adjusted until its voltage becomes equal to that of the supply source, after which the switch is closed. From the indication of the various instruments shown, the losses may easily be determined, and from this the efficiency, by a simple application of Ohm's law.

When the two machines are identical in construction, an approximate value of the efficiency of each may be obtained by measuring the current supplied by the power supply in addition to their terminal voltage. The total losses in the two machines are then equal to the potential across the circuit multiplied by the total current supplied. Thus, if the current and the back emf in the two machines are equal, the total power loss in each machine is simply EI. In actual practice, however, this is not exactly true since

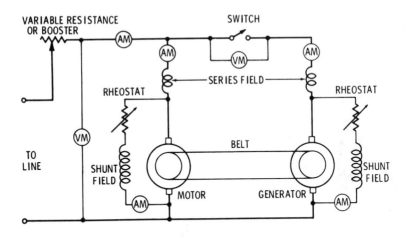

Fig. 10-9. Connections for Kapp's loading-back test.

the back emf on the motor will always be less than that on the generator. Also, the armature current of the motor will always be greater than that of the generator. Thus, the core loss of the motor will be less than the core loss of the generator, and the armature (I^2R) loss of the motor will be somewhat greater than the armature (I^2R) loss of the generator. From the foregoing, it is evident that the loss in each machine is only approximately equal to $EI/2$.

Hopkinson's loading back test differs principally from the previously described Kapp's method in that the supply losses are supplied mechanically by an auxiliary motor instead of being supplied electrically. The connections for this method are shown in Fig. 10-10. The auxiliary motor is used to supply all the losses, with the current of the armatures being circulated by weakening the field of one machine and strengthening that of the other. It should be observed, however, that the core losses in the machines are different, because the fields have different excitations. Therefore, while this method is quite satisfactory as a regulation and heating test, it is not satisfactory for efficiency tests.

Saturation Tests

In order to ascertain the characteristics of the magnetic circuit, a test known as a saturation test is made. The characteristic curve

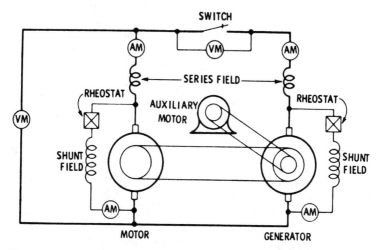

Fig. 10-10. Connections for Hopkinson's loading-back test.

may be obtained by either the generator-saturation method or the motor-saturation method.

To obtain a saturation curve by the generator-saturation method, the machine is driven at normal speed as a generator. The brushes of DC machines should always be set on neutral, and the machine run at its normal rated speed. In taking the saturation curve on polyphase AC generators, a reading of the voltage across each phase must be taken at normal field current to see if all phases are properly balanced. If they do not balance, they must be made to do so. With synchronous converters, careful reading must be taken of the DC voltage as well as the AC voltage between all phases, with the field excitation giving a normal voltage. The phase voltages must also be closely balanced.

The usual method of taking the generator saturation curve is to hold the speed constant, and then increase the field current step by step until the full field-excitation voltage has been reached, taking simultaneous readings at each step of the armature voltage, field voltage, and field amperage. About twelve points are sufficient. Four readings should be taken from zero field to 90 percent of the normal rated voltage of the machine. Five points should be taken between 90 and 100 percent of the normal rated voltage, taking one of the points of the normal rated voltage of the machine. Three points should be taken between 110 percent normal rated voltage and the full field current on the machine.

After reaching the maximum value of the field current, and without opening the field, reduce the current gradually in four or five steps, and again take readings to determine the value of residual magnetism at various points along the curve. Special care must be taken to ensure accurate readings at and above normal voltage, for in AC generators this is the portion of the curve used for calculating the regulation load. On separately excited high-voltage DC generators, the saturation curve should be carried only to 25 percent above the normal rated voltage.

When it is inconvenient or impossible to drive the machine as a generator, a motor-saturation curve may be made. For this method, the machine is operated as a free-running motor. The driving power must be furnished from a variable-voltage circuit. A certain voltage is impressed upon the armature and the motor field increased or decreased (in the case of DC machines) to give normal speed, and a record made of the armature voltage, armature amperage, field amperage, field voltage, and speed. The starting voltage should be at least 50 percent lower than the normal rated voltage of the apparatus. The applied voltage of the armature should be increased by steps to 25 percent above the normal value, and the field increased correspondingly to keep the speed constant, the same readings being recorded at the various steps as before. Readings should also be taken at three or four points as the impressed voltage and field current are lowered to approximately the values at the beginning of the test.

Care must be taken in testing DC apparatus, because unstable electrical conditions may develop and excessive speeds result. The circuit breaker in the armature circuit of the motor driving the machine must be accessible to the tester reading the speed.

On AC apparatus, the machine is run as a motor and the impressed voltage varied as already described. The speed is independent of the motor field in this case, and instead of regulating the motor field for speed, it should be regulated to give minimum input current at each voltage. Readings should be taken of each voltage impressed, the armature amperage, field amperage, and field voltage. With induction motors, it is only necessary to impress variable voltages at a constant frequency and record the readings of the impressed armature voltage, armature amperage, and speed. A curve should be plotted, using the armature voltage as ordinates and the field amperage as abscissae. Fig. 10-11 shows a typical saturation curve.

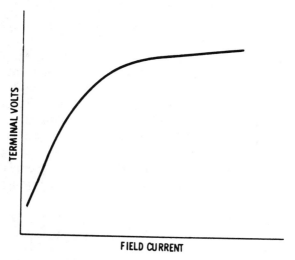

Fig. 10-11. Saturation curve of a typical shunt motor.

Core Loss

There are two methods commonly used to measure the core losses of rotating DC machines and AC synchronous apparatus: They are the *running-light* and *belted* methods.

For DC machines, the following conditions must be met in order to give satisfactory results: The brushes must be shifted on the commutator to the mechanical neutral point. The driving power must be supplied from a variable-voltage source that is not subject to sudden fluctuation. Readings must not be taken when the rotating parts are accelerating or decelerating.

In the *running-light* method, this test is made by running the machine free as a motor. It is made on most DC generators and motors that are given a running test, and occasionally on AC synchronous apparatus. When running any DC machine as a motor, the following procedure should be followed: set the brushes on mechanical neutral. Be sure that the water box (or other variable resistance) for starting is plugged across the main switch, that the main switch is open, and that the blade of the water box is nearly all the way up (Fig. 10-12). Plug the power to the armature, bring the supply voltage up, close the breaker, and lower the blade of the water rheostat slowly, watching the reading on the line amme- ter. If the machine does not start to rotate at a value of current below full load, turn the power off and investigate the cause. When the machine is running at rated voltage, close the line switch and raise the blade of the water box again. If the machine runs too fast at normal voltage, turn off the power. The connec- tions should be carefully checked to see that the field is wired properly. It may be that the field is connected directly across the main switch. If such is the case, the field current will fall rapidly as the starting resistance is cut out, and the motor will speed up. To test for incorrect wiring in the field, observe the field voltage during starting, because it will drop if the field is incorrectly connected.

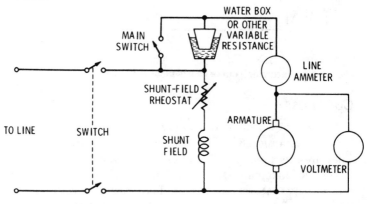

Fig. 10-12. Connections for a running-light test on a shunt motor.

When a running-light test is made on generators, the observa- tions must be made with a full-load field flux. The potential ap- plied to the armature must be equal to the normal rated voltage of the generator, increased by the IR drop under full load. The IR drop is taken as 4 percent of the normal voltage for both motors and generators. With this voltage impressed, the field current is varied until normal speed is obtained, when careful readings must be made of the armature current, armature voltage, field current, field voltage, and speed. If the machine under test is a DC motor, the voltage applied to the armature should be equal to the normal rated voltage of the motor, less the IR drop under full load. The field current is then adjusted to give normal speed, and electrical readings taken as previously outlined for DC generators.

The power supplied to machines running free will equal that absorbed in bearing, friction, brush friction, windage, and core loss, when the armature I^2R losses have been subtracted.

The record of these tests must clearly show whether the running-light current consists of the armature current plus the shunt-field current or whether it is the armature current alone. To obtain the running-light core-loss test, the machine should operate as a shunt motor.

For series-wound motors, the field should be separately excited and extreme care should be taken to see that the motor does not lose its field excitation.

To obtain running-light core loss in AC synchronous motors, they should be operated at the proper frequency and rated voltage. For the best results, both frequency and voltage must have steady value.

With normal voltage on the armature, the DC field should then be varied until minimum armature current is obtained. Readings should then be taken of the current and voltage of all phases. At minimum input current, unity power factor is obtained and, therefore, the power to drive such machines will be the volt-ampere input. Wattmeters may be used, in addition, to check the volt-ampere reading. This measurement includes friction and windage losses, together with open-circuit core losses, plus the I^2R loss in the armature. If the value of the core loss need not be separated from the other losses, the test is useful for checking load efficiency.

On motor-generator sets, unless otherwise specified in the instruction, the following running-light tests should be made: with the DC brushes down, running at the rated speed from the DC end with the line voltage 4 percent high as for any generator, and with no field on the synchronous motor, record the readings of the line voltage and current, field voltage and current, and speed.

Continuing to hold the line voltage and speed constant, bring the field excitation up on the synchronous motor until the normal voltage on the armature is obtained. Take and record the readings of the synchronous-motor armature voltage, field voltage, and amperes, together with another set of input readings on the generator as before. The difference between these two sets of input readings will be the core loss of the synchronous motor.

By means of the *belted core-loss* method, the core loss can be separated from the bearing friction, brush friction, and windage. A small DC motor is used to drive the machine under test as a generator at its rated speed (Fig. 10-13). A belt drive between these machines is most commonly used, but whenever great accuracy or a high speed is necessary, a direct drive by means of a coupling is often used.

The driving motor for this test should be such that good commutation is obtained with a fixed setting of the brushes for all the loads required by the core-loss test. With the maximum voltage on the machine under test, it should not carry more than 100 percent of its normal rated capacity.

A good rule to follow is to select a motor that has a rated capacity of approximately 10 percent of the rated output of the machine under test. When the brush setting to give the best possible commutation at all loads has been obtained, the brushes should be left in a position that will give the motor a slightly falling speed characteristic. With the brushes in this position, the voltage applied at the armature will have to be increased as the load is applied in order to maintain constant speed. The commutator surface should be in first-class condition and should have the brushes closely fitted to it.

The belt should be of minimum width and weight to carry the load without slipping. When testing motor-generator sets, synchronous converters, and other machines that do not require belts in practice, the tension of the belt must be kept as low as practical so that the bearing friction is not increased by the belt pull. Endless belts should always be used in preference to laced belts. The diameter of the pulleys should be selected so that the driving motor will run at or near its normal rate of speed when the motor under test is running at its normal speed.

The driving motor should have its field separately excited from

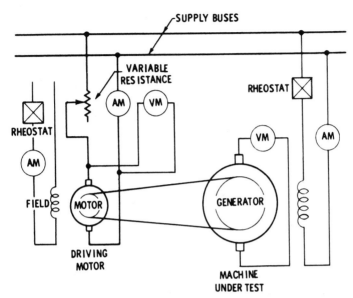

Fig. 10-13. Connections for a belted core-loss and friction test.

a constant source, and other wiring so arranged that readings may be taken of the armature current, armature voltage, field current, and speed. The test wires should be firmly attached to adjacent brushes, which should be insulated from the brush holders so that the true armature voltage may be obtained.

The machine under test should be wired as a separately excited generator with provision for reading the armature voltage, field voltage, field current, and speed. The tests should be carried out as follows: the field of the driving motor should be adjusted to about normal value and held constant, and the speed regulated by varying the voltage applied to the armature terminals. Careful readings taking simultaneous readings of the speed of both the driving motor and the machine under test: (a) with no field on the machine under test, (b) with normal field excitation. The two readings of speed should be identical. The machine should be run a length of time sufficient to allow the friction to become constant. This will be the case when the input to the driving motor remains constant when the machine under test is driven without any field excitation.

Throughout the entire test, readings must be taken at absolutely constant speed when the rotating parts are neither accelerating nor decelerating. The readings should be taken as follows: (a) take the input of the driving motor, with no field on the machine under test and with all brushes down on the commutator; (b) take the input, with all brushes raised and with no field on the test machine. The difference between these two readings is the brush friction. Then, starting with zero field on the machine under test, and with the brushes raised from the commutator, observations of input to the driving motor should be recorded at various values of field up to that which will give 125 percent of normal voltage; at least half of the readings should be taken between 90 and 110 percent of normal voltage.

The friction reading with zero excitation on the machine under test should be repeated at least three times during the test, namely, at the beginning, again near the midpoint of the curve, and finally at the end of the test.

The driving motor should then be disconnected and a running-light reading taken as follows: without changing the brush life, hold the same field current that was held during the core-loss test, and take a reading of the input to the motor to give the same speed that was read on the driving motor at the beginning of the test. The armature voltage should be lower than for any other reading taken during the core-loss test.

To check the results of the core loss as the test proceeds, the power input to the driving motor required by the core loss at a given excitation should be plotted against the voltage generated by the armature.

Correct the motor input at various field strengths by deducting the I^2R loss in the armature of the driving motor, and subtracting the power input to the driving motor with zero field on the machine under test. The core losses left correspond to the various field strengths. By subtracting the running-light input to the driving motor from the input at zero field on the machine under test, the bearing friction and windage losses of the machine are obtained.

On series motors, core-loss tests should be taken at several different speeds covering the range of the speed curve. The method used is identical with that previously described.

Synchronous AC machines generally have loss measurements as previously outlined, on open circuit, and also with the armature of the machine under test short-circuited. In the latter case, the increase in power supplied by the driving motor over that required by the friction loss is plotted as ordinates against the armature amperage as abscissae, or the open-circuit armature voltage due to a given excitation. A curve is obtained similar in character to the open-circuit core-loss curve. Such a test is commonly known as the short-circuited core loss. In making this test, careful measurements must be made of the resistance of the short-circuited armature, including all leads. Observations should be made starting at 200 percent normal current value and reducing by steps to zero current. Several thermometers should be placed on the stator winding and readings taken on the temperature after each step on the core-loss curve.

Heat-Run Tests

Heat runs are taken primarily to determine the amount of temperature rise on the different parts of the machine while running under a specified load. This rise in temperature is measured by the rise in resistance of the current-carrying part, by means of a thermometer, or by both. The results obtained by the rise in resistance, as a general rule, are used only as a check on the results obtained by reading the thermometers placed on the different parts.

The average temperature of the winding is obtained from the resistance measurements by using the following equation:

$$T = \frac{R}{r}(234.5 + t) - 234.5$$

where

T is the hot temperature in degrees C.
R is the hot resistance.
r is the cold resistance at temperature t,
t is the cold temperature of the winding in degrees C.

Example — Assume that the cold resistance is 0.50 ohm at 25°C and the hot resistance is 0.61 ohms. Then:

$$T = \frac{0.61}{0.50}(234.5 + 25) - 234.5 = 82°$$

The temperatures measured by detectors or by the change in resistance are generally higher than the thermometer measurements and are closer to the true hottest-spot temperature in the machine. For this reason, the standards permit high observable temperature rises when measured in this manner. Temperature conversions are made by using Table 10-1.

Special copper leaf brushes are commonly used for measuring the voltage of the fields of alternators and synchronous motors by the volt-ampere method. The resistance by the volt-ampere method must be made to check within 1 percent of the Wheatstone-bridge resistance measurement before the machine is started the first time. Three readings of volt-amperes should be taken at 20, 25, and 30 percent normal field voltage, with thermometers on two poles to be read at each of the foregoing voltage readings.

Self-excited DC machines must be checked, as described previously, immediately after the voltage is brought up the first time. As the heat run progresses, the temperature rise of the field should be calculated and recorded at the same time the thermometer readings are taken. All meters to be used on a heat run should be checked against other meters before the run is started. The same meters must be used throughout any given test.

Thermometers should be carefully examined for a broken mercury column before being used and should not be inverted or placed on a machine so that the bulb is at a higher level than the other end. Before starting a heat run, thermometers should be placed on the stationary accessible parts of the machine. Each thermometer should be attached with the bulb in contact with the part of which the temperature is required, and should have the bulb covered with an amount of putty sufficient to secure it to the machine and to shield it from the surrounding air. Extreme care must be exercised regarding the amount of putty used, as too much putty is as bad as too little. Just enough should be used to do the work required. There should be no restriction of the natural windage of the machine, and no radiation from the coil of which the temperature is being measured.

Table 10-1. Temperature Conversion Table

°F ⇌ °C

The numbers in italics in the center column refer to the temperature, either in Celsius or Fahrenheit, which is to be converted to the other scale. If converting Fahrenheit to Celsius, the equivalent temperature will be found in the left column. If converting Celsius to Fahrenheit, the equivalent temperature will be found in the column on the right.

-100 TO 30			31 TO 71			72 TO 212			213 TO 620			621 TO 1000		
C		F	C		F	C		F	C		F	C		F
-73	*(-)100*	-148	-0.6	*31*	87.8	22.2	*72*	161.6	104	*220*	428	332	*630*	1166
-68	*(-)90*	-130	0	*32*	89.6	22.8	*73*	163.4	110	*230*	446	338	*640*	1184
-62	*(-)80*	-112	0.6	*33*	91.4	23.3	*74*	165.2	116	*240*	464	343	*650*	1202
-57	*(-)70*	-94	1.1	*34*	93.2	23.9	*75*	167.0	121	*250*	482	349	*660*	1220
-51	*(-)60*	-76	1.7	*35*	95.0	24.4	*76*	168.8	127	*260*	500	354	*670*	1238
-46	*(-)50*	-58	2.2	*36*	96.8	25.0	*77*	170.6	132	*270*	518	360	*680*	1256
-40	*(-)40*	-40	2.8	*37*	98.6	25.6	*78*	172.4	138	*280*	536	366	*690*	1274
-34.4	*(-)30*	-22	3.3	*38*	100.4	26.1	*79*	174.2	143	*290*	554	371	*700*	1292
-28.9	*(-)20*	-4	3.9	*39*	102.2	26.7	*80*	176.0	149	*300*	572	377	*710*	1310
-23.3	*(-)10*	14	4.4	*40*	104.0	27.2	*81*	177.8	154	*310*	590	382	*720*	1328
-17.8	*0*	32	5.0	*41*	105.8	27.8	*82*	179.6	160	*320*	608	388	*730*	1346
-17.2	*1*	33.8	5.6	*42*	107.6	28.3	*83*	181.4	166	*330*	626	393	*740*	1364
-16.7	*2*	35.6	6.1	*43*	109.4	28.9	*84*	183.2	171	*340*	644	399	*750*	1382
-16.1	*3*	37.4	6.7	*44*	111.2	29.4	*85*	185.0	177	*350*	662	404	*760*	1400
-15.6	*4*	39.2	7.2	*45*	113.0	30.0	*86*	186.8	182	*360*	680	410	*770*	1418
-15.0	*5*	41.0	7.8	*46*	114.8	30.6	*87*	188.6	188	*370*	698	416	*780*	1436
-14.4	*6*	42.8	8.3	*47*	116.6	31.1	*88*	190.4	193	*380*	716	421	*790*	1454
-13.9	*7*	44.6	8.9	*48*	118.4	31.7	*89*	192.2	199	*390*	734	427	*800*	1472
-13.3	*8*	46.4	9.4	*49*	120.0	32.2	*90*	194.0	204	*400*	752	432	*810*	1490
-12.8	*9*	48.2	10.0	*50*	122.0	32.8	*91*	195.8	210	*410*	770	438	*820*	1508
-12.2	*10*	50.0	10.6	*51*	123.8	33.3	*92*	197.6	216	*420*	788	443	*830*	1526
-11.7	*11*	51.8	11.1	*52*	125.6	33.9	*93*	199.4	221	*430*	806	449	*840*	1544
-11.1	*12*	53.6	11.7	*53*	127.4	34.4	*94*	201.2	227	*440*	824	454	*850*	1562
-10.6	*13*	55.4	12.2	*54*	129.2	35.0	*95*	203.0	232	*450*	842	460	*860*	1580
-10.0	*14*	57.2	12.8	*55*	131.0	35.6	*96*	204.8	238	*460*	860	466	*870*	1598
-9.4	*15*	59.0	13.3	*56*	132.8	36.1	*97*	206.6	243	*470*	878	471	*880*	1616
-8.9	*16*	60.8	13.9	*57*	134.6	36.7	*98*	208.4	249	*480*	896	477	*890*	1634
-8.3	*17*	62.6	14.4	*58*	136.4	37.2	*99*	210.2	254	*490*	914	482	*900*	1652
-7.8	*18*	64.4	15.0	*59*	138.2	37.8	*100*	212.0	260	*500*	932	488	*910*	1670
-7.2	*19*	66.2	15.6	*60*	140.0	43	*110*	230	266	*510*	950	493	*920*	1688
-6.7	*20*	68.0	16.1	*61*	141.8	49	*120*	248	271	*520*	968	499	*930*	1706
-6.1	*21*	69.8	16.7	*62*	143.6	54	*130*	266	277	*530*	986	504	*940*	1724
-5.6	*22*	71.6	17.2	*63*	145.4	60	*140*	284	282	*540*	1004	510	*950*	1742
-5.0	*23*	73.4	17.8	*64*	147.2	66	*150*	302	288	*550*	1022	516	*960*	1760
-4.4	*24*	75.2	18.3	*65*	149.0	71	*160*	320	293	*560*	1040	521	*970*	1778
-3.9	*25*	77.0	18.9	*66*	150.8	77	*170*	338	299	*570*	1058	527	*980*	1796
-3.3	*26*	78.8	19.4	*67*	152.6	82	*180*	356	304	*580*	1076	532	*990*	1814
-2.8	*27*	80.6	20.0	*68*	154.4	88	*190*	374	310	*590*	1094	538	*1000*	1832
-2.2	*28*	82.4	20.6	*69*	156.2	93	*200*	392	316	*600*	1112			
-1.7	*29*	84.2	21.1	*70*	158.0	99	*210*	410	321	*610*	1130			
-1.1	*30*	86.0	21.7	*71*	159.8	100	*212*	414	327	*620*	1148			

(Courtesy Bodine)

Thermometers that are to register the temperature of the air ducts should be placed so that the bulbs cannot make contact with the iron laminations while the machine is running. Thermometers that are liable to be shaken off by continued action of windage or vibration should be securely fastened to the machine. When placing thermometers on field coils, care should be taken to see that they are not placed on the fiber strips protecting the outside terminals. These fiber strips run from one terminal to the other and from a nonconducting wall between the coil and its outside insulation, and thus do not represent the true temperature of the coil. Coils above the horizontal centerline of the machine should be used as the top of the machine. They are usually somewhat hotter than the bottom coils. On small machines, two thermometers will be sufficient on the coils, but larger machines should have at least four.

One thermometer will be sufficient on the frame of small machines, but two should be used on large units. At least two thermometers should be used on the laminations and ducts of small machines, and four or more used on the larger machines.

Any large machine requiring considerable floor space should have the room temperature taken at four or more nearby points. These thermomenters should be place about 6 feet (1.829 m) from the base and at the elevation of the centerline of the machine. These thermometers should be placed two together; one set should be placed in an oil cup designed for this purpose, and the other set should hang in the air to record the air temperature. All of the room-temperature thermometers, whether they are placed in air or oil, should be so located that they are not affected by the windage and radiation of the machine under test.

The machine should be shielded from currents or air coming from adjacent pulleys, belts, and other machines. A very slight current of air will cause great discrepancies in the heating results. Consequently, a suitable canvas screen should be used to protect the machine under test when necessary. Great care must be used, however, to see that the screen does not interfere with the natural ventilation of the machine under test. Care should always be taken to see that sufficient floor space is left between machines to allow free circulation of air. Under ordinary conditions, a distance of 6 feet (1.829 m) is considered to be sufficient.

Immediately after the heat run is started, various checks should be made. On all heat runs where meters are connected in both the output and input circuits, especially motor-generated sets and converters, the power-output readings plus the losses of the apparatus under test must equal the power-input readings. On all synchronous machines, the power factor specified in the instructions should be held and checked against the phase characteristics on motors and against saturation on generators. In such cases, the field current held during the heat run should be greater than the unity-power-factor field current.

Summary

Electric motors are tested to make sure that they have been constructed properly, that they meet the manufacturer's guarantee, and that they can perform their assigned duty with safety under normal conditions of operation. The tests can be applied to AC machines as well as to DC machines.

Prior to making a test, it is important that the necessary instruments and accessories are available, in addition to the proper source of power. Instruments and accessories used most commonly are tachometers, frequency indicators, meggers, megohmmeters, ohmmeters, voltmeters, ammeters, shunts, graphic recording instruments, wattmeters, temperature meters, potential transformers, current transformers, etc.

Careful inspection of the machine should be made prior to any test to see that each point connected to the field, armature, commutator, brushes, brush rigging, terminal block, etc., conforms with the manufacturer's specifications. The number of terminals on each coil indicates the form of field-winding — *series, shunt,* or *compound.*

Various resistance tests are made on commutating machines. These tests are classified as armature resistance, location of defects, field-winding resistance, insulation resistance, and dielectric strength.

The availability of a motor for a particular duty is determined almost entirely by two factors — variation of *torque with load* and variation of *speed with load.* The more common motor characteristics that can be obtained by testing are:

1. Efficiency, torque, and speed as a function of current.
2. Saturation curves.
3. Speed-torque curves (series, shunt, cumulative, and differential compound).

Heat-run tests are made primarily to determine the amount of temperature rise on the various parts of the machine while running under a specified load. The average temperature of the winding is obtained from the resistance measurements by the equation:

$$T = \frac{R}{r} (234.5 + t) - 234.5$$

in which

  T is the hot temperature in degrees C,
  R is the hot resistance,
  r is the cold resistance at temperature t,
  t is the cold temperature of the winding in degrees C.

Review Questions

1. What is the purpose of testing electric motors?
2. Why should the motor be inspected before testing is begun?
3. What are the various resistance tests made on commutating machines?
4. What motor characteristics can be obtained from tests?
5. Why are heat-run tests made on electric motors?
6. What is the purpose of a lubricant in ball bearings?
7. What is the purpose of a lubricant in sleeve bearings?
8. What is viscosity?
9. How often should a motor be lubricated?
10. What type of lubricant do you use for a sleeve bearing?

CHAPTER 11

Motor Maintenance

To ensure the best operation, a systematic motor inspection should be made at regular intervals, the frequency depending on the type of service and operating conditions. There are five important procedures that should be followed in the proper maintenance of motors. They are:

1. Schedule the shutdown time of machines and notify the electrical department whenever a machine is, or will be, available for mechanical work so that the electric equipment can be checked simultaneously.
2. Build up and maintain a reserve of units, parts, and sub-assemblies so that quick changes can be made.
3. Keep spare units and renewal parts clean, dry, and in good order for quick application.
4. Keep records of troubles and remedies as a means for running down chronic cases and finding the causes.

5. Make inspections as regularly and thoroughly as circumstances will permit.

Check the installations of new motors, controls, and replacement after putting them into service. Some maladjustment may be discovered that might cause considerable trouble if the scheduled inspection date is far ahead. Dampness or excessive temperature and humidity will, in time, deteriorate the insulation of the motor windings, resulting in grounds or short circuits. Particular care must be given to frequent inspection, maintenance, and repair of motors operating under such adverse conditions.

Keep a card file of the equipment. Set up a schedule of inspection that can be met with a reasonable man-hour demand. Find out what an increased department output will mean to the motors, and be sure that the electrical loads are not stepped up to a point where failures will curtail production.

All types of motors have many elements in common for which maintenance work is fairly uniform. Insulated windings, brushes, collector rings, commutators, and bearings are treated independently of the machines in which they are used. The same is true of the disassembly and assembly, and also of external inspection.

Inspection

Advance notice of trouble is often provided by inspection. Careful inspections by a conscientious and observant person are much more valuable than careless routine inspections at more frequent intervals. Since a lubrication person should visit each motor at fairly regular intervals, he or she should be able to check for noise, heat, changes in surroundings, or other abnormal conditions. Regular and more detailed inspections can be scheduled, but a system should be established that can be continued in a consistent manner.

The frequency of inspections that should be made will vary between plants and between departments in a given plant. It will also vary with the types of motors involved. While inspections are normally scheduled for definite intervals, the intervals are only relative and, in many cases, can be somewhat extended, depending on local conditions.

Weekly Inspection

Sleeve Bearings — If an oil gauge is used on sleeve-bearing motors, check the oil level and fill to the line, if necessary. Where the journal size is less than 2 inches, stop the motor before attempting to check the oil level.

For special lubricating systems, such as wool-packed, forced-flood, and disk, refer to the manufacturer's recommendations.

Add oil to the bearing housing if needed, but only while the motor is stopped. Check to see if oil is creeping along the shaft toward the windings. If oil creep is present, report the condition immediately.

Ball or Roller Bearings — Feel the bearing housing for evidence of vibration, any unusual noise, or excessive temperature. Inspect for creepage of grease on inside of motor.

Brushes, Commutator, and Rings — Check for sparking at the brushes. If the motor is on a cycle cuty, observe it through several cycles. Note the color and condition of the commutator or rings. Inspect all brushes for wear, and check pigtail connections for looseness. Check commutators or rings for burned spots or roughness. Clean the surfaces with a commutator cleaning stick.

Rotors and Armatures — Check the air gap on sleeve-bearing motors, especially if recently overhauled. If new bearings have been installed, make sure that the average air gap is within 10 percent if the gap is normally less than 20 mils (0.5 mm). See if the air passages through the punching are clogged with foreign matter. If needed, clean by vacuuming or mild blowing. Make sure the machine is stopped and the electricity off. Wipe off the dust with a dry cloth. Check for moisture and the accumulation of water in the bottom of the frame. See if any oil or grease has worked its way up to the rotor or armature windings. If so, clean with a degreaser in a well-ventilated room.

Mechanical Condition — Check belts for suitable slack and surface condition. Look at all gears for signs of uneven wear. Check flexible couplings. If they are properly lined up, there will be no noise or evidence of excessive use of the flexible factor. See if chains clear the housings. Check the chain lubrication, especially for excess grease and oil in the bottom of the housing. See if sprockets show even wear across the teeth. Check the motor starting to see that the motor comes up to proper speed each time power is applied. Report any abnormal action for a control checkup.

Monthly Inspection

Winding — Check shunt, series, and commutating fields for tightness. Try to move the field spools on the poles; drying out may cause some play. Check the machine cable connections for tightness.

Brushes — Check the brushes in their holders for fit and free play. Check the brush-spring pressure. Tighten the brush studs to take up any slack caused by the washers' drying out. Be sure the studs are not displaced on DC units. Replace all brushes that are worn down almost to the brush rivet. Examine the brush faces for chipped toes or heels, and for heat cracks. Replace any that are damaged.

Commutators — Examine commutator surfaces for high bars and high mica, or for evidence of scratches or roughness. See that the risers are clean and have not been damaged in any way.

Ball or Roller Bearings — On hard-driven twenty-four-hour-service motors, purge out all old grease through the drain hole and apply new grease. Check to make sure that grease or oil is not leaking out of the bearing housing. If any leakage is present, correct the condition before continuing to operate.

Sleeve Bearings — Look for bearing wear, including end play at the bearing surfaces. Clean out oil wells if there is evidence of dirt or sludge. Flush well with lighter oil before refilling.

Enclosed Gears — Open the drain plug and check the oil flow for the presence of metal scale, sand, or water. If the condition of the oil is bad, drain, flush, and refill as directed. Rock the rotor to see if the slack or backlash is increasing.

Couplings and Drive Units — Note if the belt-tightening adjustment is all taken up. Shorten the belt if this condition exists. See if the belt runs steadily and close to the motor edge of the pulley. Check chains for evidence of wear and stretch. Clean the inside of chain housings. Check the chain-lubricating system. Note the incline of slanting bases to make sure they do not cause the oil rings to rub on the housing.

Loads — Check loads for changed conditions, bad adjustment, poor handling, or control trouble.

Yearly Inspection

Windings — Check the insulation resistance. Also check the insulation surfaces for dry cracks and other evidence of the need for coatings of insulating material. Clean all surfaces and ventilation passages thoroughly if an inspection shows an accumulation of dust. Check for moisture, mold, or water standing in motors to see if varnishing and baking are needed after drying out the windings.

Air Gaps and Bearings — Check all air gaps to make sure that the average reading is within 10 percent, providing the reading should be less than 20 mils (0.5 mm). All bearings should be thoroughly checked and the defective ones replaced. Waste-packed and wick-oiled bearings should have the waste or wicks renewed if they have become glazed or filled with metal or dirt. See that the new waste bears snugly against the shaft.

Squirrel-Cage Rotors — Check for broken or loose bars and evidence of local heating. If fan blades are not cast in place, check for loose blades. Look for marks on the rotor surface indicating foreign matter in the air gap, or worn bearings.

Wound Rotors — Thoroughly clean around the collector rings, washers, and connections. Tighten the connections if necessary. If the rings are rough, spotted, or eccentric, send them to the repair shop for refinishing. See that all top sticks or wedges are tight. Thoroughly clean all air passages. Look for oil or grease creepage along the shaft, checking back to the bearing. Check the commutator for surface condition, high bars, high mica, and eccentricity.

Mechanical Condition — Check belts or chains for possible need of replacement. Do not wait for them to break during a production rush. Clean the inside and outside of the frames and end belts. Turn the rotor by hand to locate misalignment, a sprung shaft, or rubbing. If the motor is connected to a machine, check the turning with a lever to see if the machine imposes an excessive friction load.

Loads — Read the load on the motor with the instruments at no load, also at full load, or through a complete cycle. This will serve as a check on the mechanical condition of the driven machine.

Initial Motor Starting

Prior to starting a motor for the first time, observe the following:

1. Dry out all moisture. If the motor has been exposed to a moist atmosphere for long periods while in transit or storage (or has remained idle for a long period after being installed in a moist atmosphere), it should be dried out thoroughly before being placed in service. If possible, place the motor in an oven and bake at a temperature not exceeding 85° (185°F). Fair results can be obtained by enclosing the motor with canvas or other covering, inserting heating units or incandescent lamps to raise the temperature, and leaving a hole at the top of the enclosure to permit the escape of moisture. The motor may also be dried out by passing a current at low voltage (motor at rest) through the field windings sufficient to raise the temperature to not over 85°C (185°F). Increase the temperature gradually to this value, keeping the temperature of the entire winding as nearly uniform as possible.

2. See that the voltage on the motor and control nameplates corresponds with that of the power supply.

3. Check all connections to the motor and control with the wiring diagrams.

4. Make sure that the drain plugs are tight and that the bearings are lubricated properly.

5. If oil-ring bearings are used, make sure that the oil rings turn freely.

6. Remove all the external load, if possible, and turn the armature by hand to see that it rotates freely.

7. Before putting the motor in service, it is desirable to operate without load long enough to determine that there is no unusual localized heating.

Determination of Belt Sizes

For a given condition, the minimum belt width is determined by the horsepower to be transmitted by the speed of the belt. The rules are as follows:

Rule 1

A single belt 1 in. (25.4 mm) wide, running at 800 ft. (243.84 m) per minute, will deliver approximately 1 horsepower (0.746kW) (up to about 4,000 rpm).

Rule 2

A double belt 1 in. (25.4 mm) wide, running at 500 ft. per minute (152.4 m), will deliver approximately 1 horsepower (0.746kW) (up to about 4,000 rpm).

These rules give wide margins of safety in ordinary power transmission when the speed of the belt does not exceed 4,000 feet (1.22 km) per minute. Above this speed, the centrifugal effects due to the belt weight begin to be rather appreciable and should therefore be included.

For heavy transmission and speeds exceeding 4,000 feet (1.22 km) per minute, the foregoing rules should be checked against the manufacturer's recommendations. It is more convenient, for calculation purposes, to express the rules just given in algebraic form. Thus, for single belts:

$$B_w = \frac{hp \times 800}{V}$$

where

B_w is the belt width in inches,
V is the belt speed in feet per minute,
hp is the horsepower transmitted,

Similary, for double belts:

$$B_w = \frac{hp \times 500}{V}$$

Note — If metric measurement is used you will have to develop a different formula.

Example — If a single belt is used in the previous example, what will be its size?

Solution — Substituting numerical values, the width of the single belt is:

$$B_w = \frac{50 \times 800}{4000} = 10 \text{ inches}$$

The ratio

$$\frac{10}{6.25} = 1.6$$

indicates that a double belt is not considered twice as effective as a single belt, but only 1.6 times as effective. A useful listing giving the horsepower per inch of width transmitted by leather belts is given in Table 11-1. To obtain the horsepower transmitted by belts of any width, multiply the figure shown for the given belt speed by the width of the belt used. Thus, for example, if the belt speed is 4,000 feet (1.22 km) per minute and the width of the double belt is 6 1/4 in. (15.9 cm), multiply the figure corresponding to a speed of 4,000 feet per minute by the width of the belt to obtain the horsepower transmitted. Thus,

$$6.25 \times 7.95 = 49\,^5/_8, \text{ or approximately 50 hp}$$

Again, if the horsepower to be transmitted and the belt speed in feet per minute are known, divide the figure shown for a given belt speed by the horsepower to be transmitted to obtain the belt width.

Example — An electric motor rated at 10 hp has a full-load speed of 1,150 rpm and a pulley diameter of 8 in., and is to drive a drilling machine by means of a belt. What width of a single belt is required?

Solution — Since the speed of the motor is 1,150 rpm and its pulley diameter is 8 in., it is necessary to first find the belt speed in feet per minute. This will conveniently be found by remembering that

$$V = \pi DN$$

where

V is the belt speed,
π is 3.1416,
D is the pulley diameter in feet,
N is the rpm of motor.

$$V = \pi \times \frac{8}{12} \times 1,150 = 2407.33 \text{ feet per min.}$$

By consulting Table 11-1, the figure corresponding most nearly to a belt speed of 2,407.33 is 3.33. By dividing this figure by the horsepower transmitted to the belt, the width of the belt is finally obtained:

$$\text{belt width (single)} = \frac{10}{3.33} = 3 \text{ inches}$$

Belt-Drive Arrangements

Whenever conditions permit, horizontal drives should be used for belts. The proper procedure is to arrange the belt as shown in Fig. 11-1 — that is, with the lower side of the belt driving. The sag of the upper side of the belt tends to increase the arc of contact on the pulley. This sag should be about $1\frac{1}{2}$ inches for every 10 feet of center distance between the shafts. If too loose, the belt will have an unsteady flapping motion, which will injure both the belt and

Table 11-1. Speed and Transmitted Horsepower of Leather Belts

Belt Speed, Ft. per Min.	HP per Inch of Width	Belt Speed, Ft. per Min.	HP per Inch of Width	Belt Speed, Ft. per Min.	HP per Inch of Width	Belt Speed, Ft. per Min.	HP per Inch of Width
Single Belting							
900	1.25	2000	2.78	3100	4.31	4200	5.84
1000	1.39	2100	2.92	3200	4.45	4300	5.98
1100	1.53	2200	3.06	3300	4.59	4400	6.12
1200	1.67	2300	3.20	3400	4.72	4500	6.26
1300	1.81	2400	3.33	3500	4.86	4600	6.40
1400	1.95	2500	3.47	3600	5.00	4700	6.54
1500	2.09	2600	3.61	3700	5.14	4800	6.68
1600	2.22	2700	3.75	3800	5.28	4900	6.82
1700	2.36	2800	3.89	3900	5.42	5000	6.96
1800	2.50	2900	4.03	4000	5.56	.....	.....
1900	2.64	3000	4.17	4100	5.70	.....	.....
Double Belting							
900	1.79	2000	3.97	3100	6.16	4200	8.34
1000	1.99	2100	4.17	3200	6.36	4300	8.54
1100	2.19	2200	4.37	3300	6.55	4400	8.74
1200	2.39	2300	4.57	3400	6.75	4500	8.94
1300	2.58	2400	4.77	3500	6.95	4600	9.13
1400	2.78	2500	4.97	3600	7.15	4700	9.33
1500	2.98	2600	5.16	3700	7.35	4800	9.53
1600	3.18	2700	5.36	3800	7.55	4900	9.73
1700	3.38	2800	5.56	3900	7.74	5000	9.93
1800	3.58	2900	5.76	4000	7.94	.....	.....
1900	3.78	3000	5.96	4100	8.14	.....	.....

the machinery. If too tight, the bearings will be put under a severe strain, resulting in a probable early failure, and the belt will be quickly destroyed.

Note: To Convert to Metric

Belt speed ft/min to Belt Speed m/sec.	hp/inch width to kW/mm width
Multiply ft/min by 304.8, then divide by 60	*Multiply hp/inch by 0.746 and divide by 25.4*

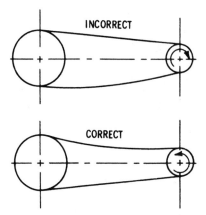

Fig. 11-1. Correct and incorrect direction of the belt for horizontal drive.

Vertical drives arranged as shown in Fig. 11-2 should be avoided whenever possible. This is particularly true where the lower pulley is the smaller. Vertical drives prove less satisfactory than horizontal drives because, in the vertical drive, the effective tension and arc contact are substantially reduced, and as a result the slip will increase in extreme cases to the point where the normal load

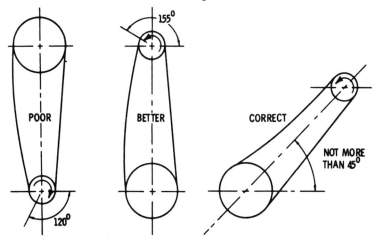

Fig. 11-2. Various arrangements of vertical drives.

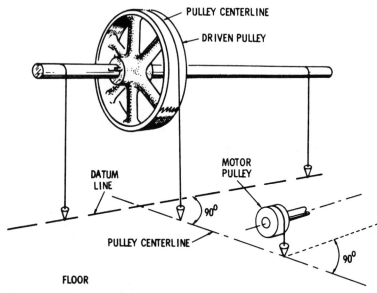

Fig. 11-3. Method of aligning pulleys.

cannot be carried. Fig. 11-2 illustrates how, in the best arrangement, the angle should not exceed 45°.

In group drives where several belts transmit power from a line shaft, it is advantageous to locate the drive shaft so that the bearing pressure can be equalized and reduced by alternating the direction of the drives, first on one side and then on the other.

Alignment

After the motor has been properly located and the right kind of mounting has been provided, the next step is to align the machine with its drive. The tools usually employed for alignment of machinery or line shafting are the square, plumb bob, and level. Fig. 11-3 shows a simple and easy method of aligning a motor pulley with the driven pulley. First, the crown or centerline of the pulleys must be on the same centerline, and second, the motor shaft must be parallel to the driven shaft.

Use a plumb bob and draw a datum line on the floor to establish operation. Next, drop a plumb line from the center of the driven pulley to the floor. With a square, draw a line perpendicular to the datum line. Next, drop a plumb line from the center of the motor pulley and move the motor up or back until the plumb bob rests on the centerline of the driven pulley. From the pulley centerline, perpendiculars may be drawn through the centers of the holes in the motor feet. A level should be used to see that the line shafting is level. If it is not, then the motor feet must be shimmed up so that the motor shaft and the line shaft will be out of level the same amount. Chain drives may be aligned in a similar manner.

With belt drives, a sliding base is nearly always used to allow for belt adjustment. Use the following procedure when aligning two pulleys:

1. Place the motor on the base so that there will be an equal amount of adjustment in either direction, and firmly fasten the motor to the base by means of the four hold-down bolts.
2. Mount the motor pulley on the motor shaft.
3. Locate the base and motor in the approximate final position, as determined by the length of the belt.
4. Stretch a string from the face of the driven pulley toward the face of the motor pulley.
5. Parallel the face of the motor pulley with this string.
6. Using a scratch pin, mark the end positions of the sliding base.
7. Extend these lines.
8. Move the base and motor away from the string an amount equal to one-half the difference in face width of the two pulleys. Use the two extended lines as a guide to keep the base in its proper position.
9. The belt should now be placed on the pulleys to see if it operates satisfactorily. If it does not, the base may be shifted slightly until proper operation is obtained.
10. Finally, firmly fasten the base to the floor, ceiling, or side walls by means of lag screws or bolts.

Belt Maintenance

It is of paramount importance that a proper maintenance schedule be observed. This consists of numerous details, the most important of which are:

1. Keeping belts tight.
2. Taking up the slack of belts.
3. Running in new belts.
4. Cleaning dirty belts.
5. Dressing belts.
6. Maintenance of belt guards.

In locations where idlers or tension-base drives are employed, the tightness of the belts is automatically adjusted to a desired value. Because of the absence of the necessary space requirement, however, this method is not always feasible, and the surplus length will have to be removed at certain intervals of time. Some belts stretch more than others under similar conditions and loads. Temperature and humidity are two factors that greatly influence belt tension. Locations of high humidity and moisture are known to cause a belt to expand more quickly.

After the elastic limit of a leather belt is reached, it may be stretched to the breaking point without becoming any tighter. The friction increases with the pressure, and the more elastic the surface, the greater the friction. To obtain the greatest amount of power from leather belts, the pulleys should also be covered with leather — this will allow the belts to run very slackly and give more wear. The leather in a belt should be pliable, of fine close fiber, solid in appearance, and have a smooth, polished surface.

Belts derive their power to transmit motion from the friction between the surface of the belt and the surface of the pulley, and from nothing else. The thickness as well as the width of belts must always be considered. Consequently, a double belt must be used where it is necessary to transmit a greater power of a belt, larger pulleys may be used, thus increasing the speed of the belt. This can often be done to advantage, provided that the belt speed is not carried above the safe limit.

Direction of Rotation

The rules in Table 11-2 govern the standard direction of rotation of generators and motors. In all cases, the observer will stand facing the designated end of the machine and in line with the shaft.

Table 11-2. Standard Direction of Motor and Generator Rotation

	Rotation	Viewed From	See Note
Generators (except gas-engine-driven generators, and in motor-generators)	Clockwise	A	1
Induction-motor-generator sets (except sets with motors not having conduit boxes)	Clockwise	C	3
Standard nonreversing DC motors	Counterclockwise	A	1
Synchronous motors with direct-connected exciters	Counterclockwise	A	1
Generators — gas-engine-driven	Counterclockwise	A	1
Single-phase motors	Counterclockwise	A	1
Synchronous-motor-generator sets and induction-motor-generator sets with motors not having conduit boxes	Counterclockwise	B	2

Note 1. *End A.* The direction of rotation of standard motors and generators will be determined by viewing the end opposite the driven end of generators, and opposite the driving end of motors.

Note 2. *End B.* The direction of rotation of synchronous-motor-generator sets, induction-motor-generator sets with motors not having conduit boxes, and frequency changers will be determined by the direction of rotation of the driving motor. When there are one or more generators on each end of the driving motor, the direction of rotation will be determined by viewing the motor from the connection end of the stator coils.

Note 3. *End C.* The direction of rotation of induction-motor-generator sets will be determined by that of the generator. When there are one or more generators on each end of the driving motor, the direction will be determined by viewing the set from the end which places the conduit boxes on the left-hand side.

Polyphase induction motors, except certain high-speed motors equipped with unidirectional fans, are suitable for either direction of rotation, and can be reversed by interchanging any two leads. High-speed motors with unidirectional fans must be ordered for the desired direction of rotation.

Summary

Motor maintenance should be performed on a systematic schedule to minimize down time. Adequate spare parts should be stocked for all motors and a record of troubles and cures maintained. Inspections should be made as regularly and thoroughly as circumstances permit.

Care should be taken when starting a new motor for the first time. It should be dry and the bearings properly lubricated. The armature should be turned by hand to make sure it rotates freely. All electrical connections should be double-checked. If possible, run the motor without load long enough to determine that no unusual localized heating is occurring.

Correct belt sizes are important to ensure proper operation of the equipment. Determine the correct type and size of belt by consulting the manufacturer's instructions or by calculations by means of one of the formulas listed. Belts must be properly aligned to prevent excessive wear, loss of power, and premature failure. A regular maintenance schedule for belts should also be followed.

Review Questions

1. List five important procedures that should be followed for the proper maintenance of motors.
2. List five points that should be checked before placing a new motor in operation.
3. What size of double belt will be required to transmit 25 horsepower at a belt speed of 3,000 feet per minute?
4. What characteristic of a leather belt must be especially considered when it is to be installed in a location where the humidity is high?
5. What is the standard direction of rotation of single-phase motors?
6. What size belt will be required to transmit 18.56 kW at a belt speed of 15.24 m/sec?

CHAPTER 12

Motor Calculations

The following simple calculations for motor circuits will serve to acquaint the reader with methods used in determining voltage drops, currents, and efficiency of DC and AC motors.

Problem — A 115-volt DC motor draws a current of 200 amperes and is located 1,000 ft. from the supply source. If the copper transmission wire has a diameter of 0.45 inch, what must be the voltage of the supply source (Fig. 12-1)?

Solution — The resistance of any conductor varies directly with its length and inversely as its cross-sectional area. Therefore, the formula is

$$R = \frac{P \times L}{A}$$

where

R is the resistance in ohms

P is the specific resistance of one mil. ft. of copper wire (10.4),
L is the length in feet,
A is the arc in circular mils (diameter of wire in thousandths)

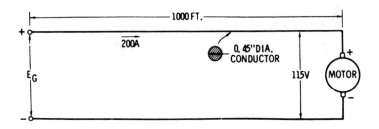

Fig. 12-1. Voltage-drop calculation in a motor feeder line.

Note — The cross-sectional area of a circular conductor can be expressed in circular mils by squaring the diameter of the conductor expressed in thousandths. For example, a wire 0.625 inch in diameter has a circular-mil area of 625 × 625 = 390,625 circular mils.

Thus, the resistance of the feed lines is

$$R = \frac{10.4 \times 2000^*}{450^2} = 0.103 \text{ ohm}$$

*Indicates the length of 2 conductors.

and

$$E_G = E_R + I_L R = 115 + (200 \times 0.103) = 135.6 \text{ volts}$$

Problem — A 10-hp, 230-volt DC motor of 84 percent full-load efficiency is located 500 ft. from the supply mains. If the motor-starting current is 1.5 times the full-load current, what is the smallest cross-sectional area of copper wire required when the allowable voltage drop in the feeder at starting is 24 volts (Fig. 12-2)?

Solution — The motor full-load current is

$$I_L = \frac{\text{hp} \times 746}{E \times \text{efficiency}} = \frac{10 \times 746}{230 \times 0.84} = 38.6 \text{ amperes}$$

where

Efficiency is percent in decimal form
E is voltage
I_L is full current load

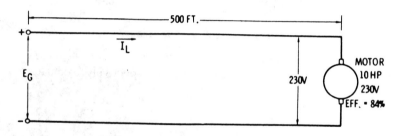

Fig. 12-2. Cross-sectional area calculation for the feeder line to a 10-hp motor.

Motor-starting current is

$$I_S = 38.6 \times 1.5 = 57.9 \text{ amperes}$$

where

I_S is starting current
R is resistance
A is cross-sectional area

Since the voltage drop in the feeder at starting is 24 volts, then, according to Ohm's law, $R = \dfrac{E}{I}$

$$R = \frac{24V}{57.9A} = 0.415 \text{ ohm}$$

The minimum cross-sectional area is therefore

$$A = \frac{10.4 \times 2 \times 500}{0.415} = 25,060.241 \text{ circular mils}$$

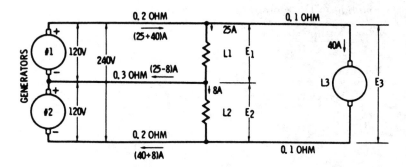

Fig. 12-3. A three-wire system supplying three loads.

Problem — In the circuit of Fig. 12-3, with loads L_1, L_2, L_3 drawing currents of 25, 8, and 40 amperes respectively, calculate:

(a) The power supplied by each generator.

(b) The voltages E_1, E_2, and E_3.

Solution — By inspection, the currents supplied by generators Nos. 1 and 2 are 65 and 48 amperes, respectively.

(a)

$$P_{G1} = 120 \times 65 = 7800 \text{ watts}$$
$$P_{G2} = 120 \times 48 = 5760 \text{ watts}$$

where

P_{G1} is Power (Generator 1)
P_{G2} is Power (Generator 2)

(b) According to Kirchhoff's law,

$$E_1 = 120 - (65 \times 0.2) - (17 \times 0.3) = 101.9 \text{ volts}$$
$$E_2 = 120 + (17 \times 0.3) - (48 \times 0.2) = 115.5 \text{ volts}$$

and

$$E_1 + E_2 = 217.4 \text{ volts}$$

Similarly,

$$E_3 = 217.4 - (0.2 \times 40) = 209.4 \text{ volts}$$

where

E_1 is Voltage (Generator 1)
E_2 is Voltage (Generator 2)
E_3 is Voltage output across both generators

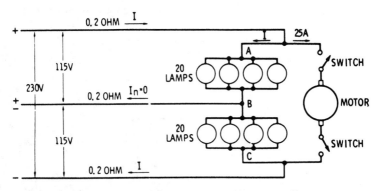

Fig. 12-4. A three-wire system supplying a lamp and motor load.

Problem — A three-wire system supplies the load shown in Fig. 12-4. If the resistance of each lamp is 110 ohms and the motor takes a current of 25 amperes, calculate the voltage across each group of lamps:

(a) When the motor is disconnected.
(b) When the motor is operating.

Solution — the combined resistance of a group of 20 lamps, each having a resistance of 110 ohms, is

$$R = \frac{110}{20} = 5.5 \text{ ohms}$$

Since the load is a balanced one, it is evident that the current in the neutral is zero.

The current through the lamps with the motor disconnected is

$$I = \frac{230}{11.4} = 20.2 \text{ amperes}$$

(a) Voltage $E_{AB} = E_{BC} = \frac{20.2 \times 11}{2} = 111.1 \text{ volts}$

The current through the lamps with the motor operating and drawing 25 amperes can be obtained if Kirchhoff's law is applied to the circuit, remembering that the current flowing in the line is now $(I + 25)$ amperes.

$$230 = 0.4(I + 25) + 11.4I + 10$$

and

$$I = \frac{220}{11.4} = 19.3 \text{ amperes}$$

(b) Voltage $E_{AB} = E_{BC} = \frac{19.3 \times 11}{2} = 106.15 \text{ volts}$

where

I is current
E_{AB} is voltage across points A to B
E_{BC} is voltage across points B to C

Thus, when the motor is thrown on the line, the voltage across the lamps will fall from 111.10 to 106.15 volts.

Problem — The field winding of a shunt motor has a resistance of 110 ohms, and the voltage applied to it is 220 volts. What is the amount of power expended in the field excitation?

Solution — The current through the field is

$$I_f = \frac{E_s}{R_f} = \frac{220}{110} = 2 \text{ amperes}$$

Power expended is

$$E_s I_f = 220 \times 2 = 440 \text{ watts}$$

The same results will also be obtained by using the equation

$$\frac{E_s^2}{R_f} = P_f = \frac{220^2}{110} = 440 \text{ watts}$$

where

> I_f is field current
> E_s is source voltage
> R_f is field resistance
> P_f is power expended in field

Problem — Assume that the DC motor shown in Fig. 12-5 draws a current of 10 amperes from the line with a supply voltage of 100 volts. If the total mechanical loss (friction, windage, etc.) is 90 watts, calculate the:

(a) Copper losses in the field.
(b) Armature current.
(c) Copper losses in the armature.
(d) Total loss.
(e) Motor input.
(f) Efficiency.

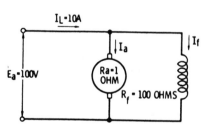

Fig. 12-5. A DC-motor circuit.

Solution — From the foregoing data we obtain:

(a) The copper losses in the field are

$$I_f^2 \times R_f = 1^2 \times 100 = 100 \text{ watts}$$

(b) The armature current is

$$10 - 1 = 9 \text{ amperes}$$

(c) Copper losses in the armature are

$$I_a^2 \times R_a = 9^2 \times 1 = 81 \text{ watts}$$

(d) The total loss is

$$100 + 81 + 90 = 271 \text{ watts}$$

(e) Motor input is

$$E_a \times I_L = 100 \times 10 = 1000 \text{ watts}$$

(f) Motor output is equal to

$$\text{input} - \text{losses} = 1000 - 271 = 729 \text{ watts}$$

(g) The efficiency is equal to

$$\frac{\text{output}}{\text{input}} \times 100 = \frac{729}{1000} \times 100 = 72.9 \text{ percent}$$

where

I_f is field current
R_f is field resistance
I_a is armature current
I_L is load current

Problem — A 50-hp, 500-volt shunt motor draws a line current of 4.5 amperes at no load. The shunt field resistance is 250 ohms and the armature resistance, exclusive of brushes, is 0.3 ohm. The brush drop is 2 volts. The full-load line current is 84 amperes. What is the horsepower output and efficiency?

Solution — From the data supplied, we obtain:

Full-load armature current

$$I_a = I_L - I_f = 84 - \frac{500}{250} = 82 \text{ amperes}$$

No-load armature current

$$I_a = I_L - I_f = 4.5 - \frac{500}{250} = 2.5 \text{ amperes}$$

Stray power loss

$$P_{sp} = I_L \times E = 2.5 \times 500 = 1250 \text{ watts}$$

$$\text{Brush loss} = 2 \times I_a = 2 \times 82 = 164 \text{ watts}$$

The efficiency of the motor

$$= \frac{500 \times 84 - [(82^2 \times 0.3) + (500 \times 2) + 164 + 1250]}{500 \times 84}$$

where

I_a is armature current
I_L is load current
I_a is field current
P_{sp} is stray power loss
E is voltage

Motor efficiency $= \dfrac{42,000 - 4431}{42,000} \quad \dfrac{37,569}{42,000} = 0.8945, \text{ or } 89.45\%$

Horsepower output $= \dfrac{37,569}{746} = 50.36 \text{ hp}$

Problem — Calculate the horsepower output, torque, and efficiency of a shunt motor from the following data:

$$I_1 = 19.8 \text{ amperes}$$

$$E_1 = 230 \text{ volts}$$

Balance reading = 12 lbs. corrected for zero reading

Brake arm = 2 ft.

Speed = 1,100 rpm

where

I_1 is current
E_1 is voltage

Solution — The horsepower output of the motor is

$$\text{hp} = \frac{2\pi \times 2 \times 12 \times 1100}{33,000} = 5.03 \text{ hp}$$

Torque $= FR = 12 \times 2 = 24$ lb. ft.

$$\text{Eff} = \frac{\text{output}}{\text{input}} = \frac{5.03 \times 746}{230 \times 19.8} = 0.824, \text{ or } 82.4\%$$

Problem — A shunt motor with an armature and field resistance of 0.055 and 32 ohms, respectively, is to be tested for its mechanical efficiency by means of a rope brake. When the motor is running at 1,400 rpm, the longitudinal pull on the 6-inch-diameter pulley is 57 lbs. Simultaneous readings of the line voltmeter and ammeter are 105 and 35, respectively. Calculate the:

(a) Counter emf.
(b) Copper losses.
(c) Efficiency.

Solution — From the foregoing data:

$$I_a = I_L - I_F = 35 - \frac{105}{32} = 31.7 \text{ amperes}$$

where

I_s is current of armature
I_L is current of line
I_f is current of field

(a) The counter emf is

$$(E_{armature} = E_{line} - (I_{arm} \times R_{arm})$$

$$E_a = E_1 - (I_a R_a) = 105 - (31.7 \times 0.055) = 103.26 \text{ volts}$$

(b) The copper losses are

$$\text{Power loss for copper} = (I^2 \times R_f) + (I_a^2 \times R_a)$$

$$P_c = I^2 R_f + I_a^2 R_a =$$
$$(3.3^2 \times 32) + (31.7^2 \times 0.055) = 404 \text{ watts}$$

(c) The efficiency is

$$\text{output} = \frac{2\pi \times \text{rpm} \times \text{radius of pulley in ft.} \times \text{pull on pulley in lbs.}}{33,000}$$

$$\text{Output} = \frac{2\pi \times 1400 \times 3/12 \times 57}{33,000} = 3.8 \text{ hp}$$

$$\text{Input} = \frac{\text{Voltage of line} \times \text{Current of line}}{746 \text{ watts}}$$

$$\text{Input} = \frac{105 \times 35}{746} = 4.93 \text{ hp}$$

or

$$\text{Efficiency} = \frac{\text{output}}{\text{input}}$$

$$\text{Eff.} = \frac{3.8}{4.93} = 0.771, \text{ or } 77\% \text{ (approx.)}$$

Problem — A DC motor requires 10 kilowatts to enable it to supply its full capacity of 10 horsepower to its pulley. What is its full-load efficiency?

Solution —

$$\text{Efficiency} = \frac{\text{output}}{\text{input}} = \frac{10 \times 746}{10,000} = 0.746, \text{ or } 74.6\%$$

Problem — A 7-hp motor takes 6.3 kilowatts at full load. What is its efficiency?

Solution —

$$\text{Efficiency} = \frac{\text{output}}{\text{input}}$$

$$\text{Output} = 7 \times 746 = 5,222 \text{ watts}$$

$$\text{Input} = 1000 \times 6.3 = 6,300 \text{ watts}$$

$$\text{Efficiency} = \frac{5,222}{6,300} = 0.829, \text{ or } 83\% \text{ (approx.)}$$

Problem — A certain load to be driven at 1,750 rpm requires a torque of 60 lb-ft. What horsepower will be required to drive the load?

Solution —

$$\frac{2\pi \times T \times N}{33,000}$$

$$= \frac{2\pi \times 60 \times 1,750}{33,000} = 20 \text{ hp}$$

Problem — A 15-hp, 220-volt, 1,800 rpm shunt motor has an efficiency of 87 percent at full load. The resistance of the field is 440 ohms. Calculate the:

(a) Full-load armature current.
(b) Torque of the machine.

Solution —

(a) The armature current is

$$I_L = \frac{15 \times 746}{0.87 \times 220}$$

$$I_f = \frac{220V}{440\Omega} = 0.5$$

$$I_a = I_L - I_f = \frac{15 \times 746}{0.87 \times 220} - \frac{220}{440}$$

$$= 58.46 - 0.5 = 57.96 \text{ amps}$$

where

I_L is current of line
I_f is current of field
N is speed of rotation in rpm

(b) The torque is

$$T = \frac{\text{hp} \times 5{,}252}{N}$$

$$= \frac{15 \times 5{,}252}{1{,}800} = 43.77 \text{ lb-ft.}$$

Problem — When the field rheostat is cut out, a 230-volt shunt motor generates a counter emf of 220 volts at no load. The resistance of the armature is 2.3 ohms and that of the field is 115 ohms. Calculate the:

(a) Current through the armature when the field rheostat is cut out.

(b) Current through the armature when sufficient external resistance has been inserted in the field circuit to make the field current one-half as great.

Solution —

(a) The armature current when the field rheostat is cut out is

$$I_a = \frac{E_s - E_a}{R_a} = \frac{230 - 220}{2.3} = 4.35 \text{ amperes}$$

(b) The current through the field without external resistance is

$$\frac{230}{115} = 2 \text{ amperes}$$

When the field current has been made half as great by inserting external resistance, the field flux and therefore the counter emf will become half as great, or 110 volts. The armature current in this particular case is therefore

$$I_a = \frac{230 - 110}{2.3} = 52.2 \text{ amperes}$$

Problem — As shown in Fig. 12-6, a resistance of 130 ohms and a capacitance of 30 microfarads are connected in parallel across a 230-volt, 50-hertz supply. Find the following:

(a) Current in each circuit.
(b) Total current.
(c) Phase difference between the total current and the applied voltage.
(d) Power consumed.
(e) Power factor.

Solution — The capacitive reactance (X_C) of the circuit is

$$X_C = \frac{10^6}{2\pi \times 50 \times 30} = 106 \text{ ohms}$$

where

X_C is capacitive reactance, $\dfrac{1}{2\pi FC}$

2π is 6.28; 10^6 converts μF to F
I_R is current through resistance
F is frequency (Hz)
C is capacitance (F)

(a) Current through the resistance is

$$I_R = \frac{E}{R} = \frac{230}{130} = 1.77 \text{ amperes}$$

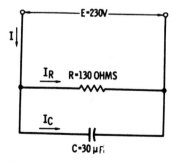
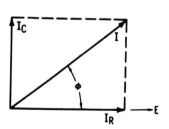

Fig. 12-6. Resistance and capacitance in parallel.

Current through the capacitance is

$$I_C = \frac{E}{X_C} = \frac{230}{106} = 2.17 \text{ amperes}$$

(b) Total current is

$$I = \sqrt{I_R{}^2 + I_C{}^2} = \sqrt{1.77^2 + 2.17^2} = 2.8 \text{ amperes}$$

(c) Phase difference is

$$\cos \angle\theta = \frac{I_R}{I} = \frac{1.77}{2.8} = 0.632$$

$$\angle\theta = 50.8° \text{ (angle of lead)}$$

(d) Power consumed is

$$P = I_R{}^2R = 1.77^2 \times 130 = 407 \text{ watts}$$

(e) Power factor according to (c) is 0.632, or 63.2%.

Problem—Two circuits, Fig. 12–7, are connected in parallel as shown. If the voltage of the source is 120, calculate the:

(a) Phase displacement.
(b) Power factor the circuit.
(c) Total current.

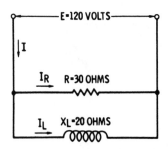
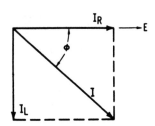

Fig. 12–7. Resistance and Inductance in parallel.

Solution — With reference to the vector diagram in Fig. 7, the current through the ohmic resistance is

$$I_R = \frac{E}{R} = \frac{120}{30} = 4 \text{ amperes}$$

Current through the inductive reactance is

$$I_L = \frac{E}{X_L} = \frac{120}{20} = 6 \text{ amperes}$$

(a) Phase displacement is

$$\tan \theta = \frac{I_L}{I_R} = \frac{6}{4} = 1.5$$

$$\theta = 56.3°$$

(b) Power factor is

$$\cos \angle \theta = 0.555, \text{ or approximately 56\% (lagging)}$$

(c) Total current is

$$I = \sqrt{4^2 + 6^2} = 7.2 \text{ amperes}$$

Problem—In two circuits (Fig. 12-8) in parallel, one branch consists of a resistance of 15 ohms and the other of an inductive reactance of 10 ohms. When the impressed voltage is 110, find the:

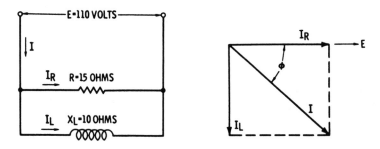

Fig. 12-8. Resistance and Inductance in parallel.

(a) Current through the ohmic resistance (I_R).
(b) Current through the inductive reactance (I_L).
(c) Line current (I_T).
(d) Power factor ($\cos \angle\theta$).
(e) Angle of lag of the line current ($\angle\theta$).

Solution — With reference to the vector diagram in Fig. 12-8, the

(a) Current through the ohmic resistance is

$$I_R = \frac{E}{R} = \frac{110}{15} = 7.34 \text{ amperes}$$

(b) Current through the inductive reactance is

$$I_L = \frac{E}{X_L} = \frac{110}{10} = 11 \text{ amperes}$$

(c) Total line current (I_T) is

$$I = \sqrt{I_R^2 + I_L^2} = \sqrt{7.34^2 + 11^2} = 13.2 \text{ amperes}$$

(d) Power factor is

$$\cos \angle\theta = \frac{I_R}{I_T} = \frac{7.34}{13.2} = 0.556$$

(e) Angle of lag of the line current is

$$\angle\theta = 56.2°$$

where

I_R is resistance current
E is voltage applied
R is resistance
I_L is current through coil
X_L is inductive reactance
I_T is total current

Problem — As shown in Fig. 12–9, a resistance of 10 ohms, an inductance of 0.6 henry, and a capacitance of 300 microfarads are connected in parallel across a 100-volt, 25-hertz supply. Find the:

(a) Current in each circuit.
(b) Total current.
(c) Power factor.
(d) Power consumption.

Solution — In this example, it is first necessary to find the inductive and capacitative reactances. They are, respectively,

$$X_L = 2\pi \ FL$$

$$X_L = 2\pi \times 25 \times 0.06 = 9.425 \text{ ohms}$$

$$X_C = \frac{1}{2\pi FC}$$

$$X_C = \frac{10^6}{2\pi \times 25 \times 300} = 21.2 \text{ ohms}$$

(a) Current through the resistance is

$$I_R = \frac{E}{R} = \frac{100}{10} = 10 \text{ amperes}$$

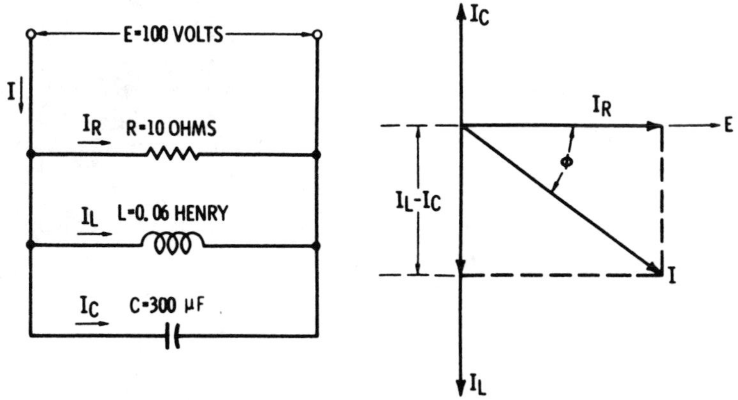

Fig. 12–9. Resistance, inductance, and capacitance in parallel.

Current through the inductive reactance is

$$I_L = \frac{E}{X_L} = \frac{100}{9.425} = 10.61 \text{ amperes}$$

Current through the capacitive reactance is

$$I_C = \frac{E}{X_C} = \frac{100}{21.2} = 4.72 \text{ amperes}$$

(b) Total current is

$$I = \sqrt{I_R^2 + (L_L - I_C)^2}$$

$$= \sqrt{10^2 + (10.61 - 4.72)^2}$$

$$= 11.61 \text{ amperes}$$

(c) Power factor is

$$\cos \angle\theta = \frac{I_R}{I_T} = \frac{10}{11.61} = 0.861 \text{ (lagging)}$$

$$\angle\theta = 30.53°$$

(d) Power consumption is

$$P = EI \cos\angle\theta = 100 \times 11.61 \times 0.861 = 1000 \text{ watts}$$

where

X_L is inductive reactance
2π is constant 6.28
F is frequency (Hz)
L is inductance (H)
X_C is capacitive reactance
I_R is current through resistance
E is applied voltage
R is resistance

Problem — A circuit connected as shown in Fig. 12-10 contains a 10-ohm resistance and 0.5-henry inductance in parallel with a capacitor of 20 microfarads. The voltage and frequency of the source are 1,000 and 60, respectively. Find the:

(a) Current through the coil.
(b) Phase angle between the current through the coil and the potential across it.
(c) Current through the capacitator.
(d) Total current.

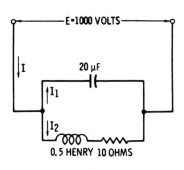

Fig. 12-10. Impedance and capacitance in parallel.

Solution —

(a) The current through the coil (I_L) is

$$I_2 = \frac{1000}{\sqrt{10^2 + (2\pi \times 60 \times 0.5)^2}} = 5.30 \text{ amperes}$$

(b) Phase angle is

$$\tan \theta = \frac{2\pi 60 + 0.5}{10} = 18.85$$

Use calculator and hit

$$[\text{arc}] \ [\text{tan}] = 86.9°$$

(tangent 18.85 is 86.9°)

(c) Current through the capacitor (I_C) is

$$I_1 = \frac{1000}{X_C} = 20 \times 10^{-6} \times 2\pi \times 60 \times 1000 = 7.54 \text{ amps}$$

With reference to the vector diagram

$$OA = I_1 = 7.54 \text{ amps, and } OB = I_2 = 5.30 \text{ amps}$$

As the current, I, is the resultant of these two vectors, it is now possible to construct the parallelogram as indicated by the dotted lines. It follows from the construction that $\beta OBC = 90°$, and from the law of cosines,

$$I = \sqrt{(5.30^2 + 7.54^2) - (2 \times 5.30 \times 7.54 \sin \angle\theta)}$$

$$I = 2.26 \text{ amperes (approximately)}$$

Problem — A series circuit consists of a 30-microfarad capacitance and a resistance of 50 ohms connected across a 110-volt, 60 hertz supply. Calculate the:

(a) Impedance of the circuit (Z).
(b) Current in the circuit (I_T).
(c) Voltage drop across the resistance (E_R).
(d) Voltage drop across the capacitance (E_C).
(e) Angle between the voltage and the current $(\angle\theta)$.
(f) Power loss (P).
(g) Power factor of the circuit (PF or $\cos \angle\theta$)

Solution —

(a) Impedance of the circuit is

$$X_C = \frac{1}{2\pi fC} = \frac{1}{2\pi \times 60 \times 0.000030} = 88.4 \text{ ohms}$$

$$Z = \sqrt{R^2 + X_C^2} = \sqrt{50^2 + 88.4^2} = 101.6 \text{ ohms}$$

(b) Current in the circuit is

$$I = \frac{E}{Z} = \frac{110}{101.6} = 1.08 \text{ amperes}$$

(c) Voltage drop across the resistance is

$$E_R = IR = 1.08 \times 50 = 54 \text{ volts}$$

(d) Voltage drop across the capacitance is

$$E_C = IX_C = 1.08 \times 88.4 = 95.5 \text{ volts}$$

(e) Angle between voltage and current is

$$\cos \theta = \frac{R}{Z} = \frac{50}{101.6} = 0.492$$

$$\angle \theta = 60.5°$$

(f) Power loss is

$$P = I^2R = 1.08^2 \times 50 = 58.3 \text{ watts}$$

(g) Power factor is cos, therefore:

$$\cos \angle \theta = 0.492, \text{ or } 49.2 \text{ percent}$$

where

X_C is capacitive reactance
2π is constant 6.28
F is frequency (Hz)
C is capacitance (F)
E is voltage
Z is impedance
I is current
E_R is voltage across resistor
E_C is voltage across capacitor

Problem — A single-phase motor is taking 20 amperes from a 400-volt, 50-hertz supply, the power factor being 80 percent lagging. What value of capacitor connected across the circuit will be necessary to raise the power factor to unity?

Solution — With reference to the vector diagram in Fig. 12-11, it is evident that

$$I_C = I_m \sin \theta$$

Since I_m = 20, and $\sin \theta = \sqrt{1 - 0.8^2}$ = 0.6

then I_C = 20 × 0.6 = 12 amperes

and

$$X_C = \frac{E}{I_C} = \frac{400}{12} = 33.3 \text{ ohms}$$

The equation for the capacitive reactance is

$$X_C = \frac{1}{2\pi f C}$$

or

$$C = \frac{1}{2\pi f X_C}$$

Inserting values and simplifying,

$$C = \frac{1}{6.28 \times 50 \times 33.3} = .0000956F \text{ or } 95.6 \ \mu F$$

$$C = 95.6 \ \mu F$$

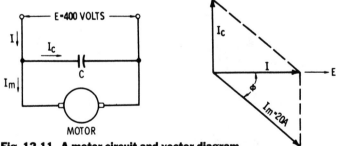

Fig. 12-11. A motor circuit and vector diagram.

Problem — A 220-volt, 60-hertz, 10-hp, single-phase induction motor operates at an efficiency of 86 percent and a power factor of 90 percent. What capacity should be placed in parallel with the motor so that the feeder supplying the motor will operate at unity power factor?

Solution —

The current taken by the motor is

$$I = \frac{10 \times 746}{220 \times 0.9 \times 0.86} = 43.81 \text{ amperes (lagging)}$$

Current taken by the capacitor is

$I_C = I \times \sin \angle\theta$

$I_C = 43.81 \times 0.4358 = 19$ amperes

$PF = \cos \angle\theta$

$PF = 90\%$

$PF = .9$ or use calculator to obtain:

[arc] [cos] $= 25.84°$

[sin] of $25.84° = 0.4358$

The capacitance reactance of the capacitor is

$$X_C = \frac{220}{19} = 11.58 \text{ ohms}$$

Capacity necessary is

$$C = \frac{10^6}{2\pi \times 60 \times 11.58} = 229 \text{ microfarads}$$

Problem — In a balanced three-phase, 208-volt circuit, the line current is 100 amperes. The power is measured by the two-wattmeter method; one meter reads 18 kW and the other zero. What is the power factor of the load? If the power factor were unity and the line current the same, what would each wattmeter read?

Solution —

The expression for power is

$$P = EI \cos \phi \sqrt{3}$$

Since one wattmeter reads zero, then

$$18,000 = 208 \times 100 \times \cos \phi \sqrt{3}$$

so

$$\cos \phi = \frac{18,000}{208 \times 100 \times \sqrt{3}} = 0.5$$

With the power factor unity and with the same line current, then

$$\tan \phi = 3 \left[\frac{W_1 - W_2}{W_1 + W_2} \right] = 0, \text{ or } W_1 = W_2$$

Also

$$W_1 + W_2 = 208 \times 100 \times \sqrt{3} = 36 \text{ kW}$$

That is, each wattmeter reads 36/2, or 18 kW.

where

P is power
E is voltage
I is current
$\left. \begin{array}{l} W_1 \\ W_2 \end{array} \right]$ are wattmeters
ϕ is one phase of 3 ϕ power

Problem — In a certain manufacturing plant, a single-phase alternating current of 110 volts, 60 hertz is applied to a parallel circuit having two branches. Branch 1 consumes 27 kW with a lagging power factor of 0.8, while Branch 2 consumes 12 kW with a leading power factor of 0.9. Calculate the:

(a) Total power supplied in kW and kVA.

(b) Power factor of the total load.

Solution —

(a) Total power supplied is

$$P = P_1 + P_2 = 27 + 12 = 39 \text{ kW}$$

The reactive-power component supplied by Branch 1 is

$$P_1 \left[\frac{\sin \phi_1}{\cos \phi_1} \right] = P_1 \tan \phi_1 = 27 \times 0.7508 = 20.27 \text{ kVA}$$

Similarly, the reactive-power component supplied by Branch 2 is

$$P_2 \left[\frac{\sin \phi_2}{\cos \phi_2} \right] = P_2 \tan \phi_2 = 12 \times 0.4834 = 5.80 \text{ kVA}$$

The sum of the reactive-power components is

$$P_1 \tan \phi_1 - P_2 \tan \phi_2 = 20.27 - 5.80 = 14.47 \text{ kVA}$$

Total power supplied, in kVA, is

$$\text{kVA} = \sqrt{39^2 + 14.47^2} = 41.6$$

(b) Power factor of the total load is

$$\cos \phi = \frac{39}{41.6} = 0.94$$

Problem — A certain load takes 40 kVA at 50 percent lagging power factor, while another load connected to the same source takes 80 kVA at 86.7 percent lagging power factor. Find the:

(a) Total effective power.
(b) Reactive power.
(c) Power factor.
(d) Apparent power.

Solution — The apparent power taken by each load, expressed with reference to the voltage, is given in Fig. 12-12. From a trigonometric table, the angle corresponding to a 50 percent power factor is 60°, and the angle corresponding to an 86.7 percent power factor is 30°. To find the resultant, it is necessary to complete the parallelogram. As shown in Fig. 12-12, the sum of the effective power is indicated along the horizontal, and is as follows:

(a) $(40 \times 0.5) + (80 \times 0.867) = 89.36 \text{ kW}$

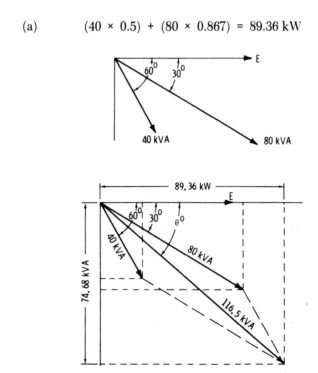

Fig. 12-12. Vector diagrams of true, reactive, and apparent power in an AC circuit.

The total reactive power is indicated along the vertical, and is

(b) $(40 \times 0.867) + (80 \times 0.5) = 74.68 \text{ kVA}$

$$\tan \theta = \frac{74.68}{89.36} = 0.836$$

(c) $\cos \theta = 0.767$

(d) $\theta = 39.9°$

The apparent power is

(e) $\dfrac{89.36}{\cos \theta} = \dfrac{89.36}{0.767} = 116.5 \text{ kVA}$

Problem — A group of induction motors takes 100 kVA at 84 percent lagging power factor. A synchronous motor connected to the same line takes 60 kVA at 70.7 percent leading power factor. Determine the:

  (a) Total effective power.
  (b) Reactive power.
  (c) Power factor.
  (d) Apparent power.

Solution — This is a simple problem on power-factor correction. Remember here that, since the synchronous motor has a leading power factor, this vector must be laid out in the proper direction or 45° above the reference or voltage vector. Thus, from Fig. 12-13

(a) $(100 \times 0.84) + (60 \times 0.707) = 126.42 \text{ kVA}$

The total reactive power is the difference between the two, since one is leading the other is lagging. Thus

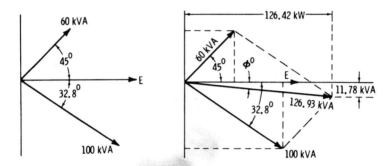

Fig. 12-13. Vector diagrams of leading and lagging loads.

(b) $\qquad (100 \times 0.542) - (60 \times 0.707) = 11.78 \text{ kVA}$

(c) $\qquad \tan \theta = \dfrac{11.78}{126.42} = 0.093$

$$\cos \theta = 0.996$$

The apparent power is

(d) $\qquad \dfrac{126.42}{\cos \theta} = \dfrac{126.42}{0.996} = 126.93 \text{ kVA}$

Table 12-1. NEMA Resistor Classification for Wound-Rotor Motors

Per Cent Full-Load Current on First Point	Starting Torque % of Full Load					Resistor Class Number					
	Series Motors	Compound Motors	Shunt Motors	Wound-Rotor Induction Motors		5 Sec. on Out of 80 Sec.	10 Sec. on Out of 80 Sec.	15 Sec. on Out of 90 Sec.	15 Sec. on Out of 60 Sec.	15 Sec. on Out of 45 Sec.	15 Sec. on Out of 30 Sec.	Cont.
				1 Ph. Stg.	3 Ph. Stg.							
25	8	12	25	15	25	111	131	141	151	161	171	91
30	30	40	50	30	50	112	132	142	152	162	172	92
70	50	60	70	40	70	113	133	143	153	163	173	93
100	100	100	100	55	100	114	134	144	154	164	174	94
150	170	160	150	85	150	115	135	145	155	165	175	95
200	250	230	200		200	116	136	146	156	166	176	96

Summary

It is often advantageous for anyone concerned with electric motors to be able to determine certain facts about the circuits by means of simple calculations. For example, it may be necessary to provide the proper size of wire to install a motor. This can be easily done by substituting the known factors in the proper formula. This chapter offers many examples of the solving of similar problems.

Review Questions

1. What is the efficiency of a 10-hp motor that requires 8 kilowatts at full load?
2. What is the formula for finding capacitance reactance?
3. How much current flows through a 10-henry inductance when connected across 220 volts, 60 hertz AC?
4. What is the circular-mil area of a conductor, the diameter of which is 0.125 inch?
5. List the three forms of Ohm's law for DC circuits.

CHAPTER 13

Meters

The more common electric meters may be roughly divided into the following classes:

According to the function performed, as:

1. Ammeters.
2. Voltmeters.
3. Wattmeters.

According to the circuit on which they are used, as:

1. Alternating current.
2. Direct current.

According to the principle of operation, as:

1. Permanent-magnet, moving coil.
2. Dynamometer.
3. Magnetic-vane.

4. Induction.
5. Digital.

The essential parts of these instruments generally include:

1. Means for providing a deflection torque (obtained by inter-action of magnetic fields).
2. A spring or other means to provide a countertorque.
3. A pointer to indicate the resultant position of the moving element of the meter.
4. Electronic circuits for driving a display.

Permanent-Magnet, Moving-Coil Meters

A permanent-magnet, moving-coil type of meter is suitable for measuring only DC, and its operation depends on the reaction between the current in a movable coil and the field of a fixed permanent magnet. The essential parts of an instrument of this type are shown in Fig. 13-1.

The permanent magnet supplies a uniform magnetic field through pole pieces and across the air gap in which a moving coil wound with fine wire is located. The moving coil (Fig. 13-2) is provided with hardened steel pivots fitted into highly polished jewels that allow the moving coil to rotate with as little friction as possible. Springs of carefully selected phosphor-bronze strip are used to lead current into and away from the moving system, and also to supply a restraining torque. The position of the moving coil is indicated by a pointer that moves over a suitably marked scale. The entire moving-coil assembly is made very light in weight to decrease the load on the sharp steel pivots as much as possible. In operation, the current through the moving coil provides a field that interacts with the field of the permanent magnet, and thus supplies a deflecting torque.

Damping is accomplished by winding the moving coil on a light aluminum frame. Eddy currents are set up in the frame because of the motion of the coil in the permanent-magnet field. The field produced by the eddy currents interacts with the permanent-

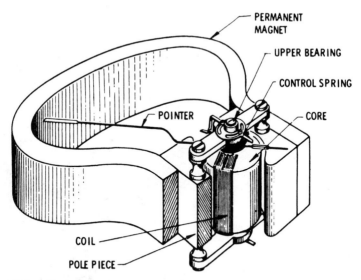

Fig. 13-1. Principal parts of a permanent-magnet, moving-coil motor movement.

magnet field in such a manner as to oppose the motion of the coil, thus causing it to come to rest quickly.

Ammeters

In general practice, it is not feasible to send more than 0.1

Fig. 13-2. A typical moving-coil assembly. Three balance weights are used to statically balance the unit.

ampere directly through the moving coil; hence, to accommodate larger currents, shunts are provided, such as those shown in Fig. 13-3. A shunt for instrument service is usually made from a material having a very low temperature coefficient of resistance and a low thermal emf to copper.

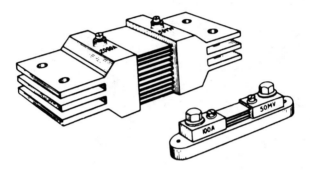

Fig. 13-3. Two types of shunts used for current measurements.

The low temperature coefficient of resistance is necessary to assure that the resistance of the shunt, and therefore the instrument indication, will not change when the shunt carries currents that are sufficiently high to cause it to become quite warm. The second requirement — low thermal emf to copper — is important, because if the shunt terminals become unequally heated, an appreciable voltage may be superimposed on the normal voltage drop of the shunt and thus cause an error.

Millivoltmeters

DC millivoltmeters for use with shunts (Fig. 13-4) generally require 10 to 15 milliamperes at 30 to 200 millivolts for full-scale deflection, depending on the class of instrument. One common type of portable instrument requires 25 milliamperes at 200 millivolts for full-scale deflection. The corresponding switchboard instruments, which need not be held to such close limits of accuracy, require approximately 20 milliamperes at 50 millivolts for full-scale deflection.

Note that the instruments just described are suitable for use only on direct current because the field supplied by the instrument magnet is unidirectional. Impressing 60-hertz current on such an

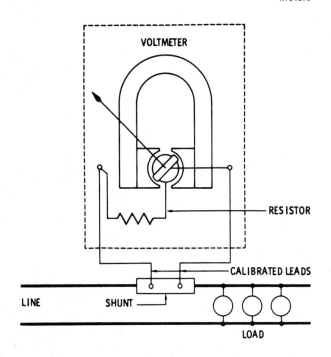

Fig. 13-4. Connections of a millivoltmeter to measure current.

element will result in a torque pulsation that tends to move the instrument pointer upscale, followed $1/120$ of a second later by a torque pulsation that tends to drive the pointer an equal distance in the opposite direction. Hence, the pointer merely vibrates a small distance on each side of the zero point on the meter scale.

Voltmeters

To make the meter movement suitable for reading DC voltage, a resistance is inserted in the instrument circuit, as shown in Fig. 13-5. This simply limits the current to such a value that full deflection of the instrument pointer is obtained when the maximum voltage that it is desired to read is applied to the instrument. For example, if an ammeter is rated at 10 milliamperes (0.01 ampere) for full-scale deflection, and it is required to make it a voltmeter having a full-scale rating of 150 volts, it would require a total resistance of 150/0.01 = 15,000 ohms.

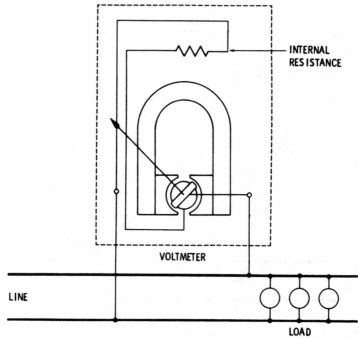

INTERNAL RESISTANCE

VOLTMETER

LINE

LOAD

Fig. 13-5. A voltmeter with internal resistance.

The milliammeter moving coil has some resistance — about 20 ohms — so only 14,980 ohms of additional resistance need be added. This resistor is made of very stable material that is not affected by ordinary changes in temperature. The amount of constant resistance added is sufficient to compensate for the effect of the moving system resistance which changes with temperature. Hence, the same current will always flow through the element when a given voltage is applied, regardless of the temperature at which the instrument operates. Several different ranges can be obtained by using separate tapped resistors (known as *multipliers*), as shown in Fig. 13-6, or by tapping off various points on the same resistor, as shown in Fig. 13-7.

It is possible to make voltmeters that require a very small amount of current for their operation. A common type of portable instrument, rated at 150 volts, requires about 10 milliamperes for full-scale deflection. It is customary to refer to DC voltmeters as

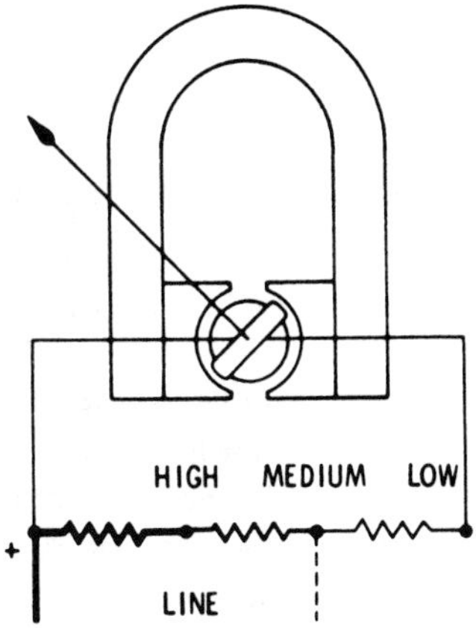

Fig. 13-6. A resistor arrangement used to provide a voltmeter with different ranges.

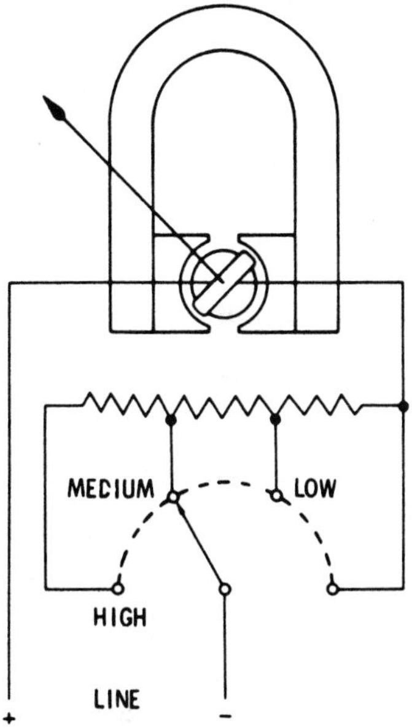

HIGH   MEDIUM   LOW

+

LINE

Fig. 13-7. A tapped resistance used to provide a voltmeter with more than one range.

MEDIUM   LOW

HIGH

LINE

+   −

possessing a certain number of *ohms-per-volt*. For example, the 150-volt instrument just mentioned required 10 milliamperes for full-scale deflection. Hence, its resistance must be 15,000 ohms. Since the resistance is 15,000 ohms and the full-scale volts are 150, the instrument has a rating of 15,000/150 or 100 ohms-per-volt. Other portable voltmeters can be made that are considerably more sensitive. Some types on the market have a resistance as high as 20,000 ohms-per-volt.

Dynamometer-Type Meters

The previously described instruments are suitable for measurements of DC only, because the field supplied by the permanent magnet is unidirectional. If the permanent magnet is replaced with stationary coils arranged as shown in Fig. 13-8, and the moving and stationary coils are connected in series, the moving coil will be deflected by an alternating current. This deflection is obtained because, when the alternating current reverses, the current in the fixed and moving coils reverses at the same instant, resulting in a pulsating torque which is always in the same direction. The amount of deflection depends directly on the amount of alternating current flowing through the coil system.

The principle of operation of this type of meter has resulted in the design of the modern dynamometer instrument shown in Fig. 13-9. Regardless of the degree of refinement used in the construction of this instrument, however, it takes a considerably larger

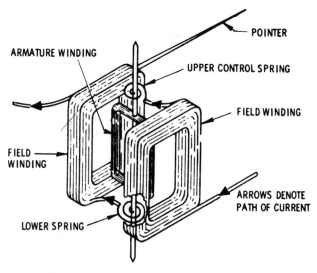

ARMATURE WINDING

POINTER

UPPER CONTROL SPRING

FIELD WINDING

FIELD WINDING

ARROWS DENOTE PATH OF CURRENT

LOWER SPRING

Fig. 13-8. Arrangement of the coils in a dynamometer-type meter.

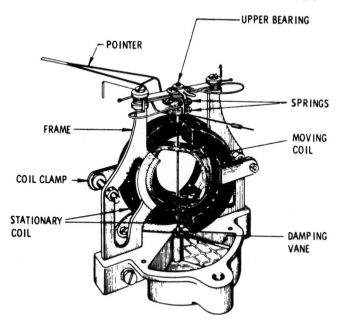

Fig. 13-9. Principal parts of a modern dynamometer mechanism.

amount of current than the permanent-magnet type employed for direct-current measurement. Here, the current being measured must not only supply the energy for the moving coil, but must also supply energy for the field winding, which in the case of the permanent-magnet instrument was supplied by the magnet.

Another important consideration in connection with this type of instrument is to keep masses of metals, as well as closed loops of wire, away from the instrument coil system. Any such material would form a path in which current could be induced, thus causing reading errors when used for AC measurements.

Because of the oscillating effect of the alternating current on the instrument pointer, the instrument is equipped with a damper. This damping is accomplished in one of two ways, either (a) by the use of a thin aluminum or copper vane that swings between the poles of a permanent magnet, or (b) by the use of a closed box in which a vane swings in a restricted space. The first method is called *magnetic damping* (Fig. 13-10), and the second is called *air damping* (Fig. 13-11). Both methods are in common use, although

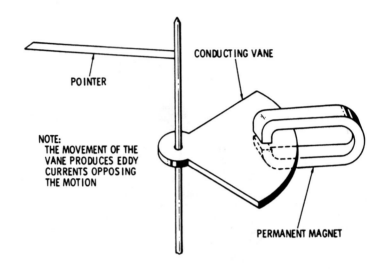

Fig. 13-10. A magnetic-damping mechanism.

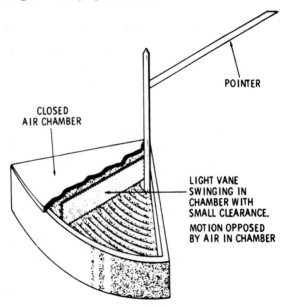

Fig. 13-11. An air-damping mechanism.

it is considerably easier to obtain high values of damping by the magnetic method.

Voltmeters

If a suitable series resistance constructed of zero-temperature coefficient material is added to this type of instrument, as shown in Fig. 13-12, then it can be used to measure AC voltage. AC voltmeters draw considerably more current from the line than DC voltmeters. In fact, it is not unusual to have a high-grade portable AC voltmeter of a 150-volt rating that requires as much as 75 milliamperes for full-scale deflection. This gives an instrument resistance of approximately 2,000 ohms and a full-scale power consumption of about 11 watts.

The series resistor in Fig. 13-12 contributes nothing to the torque of the instrument, since the torque is governed by the product of the ampere-turns of the fixed and moving coils. If the fixed and moving coils were to be wound entirely of resistance wire, the fixed resistor normally used could be dispensed with. This, however, is not practical because of the large amount of heat which would then have to be dissipated by the coil system.

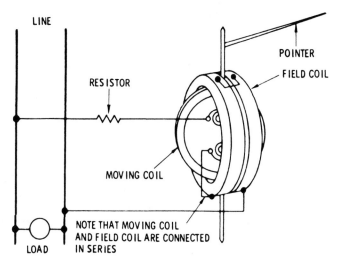

Fig. 13-12. Arrangement of a dynamometer element when used to measure voltage.

To dispose of this difficulty, the fixed and moving coils are wound with copper wire. Then, to reduce the instrument error caused by the change in the resistance of copper when the temperature is varied, it is customary to mask, or "swamp out," the error by using sufficient series resistance of negligible temperature coefficient. It has been found that a satisfactory relation between copper resistance and swamping resistance exists when the copper resistance amounts to about 10 percent of the total resistance of the instrument. In other words, in the 150-volt portable instrument previously described, the field and moving coils would have a combined resistance of about 200 ohms, and the series resistance would be approximately 1,800 ohms, making a total of 2,000 ohms.

If the full-scale voltage rating of the instrument is lowered, the resistance of the instrument, of course, must decrease. Also, it is necessary to keep the copper resistance at about the same proportion — namely 10 percent. This limits the number of turns it is possible to wind on the fixed and moving coils. Hence, to get the same torque in the instrument, more current must be used, which means that the instrument may draw a very appreciable amount of current from the line. For example, a 15-volt AC instrument of the higher-grade portable type may draw as much as 250 milliamperes from the line.

To illustrate the effect of increasing the percentage of copper resistance, an instrument could be made that would require less current for full-scale deflection by using a smaller size of copper wire on the fixed and moving coils, and by increasing the number of turns. However, the performance of the instrument under various conditions of ambient temperature might be quite poor.

Suppose that, instead of using 10 percent or 6 ohms of copper in a 15-volt, 60-ohm instrument, 30 ohms of copper and 30 ohms of zero temperature coefficient resistance were used. Copper changes 0.4 percent for each degree C change in ambient temperature; thus, if the temperature at which the instrument is used were increased 10 percent, the resistance of the copper circuit would increase to 31.2 ohms, giving an instrument resistance of 61.2 ohms at this temperature. The instrument would then read 60/61.2 or only 98 percent of the correct value, when 15 volts were applied.

The foregoing assumption as to instrument indication does not take into account the change in elastic modulus of the instrument springs at this higher temperature. As a matter of fact, in most instruments, the elastic property of the phosphor-bronze spring material changes in such a direction as to compensate partly for the error in indication caused by the change in the copper resistance of the instrument.

Ammeters

Instead of changing the milliammeter into a voltmeter by providing it with a series resistance, it can be used to measure higher values of alternating current by equipping it with a shunt. However, there would be some practical difficulties in applying this scheme if the shunt were connected as shown in Fig. 13-13. The shunt would have to have a very high voltage drop at the rated current to divert sufficient current through the ammeter element to obtain full-scale deflection. The meter element would also have to consist principally of copper. As the ambient temperature increased, the amount of current flowing in the instrument element for any given line current would be lower. The inductance of the coils must also be considered.

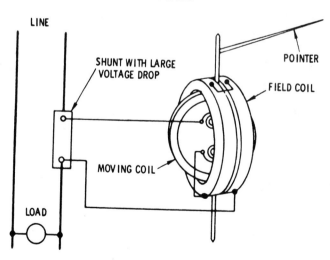

Fig. 13-13. Arrangement of a dynamometer element when used to measure current.

In order to overcome these difficulties, it is customary to send the full line current through the fixed coils of the meter and to allow only the current for the moving element to be taken off the shunt, as shown in Fig. 13-14. This arrangement permits the use of some series resistance and materially improves both the temperature coefficient and the time constant of the instrument. For lower ratings, the shunt can be located within the instrument case.

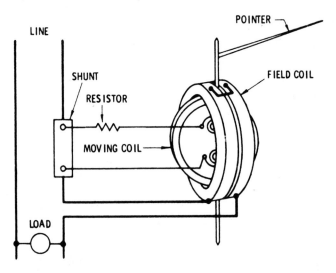

Fig. 13-14. Connection of a dynamometer to measure current. In this arrangement, the field coil carries all of the line current, whereas the moving coil carries only a fraction.

Wattmeters

Assume the ammeter arrangement shown in Fig. 13-14 is connected as shown in Fig. 13-15, omitting the shunt and adding resistance in series with the moving element instead. On alternating current, the deflection of the instrument pointer would depend on the product of the instantaneous current flowing in the fixed and moving coils. Since the current in the moving element is proportional to the line voltage, the instrument indication will be proportional to the product of the instantaneous voltage times the instantaneous line current. In other words, the instrument will read watts, and the scale can be so marked.

LINE

POTENTIAL
CIRCUIT
RESISTOR

FIELD COIL
TO CARRY
LINE CURRENT

ARMATURE COIL TAKES
CURRENT PROPORTIONAL
TO LINE VOLTAGE

LOAD

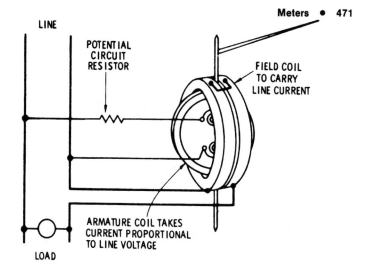

Fig. 13-15. Connection of a dynamometer to measure power consumption.

If the maximum and other corresponding values of current and voltage occur at the same instant, as shown in Fig. 13-16, the load is said to have a power factor of unity, since cos θ (θ is the angle between current and voltage) is equal to 1. If the current lags 30° behind the voltage, as in Fig. 13-17, the power factor is 0.866 lagging.

Note how the instrument in Fig. 13-15 will respond to this condition, remembering that it always reads the quantity proportional to the instantaneous product of the volts and amperes. To take a specific example: Draw lines corresponding to the voltage and current (values E and I) at a given instant, t, as shown in Fig. 13-18, and take their product. This product ($e \times i$ in Fig. 13-19) will be indicative of the torque on the moving element at that particular instant. Repeating this at suitable intervals, a torque curve that approximates that shown in Fig. 13-19 will be obtained.

Note that the torque is positive in direction at certain intervals during the cycle, and negative in others. The instrument pointer will, of course, assume a position dependent on the average value of the torque over a period of time, since the instrument element is not light enough to respond to each impulse of a 60-hertz source. Had a power factor of unity been selected, the torque would have been positive in direction at every instant.

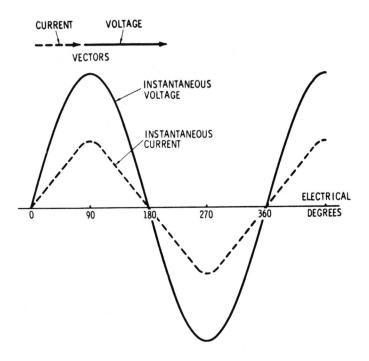

Fig. 13-16. Relation of current and voltage in an AC circuit when the power factor is unity.

While pursuing this analysis, apply it to the voltmeter (which would be connected across the same line), plotting the torque curve over one cycle. Since in a voltmeter the field and moving coils are connected in series, the instantaneous currents in each element are always in phase, regardless of the power factor of the line. To obtain the torque curve, simply square the ordinates on the current curve. Because of the mass of the moving element, the instrument pointer will again assume an average position with respect to the double-frequency torque pulsations. These conditions and the procedure are indicated in Fig. 13-20.

The dynamometer-ammeter element that was shown in Fig. 13-15 will also be subjected to the same continuous positive pulsating torque, since the current in the movable coil is always in phase with the line current which flows in the current coil.

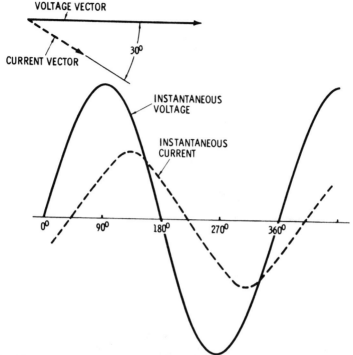

Fig. 13-17. Relation of current and voltage when the current is lagging at an angle of 30°.

It should always be remembered that:

1. The dynamometer voltmeter indicates *root-mean-square (rms)* voltage.
2. The dynamometer ammeter indicates *root-mean-square (rms)* current.
3. The dynamometer wattmeter indicates *average* power.

A wattmeter may be connected in a circuit as shown in Fig. 13-21 or as in Fig. 13-22. Note that, in one case, the wattmeter potential-circuit loss is being measured, and in the other case, the current-coil loss. The potential-circuit loss of a typical portable 500-watt wattmeter would be in the order of 2 watts at 100 volts. Hence, if this instrument were connected as in Fig. 13-21, the reading would be 2 watts higher than the true value consumed in the load. The current coil for this particular wattmeter requires

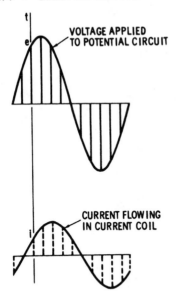

Fig. 13-18. Relation of current and voltage in wattmeter.

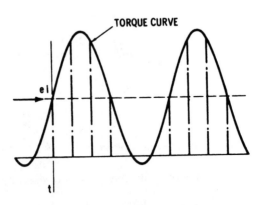

Fig. 13-19. Torque curve, which is the product of the current and voltage in Fig. 13-18. Notice that the torque reverses during each cycle.

about 2 watts at about 5 amperes. Since the voltage is usually fairly constant, but the current varies, it is much easier to correct the readings for a constant-potential circuit loss than for a varying-current coil loss. Thus, the connection including the potential-circuit loss is usually preferred.

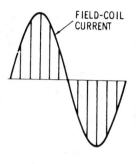

FIELD-COIL
CURRENT

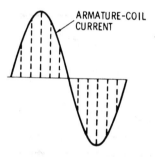

ARMATURE-COIL
CURRENT

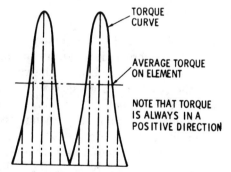

TORQUE
CURVE

AVERAGE TORQUE
ON ELEMENT

NOTE THAT TORQUE
IS ALWAYS IN A
POSITIVE DIRECTION

Fig. 13-20. Field, armature, and torque curves obtained by wattmeter measurements.

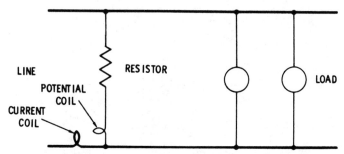

Fig. 13-21. Connection of a wattmeter in which the current coil is connected ahead of the potential coil.

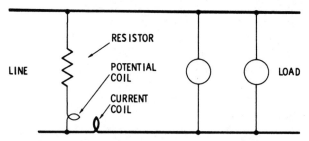

Fig. 13-22. Connection of a wattmeter in which the current coil is connected back of the potential coil.

Magnetic-Vane Meters

This type of AC and DC instrument, sometimes called a moving-iron instrument or a moving-iron-vane instrument, is often used for ammeters and voltmeters. This instrument depends for its operation on the reactions resulting from the current in one or more fixed coils acting upon one or more pieces of soft iron or magnetically similar material in the moving system. The moving-iron principle is illustrated in Fig. 13-23.

If two similar and adjacent iron bars are similarly magnetized, a repelling force is developed between them that tends to move them apart. In the moving-iron instrument, this principle is used by having one bar fixed in space and by pivoting the second so that it will tend to rotate when the magnetizing current flows. A spring attached to the moving vane opposes the motion of the vane and permits the scale to be calibrated in terms of the amount

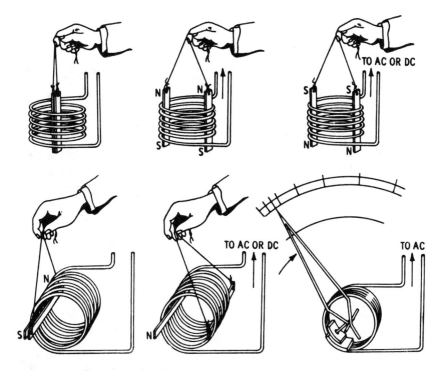

Fig. 13-23. Working principles of a magnetic-vane meter.

of current flowing. When current flows through the solenoid, the plunger is drawn into the coil and a measurable deflection of the instrument pointer is obtained. Because, however, of high power consumption, sensitivity to slight zero shifts, scale difficulties, etc., this type of movement is presently used only in less expensive instruments.

Repulsion Type

This instrument (Fig. 13-24) differs in construction principles from the previously described moving-iron-vane instrument in that the magnetic vane embedded in the side of the coil has definite N and S poles at the points shown. The moving vane will be magnetized by induction, with the polarities as shown. Notice that the north pole of the moving vane is nearest to the north pole on the piece of magnetic material embedded in the coil, and that the

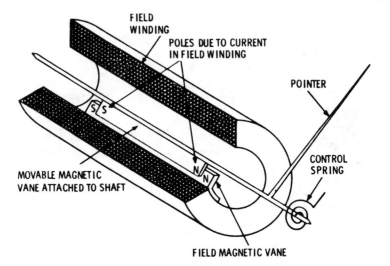

Fig. 13-24. Principle of a simple magnetic-vane, repulsion-type meter.

south poles are similarly placed. Hence, the vane tends to move away or to be repelled from the stationary magnetic piece.

If alternating current is applied to the instrument, the two vanes will simultaneously change polarities as the current varies throughout the cycle; thus, the instrument also operates on alternating current. In fact, it is in AC measurements that the instrument finds its greatest application.

Induction Type

This type of instrument depends for its operation on the reactions between a magnetic flux (or fluxes) set up by one or more currents in fixed windings, and electric currents set up by electromagnetic induction in conducting parts of the moving system. The construction of a typical induction instrument is shown in Fig. 13-25. The instrument consists primarily of a field structure with associated windings into which a rotating disc of aluminum is placed. The disc will tend to rotate when an alternating current of the proper phase relation is applied to the two coils. If a spring is inserted on the disc shaft to counteract the movement of the disc, and a pointer is added, its movement will indicate the value of the current flowing through the instrument.

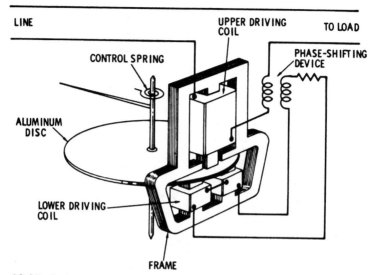

Fig. 13-25. An induction-type meter movement.

This instrument can only be used on alternating current. If a direct current is applied, the only time voltage would be induced in the disc would be at the instant the circuit is opened or closed. Thus, the disc will tend to move forward and then drop back to its initial position when a steady-state condition is reached.

The disc must be made of aluminum or copper, or of some other material of high electrical conductivity. Since all these materials possess appreciable temperature coefficients of resistance, changes in temperature will materially affect the readings unless carefully compensated for in other parts of the circuit. This has limited the usefulness of these devices to a considerable extent.

Induction Watt-Hour Meter

If the spring and pointer are omitted in the assembly shown in Fig. 13-25, and magnets placed at the edge of the disc as in Fig. 13-26, the disc will cut lines of magnetic flux from the magnets located at its periphery, and the currents induced will tend to reduce the speed of the disc.

The effect of temperature on such an arrangement is not nearly so pronounced as in the case of the spring-controlled induction

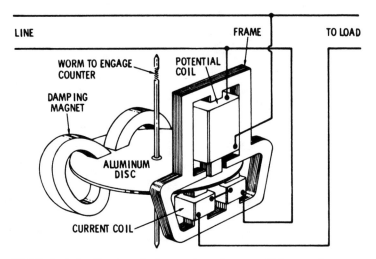

LINE   FRAME   TO LOAD

WORM TO ENGAGE
COUNTER

POTENTIAL
COIL

DAMPING
MAGNET

ALUMINUM
DISC

CURRENT COIL

Fig. 13-26. An induction-type instrument used as a watt-hour meter.

instrument. As the temperature increases, both the induced current actuating the disc and the eddy currents caused by the damping magnets tend to decrease in the same proportion. If the connections are changed and a counter connected by suitable gearing added to count the total revolutions of the disc, the basic elements of the induction watt-hour meter are obtained.

Rectifier-Type Meters

A rectified instrument is principally the combination of a meter sensitive only to direct current and a rectifying device whereby alternating currents or voltages may be measured. By inserting a rectifier in an AC circuit, as shown in Fig. 13-27, it will be possible to use the previously described permanent-magnet, moving-coil type of meter to measure the value of an alternating current.

A rectifier instrument may be equipped with any one of several types of rectifying devices. The one most commonly used is the semiconductor diode. The most common method used is to arrange four diodes in a bridge connection, as shown in Fig. 13-28. This bridge arrangement rectifies every half-cycle of the AC wave and is called a full-wave rectifier.

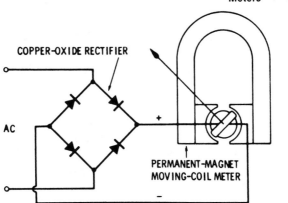

COPPER-OXIDE RECTIFIER

AC

+

PERMANENT-MAGNET
MOVING-COIL METER

−

Fig. 13-27. Rectifier circuit connected to a permanent-magnet, moving-coil meter to permit the measurement of AC voltage.

When using rectifier instruments, it should always be borne in mind what the instrument really indicates, regardless of how the scale may be marked. A permanent-magnet, moving-coil type of instrument indicates average values, shown by the lower line in Fig. 13-28. On AC circuits, however, the root-mean-square (*rms*) value is the quantity generally required. This value is shown by the upper line.

For a sine wave, the ratio between *rms* value and average value is 1:11, and the meter scale is marked directly in terms of the *rms* value. For other wave shapes this ratio is different, and hence a rectifier instrument reads *rms* quantities correctly only for the wave shape for which it is calibrated. This point must be kept in mind when using rectifier instruments to measure nonsinusoidal quantities, such as the outputs of constant-voltage transformers. Fortunately, the errors resulting from this condition are generally less than 10 percent of full-scale value. The permanent-magnet, moving-coil construction gives a higher sensitivity and a more uniform scale than is possible to secure with a moving-iron or dynamometer type.

Clamp Ammeters

Another type of instrument frequently employed in conjunction with a rectifier is a clamp ammeter. It is not, however, a precision instrument, but is used for rough approximations of current in

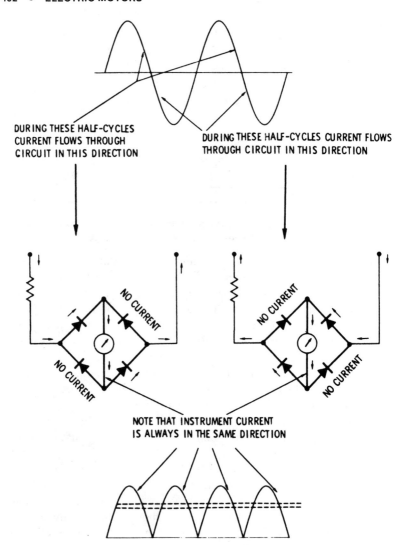

Fig. 13-28. Working principles of a rectifier-type meter.

conductors and in motor or transformer leads. Principally, the instrument consists of a current transformer of the inserted-primary
type, and a rectifier connected to a permanent-magnet, moving-
coil type of instrument. As shown in Fig. 13-29, the conductor

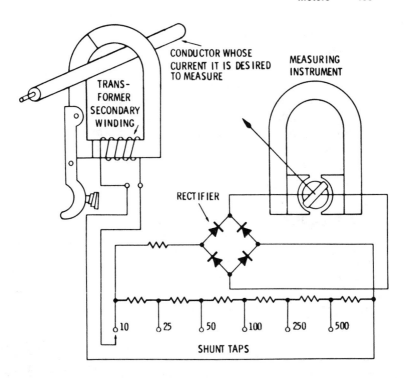

Fig. 13-29. A clamp-ammeter circuit used with a rectifier-type meter. Also see Fig. 13-35.

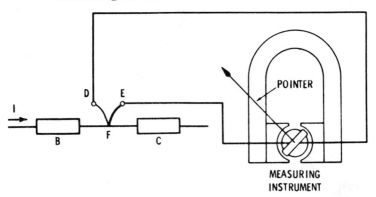

Fig. 13-30. A thermocouple added to a meter movement to permit measuring AC and DC current.

forms the transformer primary winding, which induces a current in the transformer secondary. Various ranges are obtained by using an external multiplier arrangement. When connected as shown, a small proportional current is derived from the line current through use of the hinged-core current transformer. This secondary current is then connected to a suitable tap on the multiple-range rectifier-type meter to give the reading. The rectifier instrument is made into a multiple-range device through the use of the multirange series shunt. In this combination, it is possible to measure the current in any conductor by using the proper taps on the series shunt.

Thermocouple-Type Meters

This type of instrument employs one or more thermojunctions (thermocouples), which, when heated directly or indirectly by an electric current, generate an electric current. When this generated current is made to flow through a permanent-magnet, moving-coil meter such as in Fig. 13-30, a deflection proportional to the difference in temperature of the junctions takes place.

The voltage from a thermocouple is proportional to the difference in temperature between the heated junction and the point at which the thermocouple wires are connected to the copper leads of the instrument. Therefore, any change in the temperature of this latter connection point will result in an error in instrument indication. To avoid errors caused by this condition, it is common practice to terminate the heater in rather massive blocks, and then bring the point at which the thermocouple leads join the copper circuit into good thermal contact with these blocks. This is usually done by connecting the leads to a thin copper plate, and separating them from the block by a thin mica strip. This is called *cold-junction compensation* (Fig. 13-31) and serves to maintain the connection to a fixed relation to the temperature of the heater.

The meters used with thermocouples to measure current must necessarily be more sensitive than those used with shunts, because the emf developed by a thermocouple is quite small. For example, such a thermocouple may have an output voltage of only 15 millivolts, and an internal resistance of 5 ohms. This means that thermo-

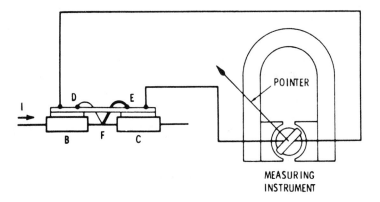

Fig. 13-31. Cold-junction compensation in a thermocouple-type meter.

couple instruments should be carefully handled, since the construction is necessarily somewhat delicate.

Hot-Wire-Type Meters

This type of instrument depends on the heat expansion of a wire carrying a current for its operation. With reference to Fig. 13-32, the instrument functions as follows: the current being measured causes a wire, through which the current flows, to heat up and thus expand or increase in length approximtaely in proportion to the square of the current flowing (I^2). The change in the length or the sag of the wire is amplified and arranged to drive the pointer.

Instruments of this type are made in different forms with various arrangements of their wires and parts. All of them, however, work on the same principle. Because this instrument functions as a result of the heating effect of the current, it can be used for measurements on both DC and AC circuits. This instrument is not employed where great accuracy is desired. This is mainly due to the instability of the wire stretch and the lack of ambient-temperature compensation. It is now being largely replaced by the more sensitive, more accurate, and better compensated permanent-magnet, moving-coil combinations previously described.

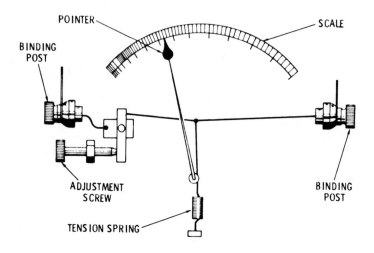

Fig. 13-32. Principal parts of a hot-wire ammeter.

Digital Multimeter

Most modern repair facilities utilize the accuracy of the digital multimeter in troubleshooting. This is a handy device that needs no power source other than the 9-volt batteries already enclosed in the case. The portable, plastic-encased meter is very useful in field testing.

The meter in Fig. 13-33 displays its readings on a liquid crystal display (LCD). A 1.8.8.8 reading indicates infinite resistance. It operates on two 9-volt batteries located under the face of the meter. When you change the selection switch to volts or millivolts, it reads 0; accuracy can be to 0.001 volt. The ohms scale reads to 200 ohms. The k-ohms scale reads up to 100,000 ohms and the M-ohms scale reads megohms. As you can see from the meter, it also measures both volts and amperes in AC and DC. Milliamperes are also available. When measuring amperes it is necessary to move the lead from the V-Ω-mA jack to the AMPS jack. This places a proper shunt in the circuit to protect the meter.

Circuitry

The internal circuitry is a bit complicated. It uses a number of modern semiconductor devices to check the voltage and current and then produce a readout on the display. In most cases the accuracy is within $\cong0.1$ percent.

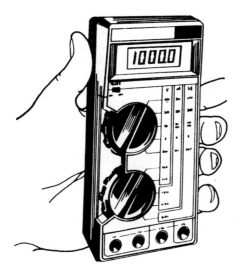

Fig. 13-33. Newer-type digital voltmeter.

Using the Meter

There is very little "learning time" with this type of meter. All you have to do is learn where the three switches are located: the on-off selector switch, the AC or DC and ohms switch, plus the movement of the lead from volts-ohms-milliamperes to amperes when necessary to measure amps. The rest is self-explanatory. The ohms switch selection has to be changed when 1.8.8.8 appears on the ohms or k-ohms scale. For greater accuracy it is necessary to change the switch from M-ohms, k-ohms to ohms to make sure you get a three-place reading. The rest of the meter is automatic. It has what is called auto-ranging, which changes the ranges automatically for volts and amps.

To extend the life of the meter, be sure to turn it to the OFF position when in use. The liquid crystal display needs ambient light to produce a reading; it does not work in the dark.

The selection of the meter is up to the individual and the job being done. However, in most instances the digital will perform better than any laboratory meter and will do it anywhere except under extreme conditions of heat and cold.

The only maintenance is changing of batteries whenever the meter indicates erratic readings.

Meter Connections

The current in an electric circuit is measured by the ammeter. Always connect an ammeter in series with the load (Fig. 13-34) whose current it is desired to measure — *never cross it*, when a clamp-on meter is used (Fig. 13-35).

Measurement of Alternating Current

If an ammeter is to be used on circuits of devices that have a high starting current, such as motors, short out the ammeter until after the starting period. Or, if the meter is to be left in circuit for some time, use a short-circuiting switch to avoid damaging the instru-

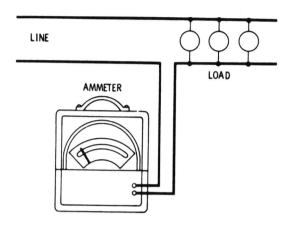

Fig. 13-34. An AC ammeter connection.

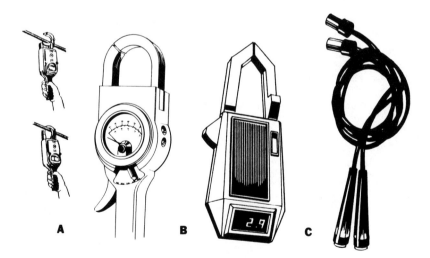

Fig. 13-35. Clamp on meters: (A) Analog type; (B) Digital type.
(C) Leads to use with voltage and resistance readings.

ment. The switch can be opened at any time when a reading is to be taken.

Measurement of Direct Current

If one side of the line is grounded, ammeters should be connected to that side. External and internal shunts are used for measurements of large current values. If the instrument reads backward, reverse the meter leads. Various DC ammeter connections are shown in Figs. 13-36, 13-37, and 13-38.

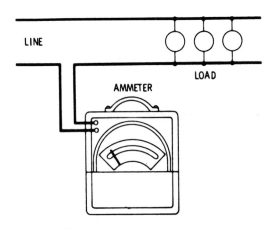

Fig. 13-36. A DC ammeter connection.

Measurement of AC Voltage

The potential or voltage across a circuit is measured with a voltmeter. Always connect a voltmeter across the load whose voltage it is designed to measure — *never in series*. To extend the range of a voltmeter, a multiplier may be used. If a higher rated voltmeter is not available, use a potential transformer. *Always ground the potential-transformer secondary.* Typical voltmeter connections are shown in Figs. 13-39 and 13-40.

Measurement of DC Voltage

If a DC voltmeter (Fig. 13-41) reads backward, reverse the instrument leads. Voltmeters are commonly equipped with an

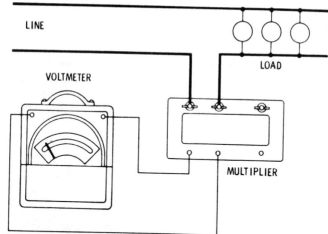

Fig. 13-37. A DC ammeter connected to a shunt for measuring DC current.

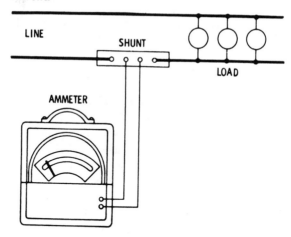

Fig. 13-38. A DC ammeter connected to a shunt to extend the range of the meter.

internal or external resistance (sometimes called a *multiplier*) to extend the range of the instrument. This is shown in Fig. 13-42.

Measurement of AC Power (Voltmeter and Ammeter Method)

The power in an electric circuit carrying a noninductive load may be measured by the combined use of a voltmeter and an ammeter, or by means of a wattmeter. An illustration showing the connections of a voltmeter and an ammeter for power measure-

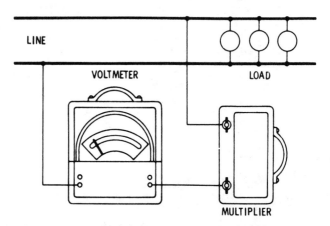

Fig. 13-39. An AC voltmeter connection using an external multiplier to extend the meter range.

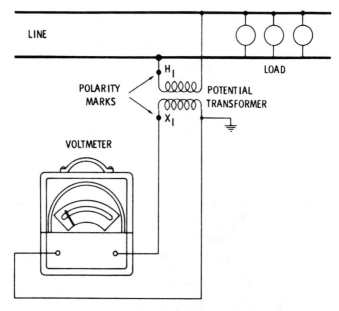

Fig. 13-40. An AC voltmeter used with a potential transformer.

ments is shown in Figs. 13-43 and 13-44. In Fig. 13-43, the voltmeter measures *line voltage*, not load voltage. The ammeter measures load current only. In Fig. 13-44, the voltmeter measures *load voltage*, not line voltage. The ammeter measures load current plus voltmeter current.

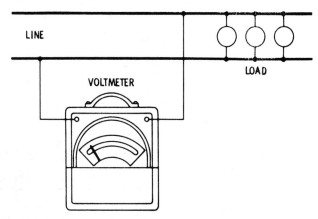

Fig. 13-41. A DC voltmeter connection.

Measurement of AC Single-Phase Power (Wattmeter Method)

Methods of connection when power in a single-phase system is to be measured by a wattmeter are shown in Figs. 13-45 through 13-50. In Fig. 13-45 the instrument is measuring load power plus the loss in its own current-coil circuit. If the instrument reads backward, reverse the current leads.

Special precautions should be taken not to overload wattmeters. Potential transformers are usually employed on circuits above 300 volts. On circuits of 750 volts and above, both potential and current transformers are generally used. *Always ground the secondary of both transformers.*

Figs. 13-49 and 13-50 illustrate the connections when it is desired to measure the voltage and power consumption of the load simultaneously. In Fig. 13-49, the wattmeter measures load power plus losses in the voltmeter and wattmeter potential circuits. In Fig. 13-50, the wattmeter measures load power plus losses in its own current-coil circuit.

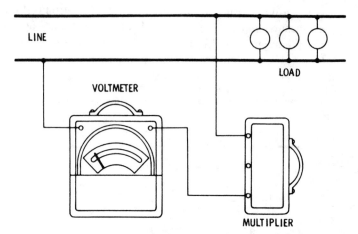

Fig. 13-42. A DC voltmeter connection with an external multiplier to extend the meter range.

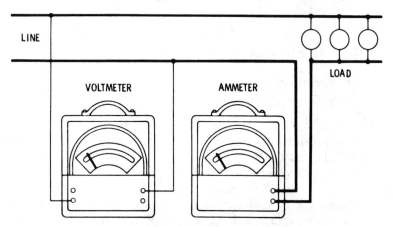

Fig. 13-43. A voltmeter and ammeter connection in which the voltmeter reads the true line voltage.

Power Factor Measurement (Indirect Method)

The power factor of a single-phase load may be determined by dividing the wattmeter reading by the product of the voltmeter and ammeter readings. When the instruments are connected as shown in Fig. 13-51, the wattmeter measures load power plus losses in the ammeter and wattmeter current-coil circuits. In Fig.

13-52, the wattmeter measures the sum of the power losses of the load, the potential circuit of the wattmeter, and the voltmeter.

Power-Factor Measurement (Direct Method)

Figs. 13-53, 13-54, and 13-55 illustrate a more convenient method

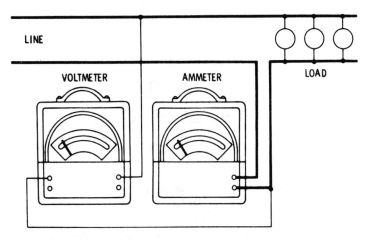

Fig. 13-44. A voltmeter and ammeter connection in which the voltmeter reads the true voltage across the load.

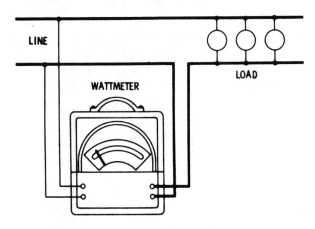

Fig. 13-45. A wattmeter connection to a single-phase circuit where the meter measures the load power plus the loss in its own current coil.

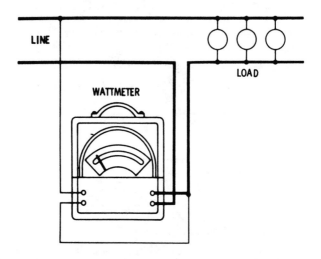

Fig. 13-46. A wattmeter connection to a single-phase circuit where the meter measures the load power plus the loss in its own potential coil.

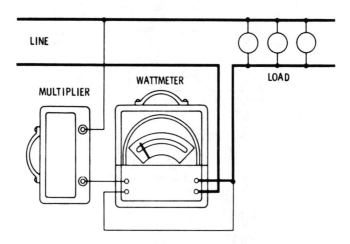

Fig. 13-47. A wattmeter connection at a single-phase circuit, using a multiplier in the meter potential circuit.

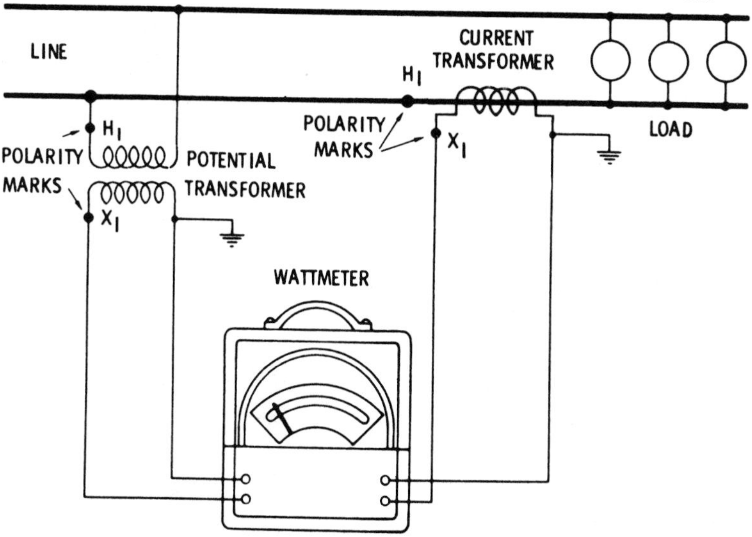

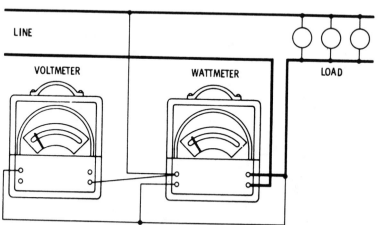

Fig. 13-48. A wattmeter connection in a single-phase circuit, and using current and potential transformers.

Fig. 13-49. A wattmeter and voltmeter connected to a single-phase circuit.

for direct reading of the power factor in single and polyphase circuits by means of power-factor meters connected as shown in Figs. 13-53, 13-54, and 13-55. These connections, when made as shown, are correct at all power factors. Polyphase power-factor meters are intended for use in balanced circuits only. Single-phase power-factor meters should be used only at the calibrated frequency.

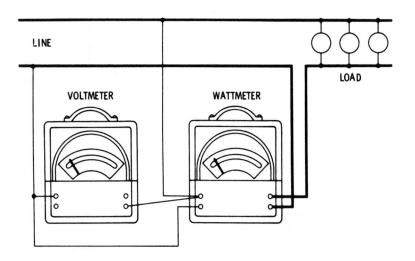

Fig. 13-50. An optional connection of a wattmeter and voltmeter to a single-phase circuit.

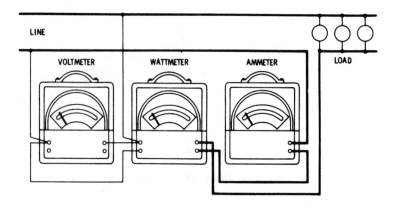

Fig. 13-51. Connection of a wattmeter, voltmeter, and ammeter to a single-phase circuit.

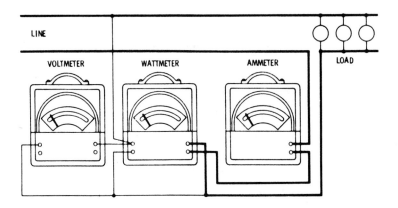

Fig. 13-52. Optional connection of a wattmeter, voltmeter, and ammeter to a single-phase circuit.

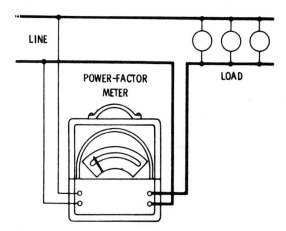

Fig. 13-53. Connection of a power-factor meter to a single-phase circuit.

Polyphase Power Measurement

Illustrations showing connections for power measurement in polyphase systems are shown in Figs. 13-56 through 13-60. In Fig. 13-56 the power of the system is three times the indication on the wattmeter. The meter indicates its own potential losses plus the power in one phase of the load. In Fig. 13-57, the two wattmeters will not indicate alike except at unity power factor. Above 50

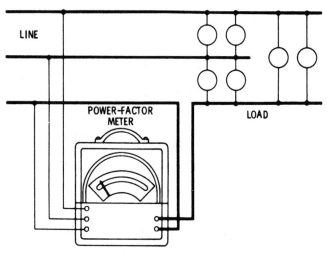

Fig. 13-54. Connection of a power-factor meter to a three-wire, three-phase circuit.

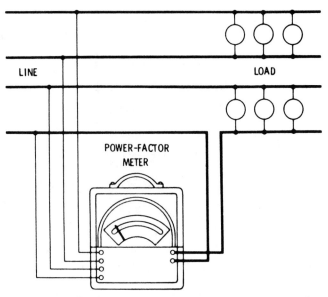

Fig. 13-55. Connection of a power-factor meter to a four-wire, three-phase circuit.

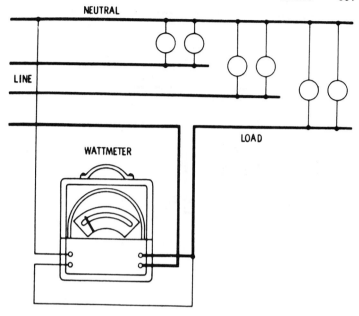

Fig. 13-56. Connection of a single-phase wattmeter to a balanced four-wire, three-phase circuit.

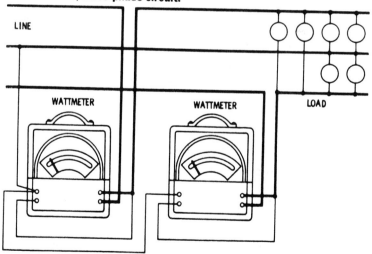

Fig. 13-57. Connection of two wattmeters to a three-phase circuit having balanced or unbalanced voltages or load.

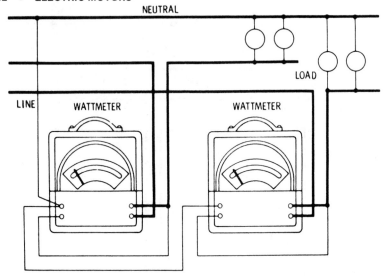

Fig. 13-58. Connection of two wattmeters to a three-wire, two-phase circuit having a balanced or unbalanced load.

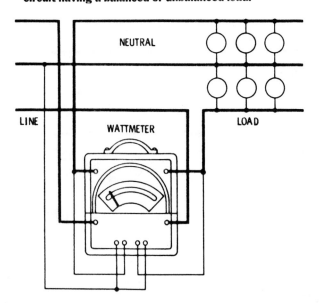

Fig. 13-59. Connection of a polyphase wattmeter to a three-wire, two-phase circuit having balanced or unbalanced voltages or load.

percent power factor, it is necessary to reverse the reading of one wattmeter (by reversing its current leads) and then take the difference between the readings of the two instruments.

Meggers

Meggers are used to test high resistances as in electric motor insulation. The handle is turned to generate a very high voltage. The scale reads out in ohms. *Do not touch the leads when the handle is turned* (Fig. 13-61).

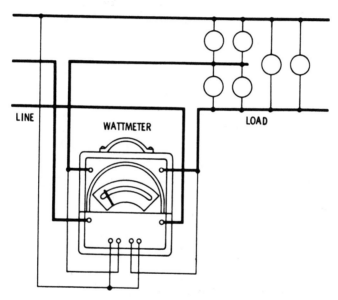

Fig. 13-60. Connection of a polyphase wattmeter to a three-wire, three-phase circuit.

Meter Maintenance

Electric meters will give good service if accorded the care that mechanisms of their delicacy deserve. If mistreated, they may fail to function properly, resulting in considerable loss in both time and money. It is therefore of the greatest importance that thoughtful attention be given to their care and use.

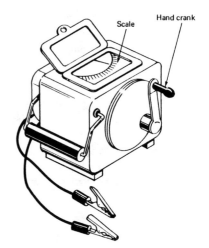

Scale   Hand crank

Fig. 13-61. Megger.

Selection

Best results in the use of instruments depend on the proper original selection. The accuracy desired and the cost are the two important factors in selection. Instruments should be chosen to give the required accuracy at the lowest cost.

The following guide for proper instrument selection is recommended:

1. What accuracy is desired?
2. Should it be AC or DC?
3. Should it be portable or for switchboard mounting?
4. Should it be indicating or recording?
5. What is the desired rating? The range should be such that the normal readings are above one-third scale.
6. If a switchboard instrument is desired, should it be of the surface-mounting or semiflush-mounting type?
7. Can a standard instrument listed in the manufacturer's catalog be utilized in order to obtain the lowest cost and the best delivery?
8. What effect will the instrument have on the circuit in which it is to be used? For example, some applications require a voltmeter with high resistance for minimum current drain. Or an ammeter with a low volt-ampere rating may be needed if it is to be used with a through-type current transformer.

9. Conditions under which the instrument is to be used are important. Indoor or outdoor humidity or other corrosive atmosphere may require special instruments.

Instrument Use

When any questions arise as to the use of a specific instrument, it is recommended that the information be obtained from the manufacturer.

The following pointers should be observed before placing *any instrument* in service:

1. The manufacturer's instruction book accompanying the instrument should be read carefully.
2. The instrument should be handled carefully to avoid any vibration or shocks that may injure the bearings in the meter movement.
3. Estimate the value of the current, voltage, etc., to be measured from the nameplate of the apparatus on which the measurement is to be made.
4. If an instrument has more than one tap or range, make connection to the highest range first to avoid damage to the instrument.
5. Instruments should always be read when in their normal operating position; that is, portables should be in a level, horizontal position; switchboard types in a vertical position.

The following pointers should be observed before placing any *portable* instrument in service. These do not apply to digital meters.

1. Metal benches should be avoided, as the instrument indication may be affected by the proximity of the metal.
2. Proper-sized leads should be used. Terminals should be clean to avoid high contact resistances, particularly in the case of meters used with shunts. Lead wires should extend away from the observer to avoid accidentally pulling the instrument off the table.
3. Whenever a shunt is used with a meter to read current, always check to make sure the leads with which the instrument was calibrated are being used.

4. If external-resistance multipliers are used, check both the serial number of the instrument and that of the multiplier to make sure of the correct combination.
5. The instrument zero setting should be checked. If the pointer is slightly off zero, reset to zero by adjusting the external-zero shifter. Never try to compensate for a bent pointer by moving the zero shifter.
6. The instrument should not be used in strong fields such as near cables or bus bars. Presence of a magnetic field that is affecting the reading of an instrument can be detected by turning the instrument (clockwise or counterclockwise) 90° at a time, and taking a reading in each position. If a difference is noted in the readings, some stray field is present.
7. Unshielded instruments should not be placed closely together, as the reading of one instrument may be affected by the magnetic field of the other.
8. When an ammeter or shunt is to be connected into a circuit, always connect it into the grounded side, if possible.
9. When a voltmeter is to be connected to a circuit, always connect the leads to the instrument first. This avoids loose leads that may be hot. Be sure that voltmeter leads are properly insulated for the voltage to be checked. (This also applies to digital meters.)
10. Instruments, shunts, and leads should always be arranged in such a way that there is no danger of knocking them off the table or tripping over the lead. If any high voltages are to be measured, the arrangement should be such that no one might accidentally touch the high-voltage equipment.
11. Whenever an instrument is to be connected to a current-transformer secondary, always short-circuit the secondary terminals before the primary is connected in the circuit to avoid dangerously high voltages. *The current-transformer secondary must be grounded.*
12. When a potential transformer is to be used, always make sure that the instrument is connected to the secondary, or low-voltage, side. *The potential-transformer secondary circuit must be grounded.*

The following pointers apply to *switchboard* instruments:

1. Switchboard instruments should be mounted on panels in accordance with manufacturer's instructions.
2. Never mount a switchboard instrument on a panel until all other work on the panel is completed, as any pounding or vibration may affect the instrument.
3. The general precautions given earlier for portable instruments also apply to switchboard instruments.

Instrument Precautions

The following pointers should be observed in the use of any electric instrument. An asterisk [°] indicates the pointer applies to digital meters also.

1. Always try to have all the instrument reading between one-half and three-fourths full scale. Never read below one-third full scale if it can be avoided.
°2. Never touch bare terminals or binding posts while an instrument is energized.
3. The instrument reading should be made from directly above the scale to avoid errors.
4. If using a split-core current transformer with an ammeter always make sure that the transformer-core joint is free from dirt, which may introduce a considerable error in the reading.
5. To avoid errors, instruments should never be used in extreme temperatures. Special temperature-compensated instruments are used for these conditions.
6. Special instruments should be used to measure motor-starting currents or welding currents.
°7. Clean the glass of an instrument by using a damp cloth. A dry cloth may induce a static charge on the glass and affect the instrument reading. If the instrument glass does become charged, the charge can be dispelled by breathing on the glass.
°8. Never slam the cover of an instrument.

*9. If an instrument is overloaded or dropped, or if for any reason the accuracy is doubted, the instrument should be checked against another instrument at several different readings before it is used again.

10. Tapping an instrument with the fingers will remove any slight stickiness. If the pointer movement is erratic, and if tapping causes appreciable movement of the pointer, the instrument bearings may require attention and should be examined by a competent instrument-repair laboratory.

11. Instrument accuracy is always expressed in terms of the percentage of error at the full-scale point, so maximum accuracy is obtainable by keeping the reading as high on the scale as possible. This calls for the selection of a properly rated instrument. The accuracy of the readings is determined as follows: Assume that there are 100 scale divisions and that the accuracy is 1 percent of full scale. A scale reading of 100 means that the quantity measured contains 100 units plus or minus 1 scale division; that is, the correct value lies between 99 and 101 units. However, while a scale reading of 20 (one-fifth of full scale) also has a margin of error of plus or minus 1 division, the error there is five times as great, or plus or minus 5 percent.

12. Wattmeters should be so connected into a circuit that the current coil and the potential coil are at the same potential. (The potential terminal, which is connected in the same side of the line as the current coil, is usually identified by a zero or + marking.) This is particularly important where multipliers are to be used, as it is possible to obtain a potential difference between the current and potential coils high enough to cause a breakdown.

13. Instruments with single ratings are preferable because if, by accident, the single-rated instrument becomes defective, the use of only one instrument is lost, whereas if one tap of a multirated instrument burns out, the entire instrument has to be repaired.

14. Where rectifier instruments are used in AC circuits to measure AC values, the shape of the waveform should be known, as rectifier instruments are normally calibrated for sine waves only.

Instrument Calibration

How often an instrument should be calibrated depends entirely on its use and the accuracy desired. If calibration standards and equipment are not available, instruments of nearly the same rating can be checked against each other. If wide discrepancies are noted, the instrument that reads incorrectly should be checked by a competent laboratory technician or returned to the manufacturer for correction.

If the meter or instrument requires a new bearing or jewel, it is advisable to replace the pivot as well, as the old one may have become damaged, especially if the meter has been operating with a damaged jewel. When inserting a new jewel, or when replacing a jewel after having removed the moving element, see that the top bearing clamping screw is loosened so that this bearing is free to move, thereby preventing the possibility of a damaged jewel. When inserting a new jewel, or when replacing a jewel after having removed the moving element, see that the top bearing clamping screw is loosened so that this bearing is free to move, thereby preventing the possibility of a damaged jewel or shaft. See that the top bearing is properly set and clamped after inserting and adjusting the lower bearing.

To replace the pivot, remove the jewel screw from the meter and insert the pivot wrench in place. Clamp the pivot in the end of the wrench and unscrew. Replace the pivot with a new one and screw it into the shaft. Apply a drop of fine watch oil to the jewel and insert the new jewel screw. Whenever the registering element of a meter is oiled, it should be wiped carefully to remove any excess oil. Only the best quality watch or instrument oil must be used for this purpose.

Summary

Meters are often a part of the control equipment for electric motors and generators. They are also used extensively in the maintenance and repair of all types of electrical equipment. For this reason, persons using meters should be thoroughly familiar with their construction, capabilities, limitations, and use.

The most common meters used in conjunction with motors and generators are *ammeters, voltmeters,* and *wattmeters.* These may be *direct-current* or *alternating-current* instruments. In addition, they may operate on a permanent-magnet, moving coil, a dynamometer, a magnetic-vane, or an induction principle.

Ammeters should never be connected *across a voltage source,* but always in series with the device of which the current is to be measured. Placing an ammeter across a voltage source will result in *permanent damage to the instrument.* The range of ammeters can be extended by means of *shunts.*

Clamp ammeters are often used where absolute current values are unnecessary. This type of instrument is useful because it measures the approximate amount of current flowing in a conductor without the necessity of disconnecting the wire and inserting the meter.

Voltmeters should always be connected across (in parallel with) the part of the circuit in which it is desired to determine the voltage.

Do not place the voltmeter in series with the device as an inaccurate reading, if any, will result. The range of voltmeters can be extended by the use of multipliers.

The digital multimeter is portable and can be used as a laboratory as well as a field measuring device. It has a 9-volt battery to power its electronic circuits. This meter uses a liquid crystal display, which needs ambient light to be read. Other digital types use light-emitting diode (LED) displays, which can be read in the dark if necessary. The digital multimeter can be calibrated to an accuracy of ± 0.1 percent.

All meters are relatively delicate instruments and should be used and handled with great care. Rough handling or improper use may result in permanent damage or inaccurate readings. An inaccurate reading is worse than no reading as it may lead to the wrong conclusions.

The proper use of a meter requires the user to be thoroughly familiar with the instrument as well as the electrical circuits and equipment on which it is used. The instruction book supplied by the manufacturer should be followed explicitly to prevent error, damage, and loss of time.

Review Questions

1. How should an ammeter be connected to measure current?
2. What is the correct method of connecting a voltmeter?
3. How can the range of an ammeter be extended?
4. How does a dynamometer-type meter read true or effective values?
5. How can a permanent-magnet, moving-coil meter be modified to allow readings to be taken on AC circuits?
6. What is a digital multimeter?
7. How accurate is a digital multimeter?
8. Where can the digital multimeter be used?
9. What type of display is there on the digital multimeter shown in Fig. 13-33?
10. Where would you use a digital multimeter?

CHAPTER 14

Wiring Diagrams

The purpose of any wiring diagram is to indicate electric circuits by means of conventional symbols, thus making it possible to show the connections between various apparatus and instruments in a diagrammatic form from which actual connections in the field may be made. In addition, wiring diagrams serve to make a permanent record of the wiring and apparatus installed, thus facilitating changes and replacement. Diagrams also assist in the location of any trouble that may develop in the circuit after installation.

Diagram Design

The general procedure in laying out a circuit or diagram, especially one that is complicated, usually consists in working out an elementary or schematic diagram first, then making the working

513

or general diagram from this layout. In an elementary diagram, no attention need be paid to the physical location of contactors or coils, but the strictest attention must be paid toward the electrical sequence of operation and to see that the equipment is correctly connected to operate properly. After a careful check of the elementary diagram has been made, the circuit is usually transferred to a working diagram where the relays, contactors, and other apparatus are laid out in a rear-view order, although the actual electrical connections conform in every detail with that of the elementary diagram. Thus, the purpose of the elementary diagram is only to simplify the circuit for construction and checking.

In addition to elementary diagrams, there are several other types, each designed for a particular purpose or installation. The *National Electrical Manufacturers Association* (NEMA) and *Institute of Electrical and Electronics Engineers* (IEEE) have given the following definitions:

Controller Wiring Diagram — A controller wiring diagram shows the electric connections among the parts comprising the controller and indicates the external connections.

External Controller Wiring Diagram — An external controller wiring diagram shows the electric connections between the controller terminals and outside points, such as the connections from the line to the motor and to auxiliary devices.

Controller Construction Diagram — A controller construction diagram indicates the physical arrangement of parts, such as wiring, busses, resistor units, etc.

Elementary Controller Wiring Diagram — An elementary controller wiring diagram uses symbols and a plan of connections to illustrate, in simple form, the scheme of control.

Control Sequence Table — A control sequence table is a tabulation of the connections that are made for each successive position of the controller.

Reading Diagrams

Since wiring diagrams consist of symbols, it is most important that these be thoroughly understood. These basic symbols, which have been generally adopted by electric equipment manufacturers, are shown in the appendix. A study and understanding of these symbols by engineers, wire workers, maintenance workers, operators, and others who work with wiring diagrams will result in a saving of time and energy when determining how control equipment should be installed and how it is intended to operate. A knowledge of the symbols will make it easy to identify the devices and apparatus represented on the print, even though there may be little resemblance between the symbol used and the actual appearance of the device. It is general practice for a diagram to show devices in their de-energized position. The power and main motor circuits are usually indicated by heavy lines, while the control circuits are indicated by light lines.

Fig. 14-1 illustrates the connections of a typical AC reversing magnetic switch, pushbutton station, and induction motor. The magnetic switch contains two three-pole contactors and a temper-

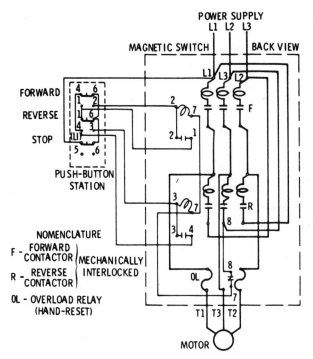

Fig. 14-1. Wiring diagram of a reversing magnetic switch.

ature overload relay. Each contactor has three normally open power contacts, a normally open electric interlock, and an operating coil. The relay includes two heaters connected in the power circuit, and a normally closed control-circuit contact. The complete symbol for the switch shows its electric parts in the same general relationships as they appear in the actual device.

When the FORWARD pushbutton is depressed (Fig. 14-1), contactor F closes and applies power to the motor. A holding circuit for the coil of F is established around the FORWARD pushbutton by auxiliary interlock F. The motor continues to run until shut down either by depressing the STOP button, by the tripping of the overload, or by a power failure. Following an overload condition, which causes relay OL to trip, it is necessary to reset the relay contact by hand before the motor can be restarted. Operation of the motor in the reverse direction is obtained by means of the REVERSE button. The back, or normally closed, contacts of the directional pushbutton units are used for electric interlocking and to prevent the coils of contactors F and R from being energized at the same time. With this arrangement, it is also possible to reverse the motor directly from the FORWARD and REVERSE buttons without first operating the STOP button.

To understand the operation of control equipment, it is necessary to have a complete idea of the circuits involved. As control systems and circuits become more complicated and require more devices, it is increasingly difficult to check the circuits and understand the operation from the conventional panel type of wiring diagram. For this reason, elementary or schematic diagrams are used. Since the main reason for this type of diagram is simplicity, the various electric elements of the devices are separated. They are shown in their respective functional position in the circuit without regard to their actual relationship in the device.

It is also customary to separate the power circuits from the control circuits. Elementary diagrams, which are often referred to as *one-line diagrams*, make it easy to visualize and understand the operation. They are particularly beneficial when troubleshooting or when changes become necessary in the operation. For example, in some cases when large equipments are involved, the elementary circuits are shown on one print, and the detailed panel wiring and interconnections on separate prints. In other cases, the circuits may be shown in both elementary and detailed forms on a single print. An elementary diagram of the magnetic switch in Fig. 14-1 is illustrated in Fig. 14-2.

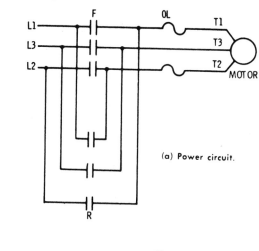

(a) Power circuit.

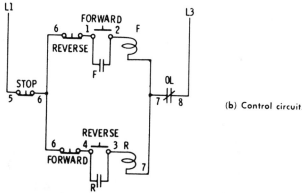

(b) Control circuit.

Fig. 14-2. Elementary diagram of the switch shown in Fig. 14-1.

In controllers of the magnetic-switch type just described, the motor power circuits are handled by magnetically operated contactors under the control of a pushbutton station or other auxiliary device. In contrast to this type, there are different kinds of controllers that are manually operated. In these, the contacts that carry the main motor circuits are closed and opened by hand. Fig.

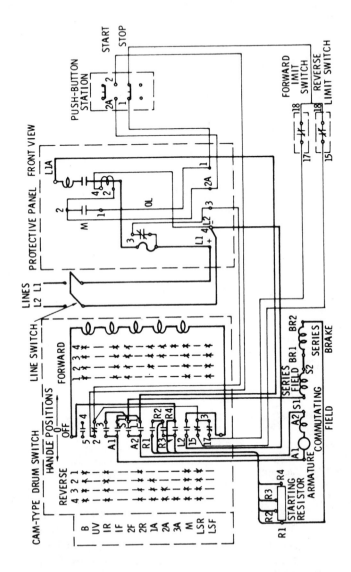

Fig. 14-3. Wiring diagram of a cam-type drum switch with protective panel and over-travel limit switches.

14-3 illustrates the connections of a manual drum switch of the cam type for use with a series-wound, direct-current motor. In this diagram, the letter *X* indicates that the contacts close when the handle moves to that particular position. Auxiliary control contacts *UV*, *LSF*, and *LSR* are closed when the switch handle is at the OFF position.

When the switch handle is moved to its first point forward, contacts *1F*, *2F*, and *M* close and apply power to the motor. When the handle is moved to the second position, contacts *1F*, *2F*, and *M* remain closed and, in addition, contact *1A* closed. In the third position, contact *2A* also closes, and in the fourth position *3A* closes to completely short-circuit the starting resistor. Returning the handle to its OFF position causes the previously mentioned contacts to open and disconnect the motor from the line. When the handle is moved to the REVERSE position, contacts *1R*, *2R*, and *M* close in the first position to energize the motor in the reverse direction.

Two control-type limit switches are shown in Fig. 14-3. The contacts of these switches are normally closed, but will assume an open position when the switches are tripped at their respective ends of travel.

The connections in Fig. 14-3 show not only the circuits of a manual cam-type drum switch, but also how undervoltage, overload, and directional over-travel protection can be obtained by the use of a protective panel, a pushbutton station, and control-type limit switches.

Abbreviations

Symbols for devices such as relays, contactors, switches, etc., are often marked to indicate the function or use of the particular device. The following abbreviations are used for device markings and designated as standards by the *National Electrical Manufacturers Association* (NEMA):

Armature accelerator...................................... A
Armature shunt... AS
Auxiliary switch (breaker), normally open .... "a"

Auxiliary switch (breaker), normally closed . . . . "b"
Balanced voltage . . . . BV
Brake . . . . BR
Compensator — running . . . . MR
Compensator — starting . . . . MS
Control . . . . CR
Door switch . . . . DS
Down . . . . D
Dynamic braking . . . . DB
Field accelerator . . . . FA
Field decelerator . . . . FD
Field discharge . . . . FD
Field dynamic brake . . . . DF
Field failure (loss of field) . . . . FL
Field forcing (decreasing on variable voltage) . . . . DF
Field forcing (increasing on variable voltage) . . . . CF
Field protective (field weakened at standstill) . . . . FP
Field reversing . . . . FR
Field weakening . . . . FW
Final limit — forward . . . . FLF
Final limit — reverse . . . . FLR
Final limit — hoist . . . . FLH
Final limit — lower . . . . FLL
Final limit — up . . . . FLU
Final limit — down . . . . FLD
Forward . . . . F
Full field . . . . FF
Generator field . . . . GF
High speed . . . . HS
Hoist . . . . H
Jam . . . . J
Kick off . . . . KO
Landing . . . . LD
Limit switch . . . . LS
Lowering . . . . LT
Low speed . . . . LS
Low torque . . . . LT
Low voltage LV
Main or line . . . . M

Terminal Markings

The purpose of applying markings to the terminals of electric power apparatus, according to a standard, is to aid in making connections to other parts of the electric power system and to avoid improper connections that may result in unsatisfactory operation or damage. The markings are placed on, or directly adjacent to, terminals to which connection must be made from outside circuits or from auxiliary devices that must be disconnected for shipment. They are not intended to be used for internal machine connections.

Although the system of terminal markings (with letters and subscript numbers) gives information and facilitates the connecting of electrical machinery, there is the possibility of finding the terminals marked without a system or according to some system other than standard (especially on old machinery or machinery of foreign manufacture). There is a further possibility that internal connections have been changed or that errors were made in markings. It is therefore advisable before connecting any apparatus to a power source to make a check test for phase rotation, phase relation, polarity, and equality of potential.

The markings consist of a capital letter of the alphabet followed by a number subscript. The letter indicates the character or function of the winding that is brought to the terminal. A terminal letter followed by the subscript number 0 designates a neutral connection. Thus, T_0 would be applied to the terminal connected to the neutral point of a stator winding.

AC Machines

Subscript number 1,2,3, etc., on AC machine terminals indicate the order of the phase succession for standard direction of rotation. *The standard direction of rotation for AC generators and synchronous motors is clockwise when facing the end of the machine opposite the drive end.* It is customary to connect the coil windings and place collector rings on this end.

Figs. 14-4, 14-5, and 14-6 show the terminal markings and connections for various types of AC motors and/or generators. The terminal letters assigned to the different windings are listed in the tables in the illustrations.

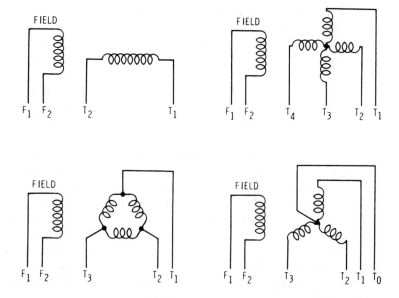

Fig. 14-4. Terminal markings for AC generators and synchronous motors.

For induction motors, subscripts *1,2,3*, etc., indicate the order of the succession of the phases. Thus, if induction-motor terminals T_1, T_2, T_3 are respectively connected to AC generator terminals T_1, T_2, T_3 the generator will cause the induction motor to turn in the same direction. In a synchronous converter, the sequence of the subscripts *1*, *2*, and *3* applied to the collector-ring leads M_1, M_2, M_3 indicates that, when the collector leads are connected to the correspondingly numbered terminals of a three-phase generator, the standard rotation of the generator (clockwise, facing the end opposite the drive) will cause a clockwise rotation when viewing the direct-current or commutator end.

DC Machines

As applied to the terminals of the DC windings of generators, motors, and synchronous converters, the subscript numbers indicate the direction of current in the windings. Thus, with a standard direction of rotation and polarity, the current in all windings will be flowing from *1* to *2;* or from a lower to a higher number, for example *3* to *4*. Figs. 14-7, 14-8, 14-9, and 14-10 show the terminal markings for various types of DC motors and generators.

The standard direction of rotation for DC generators is clockwise when facing the end of the machine opposite the drive (usually the commutator end of the machine). *The standard rotation for DC motors is counterclockwise when facing the end opposite the drive* (usually the commutator end).

Note 1 — Any DC machine can be used either as a generator or as a motor. For the desired direction of rotation, connection changes may be necessary. The conventions for current flow in combination with the standardization of opposite directions of rotation of DC generators and DC motors is such that any DC machine can be called "generator" or "motor" without changing the terminal markings.

Note 2 — A DC motor and a DC generator, by direct coupling, constitute a "motor-generator." With such coupling, the direction of rotation of the motor and generator is necessarily reversed when each is viewed from the "end opposite the drive." The standardized clockwise rotation for DC generators and counterclockwise for DC motors meets such coupling requirements with-

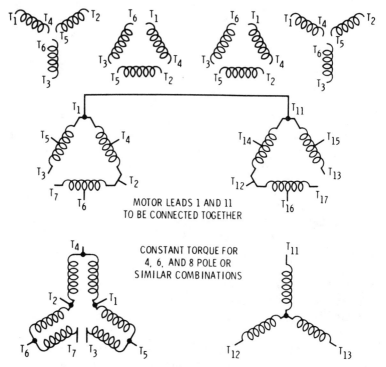

Fig. 14-5. Induction-motor stator connections.

out changing the standard connections or rotation for either machine. Similarly, a DC motor may be mechanically coupled to an AC generator without changing from the standard from either individual machine. However, the coupling of an AC motor to a DC generator cannot be made without rotation other than standard for one of the two machines. Since the rotation of the AC machine is usually easier to change, it is general practice to operate a motor-generator with a clockwise rotation, viewed from the generator end.

Transformers

On single-phase transformers, the subscript indicates the polarity relation between the terminals on the primary and secondary windings. Thus, during that part of the AC cycle when high-tension

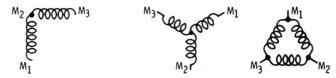

Fig. 14-6. Induction-motor rotor connections.

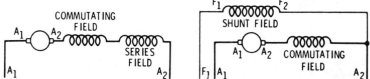

Fig. 14-7. Terminal markings on DC generators without commutating poles.

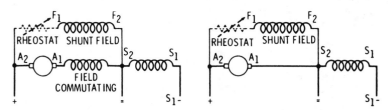

Fig. 14-8. Terminal markings on DC compound generators.

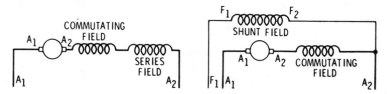

Fig. 14-9. Terminal markings on nonreversing commutating-pole types of DC motors.

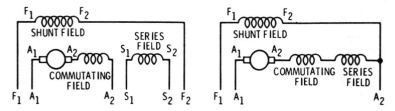

Fig. 14-10. Terminal markings on compound commutating-pole types of DC motors.

terminal H_1 is positive (+) with respect to H_2, the low tension terminal X_1 is positive with respect to X_2. The idea is further carried out on single-phase transformers having tapped windings by applying the subscripts 1, 2, 3, 4, 5, etc., to the taps so that the potential gradient follows the sequence of the subscript numbers. Figs. 14-11, 14-12, and 14-13 show the standard terminal markings of various transformers.

Note 3 — When the primary of a transformer receives energy through the connecting leads, the secondary delivers energy to its connected circuit. However, the direction of current in the winding is reversed with respect to the polarity of the voltage at the terminals. It is important to take note of the difference between the practice of applying subscripts to DC generator and motors, where the subscripts are assigned according to the direction of current, and to single-phase transformers, where subscripts are assigned according to terminal voltage.

In the case of polyphase transformers, the terminal subscripts are applied so that if the phase sequence of voltage on the high-voltage side is in the time order of H_1, H_2, H_3, etc., it is in the time order of X_1, X_2, X_3, etc., on the low-voltage side, and also in the time order of Y_1, Y_2, Y_3, etc., if there is a tertiary winding.

Note 4 — The terminal markings of polyphase transformers afford information on how phase rotation is carried through the transformer, but do not disclose completely the phase relations between correspondingly numbered primary and secondary terminals. Consequently, additional information on internal connections is required before polyphase transformers can be safely paralleled.

AC generators driven counterclockwise (clockwise is standard) will generate without change in connections, but the phases will follow the sequence of 3, 2, 1 instead of sequence 1, 2, 3.

Synchronous motors, synchronous condensers, induction motors, and synchronous converters may be operated with reversed rotations by transposing the connections so that the phase sequence of the polyphase supply is applied to the terminals in reversed order, that is, 3, 2, 1.

DC generators with connections properly made for standard rotation (clockwise) will not function if driven counterclockwise,

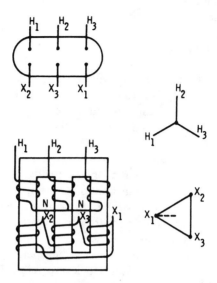

Fig. 14-11. Terminal markings for a star-delta connected three-phase transformer having additive polarity, 30° angular displacement, and standard phase rotation.

as any small current delivered by the armature tends to demagnetize the fields and thus prevent the armature from delivering current. If conditions call for reversed rotation, connections should be made with either the armature leads transposed or the field leads transposed, but not both.

The polarity of a DC generator, with the accompanying direction of current in the several windings, is determined by the N and S polarity of the residual magnetism. Accidents or special manipulations may reverse this magnetic polarity; an unforeseen change may cause a disturbance or damage when it is connected to other generators or devices.

The direction of rotation of DC motors depends not on absolute polarity, but on a relative polarity between the field and armature. As no dependence is placed on residual magnetism for determining the initial direction of rotation, it is not necessary, when connecting DC motors, to regard the polarity of the DC supply connections. With a standard direction of rotation, if current is found to be flowing from terminal *1* to *2* in one of the windings, it will be found in all the other windings as flowing from *1* to *2*. But, because of the disregard of polarity in connecting DC motors, the current is likely to be flowing from terminal *2* to terminal *1* in each of the several windings.

Reversal of rotation of a DC motor is obtained by a transposition of the two armature leads or by a transposition of the field leads (Fig. 14-14). With such reversed rotation (clockwise), when the polarity of the supply makes the direction of the current in the armature from terminal *2* to *1*, it will be flowing in the field windings from terminal *1* to *3*, and vice versa.

With synchronous converters, the practice of AC starting eliminates residual magnetism as the factor determining DC polarity. Proper polarity for connection to other apparatus is secured either by separate excitation of the field or by special manipulation of a switch that permits the converter to reverse polarity, thus correcting a start with wrong polarity.

If a synchronous converter is to be operated with reversed direction of rotation (counterclockwise, viewing the commutator end), besides transposing the AC terminal connections, it is also

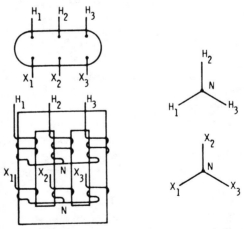

Fig. 14-12. Terminal markings for a star-star connected three-phase transformer having subtractive polarity, 0° phase displacement, and standard phase rotation.

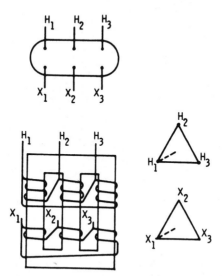

Fig. 14-13. Terminal markings for a delta-delta connected three-phase transformer having subtractive polarity, 0° phase displacement, and standard phase rotation.

necessary to make either a transposition of armature leads or a transposition of field leads.

Wiring Symbols

The symbols used on wiring diagrams to represent various electrical devices facilitate not only the reading and understanding, but also the maintenance and installation of electrical equipment of all kinds. Various fundamental symbols commonly found on wiring diagrams and employed by representative electrical manufacturers are given in the Appendix. The fundamental symbols represent component parts such as coils, connections, contacts, instruments, transformers, etc., and can be combined in various ways to represent complete electrical devices.

Table 14-1. Circuit Wire Sizes for Individual Single-Phase Motors

Horsepower of Motor	Volts	Approximate Starting Current Amperes	Approximate Full-Load Current Amperes	Feet	Length of Run in Feet (from Main Switch to Motor)							
					25	50	75	100	150	200	300	400
¹/₄	120	20	5	Wire Size	14	14	14	12	10	10	8	6
¹/₃	120	20	5.5	Wire Size	14	14	14	12	10	8	6	6
¹/₂	120	22	7	Wire Size	14	14	12	12	10	8	6	6
³/₄	120	28	9.5	Wire Size	14	12	12	10	8	6	4	4
¹/₄	240	10	2.5	Wire Size	14	14	14	14	14	14	12	12
¹/₃	240	10	3	Wire Size	14	14	14	14	14	14	12	10
¹/₂	240	11	3.5	Wire Size	14	14	14	14	14	12	12	10
³/₄	240	14	4.7	Wire Size	14	14	14	14	14	12	10	10
1	240	16	5.5	Wire Size	14	14	14	14	14	12	10	10
1¹/₂	240	22	7.6	Wire Size	14	14	14	14	12	10	8	8
2	240	30	10	Wire Size	14	14	14	12	10	10	8	6
3	240	42	14	Wire Size	14	12	12	12	10	8	6	6
5	240	69	23	Wire Size	10	10	10	8	8	6	4	4
7¹/₂	240	100	34	Wire Size	8	8	8	8	6	4	2	2
10	240	130	43	Wire Size	6	6	6	6	4	4	2	1

Circuit Wire Sizes

A motor must have the proper amount of current and voltage to operate in any prolonged condition. One way to make sure of the proper amount of power reaching the motor is to select the correct wire size for connecting the motor to the power source (Table 14-1).

Summary

Wiring diagrams are used to show the connections of electrical circuits and components in a form that is much easier to interpret than the actual wiring. Components, such as resistors, capacitors, switches, etc., are represented by standardized symbols to further simplify the diagrams. The symbols used for electrical diagrams often differ from those used for electronic diagrams. This should be kept in mind in order to avoid confusion.

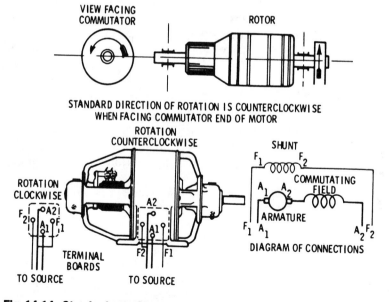

Fig. 14-14. Standard rotation, terminal markings, and terminal connections of a DC shunt motor. To reverse the direction, the field leads are interchanged as shown.

The location of the various symbols on the wiring diagram does not necessarily indicate their actual location in the electrical equipment. Instead, the correct electrical connections are the primary concern.

Symbols are often marked with an abbreviation to indicate the function of the component or device they represent. Abbreviations that have been designated as standard are used. Terminal markings for the terminals of motors, generators, and transformers have also been standardized. These markings consist of a capital and a number subscript, the combination indicating the character or function of the winding connected to the terminal. The use of terminal markings reduces the possibility of error when wiring equipment, as well as simplifying and speeding up the procedure. Terminal markings furnish other information, such as the order of phase succession, the direction of rotation, and the direction of current flow in DC machines.

The standard direction of rotation for AC generators and synchronous motors is clockwise when facing the end of the machine opposite the drive end. The standard direction of rotation for DC generators is clockwise when facing the end of the machine opposite the drive end. The rotation of DC motors is exactly opposite — counterclockwise when viewed from the end that is not driving.

Review Questions

1. Why are diagrams used to indicate electrical circuits?
2. In a wiring diagram that includes both power and control circuits, which circuits are generally drawn with heavy lines?
3. What is the purpose of the terminal marking on an AC three-phase motor?
4. How may a three-phase induction motor be reversed?
5. How may a DC motor be reversed?

CHAPTER 15

Armature Windings

With a knowledge of the methods used in rewinding DC armatures, the repairman can successfully rewind wound-rotor AC rotors. There is no difference in the winding of an DC armature and an AC rotor except that the latter is equipped with slip-rings instead of a commutator. Information of a theoretical nature is given here for an understanding of the design features of both DC armatures and AC wound rotors. Connection diagrams are furnished to facilitate the tracing out of progressive or retrogressive lap-winding connections, progressive or retrogressive wave-winding connections, winding pitch, commutator pitch, and lead swing.

Types of Windings

Armature windings are classified into two main groups: *lap* windings and *wave* windings. They differ in the manner in which the leads are connected to the commutator bars. Either group can be arranged progressively or retrogressively and connected in *simplex, duplex,* or *triplex.*

In a *simplex* lap winding, the beginning and end leads of a coil are connected to adjacent commutator bars (commutator pitch), as shown in Fig. 15-1.

A *simplex progressive lap* winding is one in which current flowing in a coil terminates in the adjacent commutator bar beyond the starting bar. Fig. 15-2 shows two coils per slot; Fig. 15-3 shows three coils per slot.

A *simplex retrogressive* lap winding terminates in the bar before and adjacent to the starting bar (Fig. 15-4). If progressive connections are changed to retrogressive connections, the armature will rotate in the opposite direction.

In a *duplex* lap winding, the beginning and end leads of a coil are connected two bars apart. The end lead of a first coil and the beginning lead of a third coil connect to the same commutator bar, the end lead of a third coil to the beginning of a fifth coil, and so on.

In a *triplex* lap winding, the beginning and end leads of a coil are

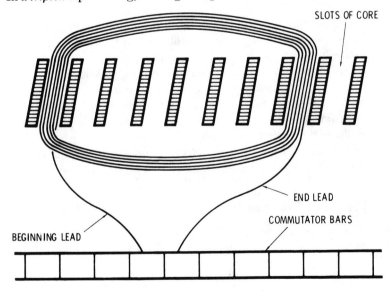

Fig. 15-1. Simplex lap winding.

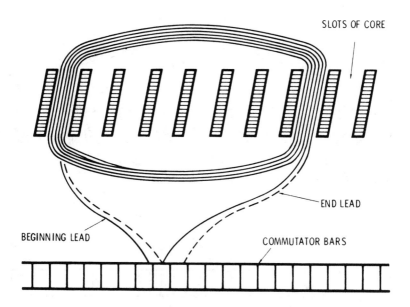

SLOTS OF CORE

END LEAD

BEGINNING LEAD

COMMUTATOR BARS

Fig. 15-2. Simplex progressive lap winding, two coils per slot.

connected three bars apart. The end lead of a first coil and the beginning lead of a fourth coil connect to the same commutator bar, the end lead of a fourth coil to the beginning of a seventh coil, and so on.

Equalizer connections, also known as *cross connections,* are used in large DC armatures to minimize circulating currents, which are due to uneven air gaps between the field poles and the armatures. The currents may be eliminated by connecting commutator bars of equal potential, depending upon the number of poles in the motor and the number of commutator bars. Equalizer connections are used mostly on repulsion type motors and only with lap windings.

Wave windings are also arranged progressively or retrogressively and connected in simplex, duplex, and triplex.

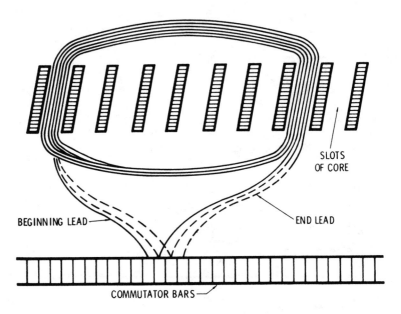

SLOTS OF CORE

BEGINNING LEAD

END LEAD

COMMUTATOR BARS

Fig. 15-3. Simplex progressive lap winding, three coils per slot.

A *simplex progressive wave* winding is one in which the current flowing through two coils in series terminates one bar beyond the starting point. A four-pole simplex progressive wave winding, one coil per slot, is shown in Fig. 15-5. A progressive wave winding, two coils per slot, is shown in Fig. 15-6. The numbers shown in these figures denote commutator bars.

A *simplex retrogressive wave* winding is one in which the current flowing through two coils in series terminates one bar before the starting point (Fig. 15-7). Numbers in this figure denote commutator bars. Wave-winding connections are made quite far apart. On a four-pole motor, the winding-end leads are connected on opposite sides of the commutator. On a six-pole motor they are connected one-third of the commutator bars apart; on an eight-pole motor one-fourth of the commutator bars apart.

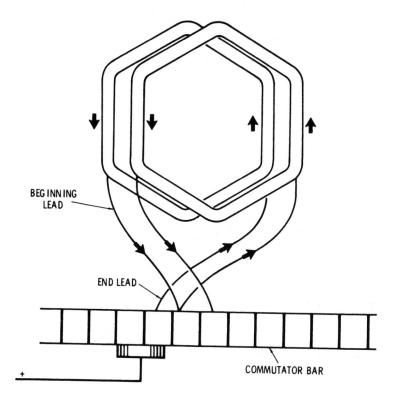

Fig. 15-4. Simple retrogressive lap winding.

Commutator Pitch

The number of bars between coil leads is termed the *commutator pitch*. For example, a commutator pitch is 24 bars if the leads are placed in bars 1 and 25. The commutator pitch is 25 bars when the leads are placed in bars 1 and 26. The formula for determining pitch commutator is

$$cp \text{ (commutator pitch)} = \frac{\text{No. of commutator bars} \pm 1}{\text{No. of pairs of poles}}$$

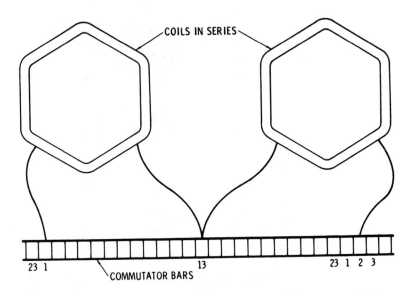

Fig. 15-5. A four-pole, simplex, progressive wave winding, one coil per slot.

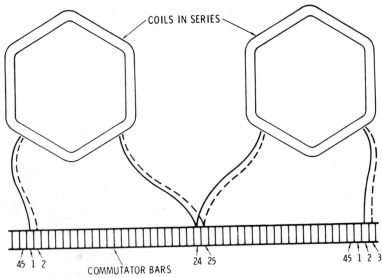

Fig. 15-6. A progressive wave winding, two coils per slot.

For example:

for a four-pole, 49-bar commutator, the cp $= \dfrac{49 \pm 1}{2} =$

24 or 25 bars spanned by the wave-winding coil leads.

Coil Pitch

The number of slots between coil sides is termed *coil pitch*. If, for example, the coil sides are placed in slots 1 and 13, the coil pitch is 12 slots. It is necessary to replace coils according to the original coil pitch so that the unit will commutate properly when assembled.

Lead Swing

Lead swing is the direction a lead takes from a coil side to its connection on a commutator bar. It is measured by the number of commutator bars from the point of alignment of a commutator bar with the coil side to the point of connection (Fig. 15-8). Lead swing can seldom be established visually and must be determined according to the alignment of the slots on the core of the armature with the centerline of the shaft. Knowledge of the correct lead swing is particularly important in armature winding in order to place the coil leads in the proper commutator bars for the correct winding. To establish correct lead swing from straight and skewed coil slots, use the procedures that follow:

Straight Slots

1. Stretch a length of cord through the center of a slot, aligning it with the centerline of the shafts (Fig. 15-9).
2. Notice whether the cord aligns with the centerline of a bar or on the mica insulation between bars.
3. Count the number of bars from the right or left of the aligned bar and coil slot to the bar in which the aligned coil side is

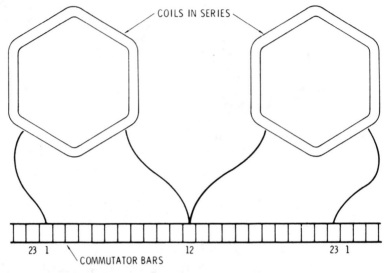

Fig. 15-7. A four-pole, simplex, retrogressive wave winding.

connected. If the alignment falls on the mica, count the bars starting with one on either side, left or right, of the mica. The lead swing count thus obtained for one coil on an armature is applicable to all coils of the armature.

Skewed Slots

1. Locate a centerline around the outer circumference of the armature core.
2. Stretch a length of cord aligned with the centerline of the shafts and the center point of a slot in line with the circumferential centerline of the armature core (Fig. 15-10).
3. Follow step 2 under *Straight Slots* instructions above.
4. Follow step 3 under *Straight Slots* instructions above.

Data Recording

One of the most necessary tasks of any rewinding operation is the accurate recording of data concerning the armature being rewound. Preliminary data is the pertinent information, furnished

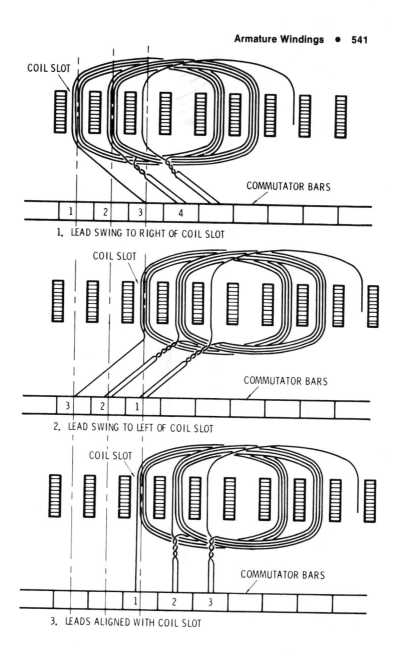

Fig. 15-8. Three conditions of lead swing.

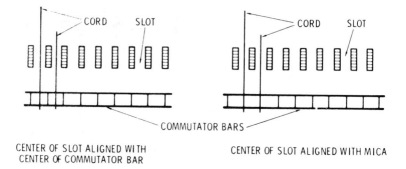

CENTER OF SLOT ALIGNED WITH
CENTER OF COMMUTATOR BAR

CENTER OF SLOT ALIGNED WITH MICA

Fig. 15-9. Determining the alignment of slot and commutator bar.

from various sources, which the repairman should have before he begins stripping the armature. Such data may be obtained from a technical manual on the machine, manufacturer's drawings, or previously collected data in the repair shop records. Without preliminary data, the repairman is responsible for recording the data he finds. A sample form and a checklist are given below to aid in recording preliminary data.

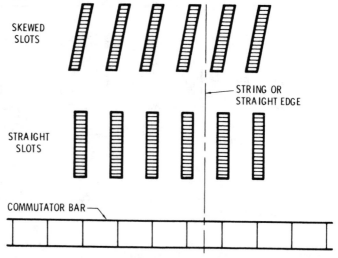

Fig. 15-10. Alignment of shaft and slots.

1. Check the recorded nameplate data against the actual nameplate data, making certain that the ratings and serial numbers are the same.
2. Record the type of winding, lap or wave.
3. Record the coil pitch (the number of slots spanned by a winding) and mark the slots as outlined in the paragraph labeled *Coil Pitch.*
4. Record the commutator pitch (the number of bars spanned by the leads of a coil) and mark the bars as outlined in the paragraph labeled *Commutator Pitch.*
5. Record the lead swing.
6. Record the end room between the ends of the windings and the back of the commutator or slip rings.
7. Record the end room between the ends of the windings and the fan or mechanism plate.
8. If possible, remove a single coil from the armature, so that the dimensions for a prewound or preformed coil may be obtained. If this is not possible, dimensions for a new coil must be ascertained by using a single conductor formed to the space which an old coil occupied.
9. Record the wire size and type of wire insulation. Remove insulation from the wire before checking the size. Enameled or silicone insulation is removed by drawing the end of the wire between 00 sandpaper. Pull the wire through only enough times to remove the insulation, otherwise the wire size will be diminished to the point of being too small for the job. Use an AWG (*American Wire Gauge*) wire gage or micrometer caliper gage to measure the wire diameter or size.

Data Taking While Stripping

The importance of observing and recording data during the stripping procedure is not to be minimized. If technical manuals, manufacturer's drawings, or shop data are not available and an armature has been completely stripped, it is almost impossible to gain information that is lost due to an oversight during the stripping procedure. The following methods and procedures can be used *before removing the commutator* if the winding is not too deteriorated and brittle and the conductors can be readily separated from the impregnated varnish:

1. Remove all bands.
2. Record the type, diameter, and number of turns of cord bands. Also, record the diameter and number of turns of the steel wire and the dimensions and thickness of clips holding the band in place. In the case of stripsteel bands, record the thickness and width.
3. Locate a top coil; that is, a coil that is exposed across the rear end of the armature from slot to slot.
4. Remove the wedges from the slots that house this coil.
5. Cut or disconnect the coil lead or leads from the commutator bar to which the coil is connected for one coil side. It may be necessary to remove more than one lead from the commutator bar to locate the top coil lead.
6. Check wire size and type of wire insulation.
7. Check the lead from the top coil by lifting the disconnected leads one at a time, until one is found that can be lifted comparatively easily along the entire length of the slot.
8. Count and record the number of turns removed by continually unwinding the coil until it is entirely removed.

ARMATURE REWINDING DATA SHEET

Job number _____ Organization _____ Horsepower _____

Volts _____ Amperes _____ RPM (R/MIN) _____

Type _____ Serial Number _____

Turns per coil _____ Coil Pitch _____ Number of slots _____

Make _____ Kind of winding _____ Amount of end room _____

Number of commutator bars _____ Wires per bar _____

Size of wire used in coils _____ Type of wire insulation _____

Location of bottom lead by considering coil in bottom of slot as slot 1 and the bar to which it is connected _____

_____

Distance commutator is pressed on shaft _____

Markings on shaft or armature core _____ Bands _____

Slot insulation _____

Remarks _____

_____

9. Count and record the number of bars on the commutator that are spanned by the top and bottom leads of the removed coil. This is the *commutator pitch.*
10. Check the thickness of the slot insulation with a micrometer caliper gage.
11. Record the size of the wire and the type of insulation.

Slot and Bar Marking

Slot and bar marking is an important operation, since incorrect lead throw causes sparking and poor operation. Markings on the core indicate the slots in which the top half and bottom half of a coil are inserted. They also prevent winding a right-hand coil when a left-handed winding is indicated, or vice versa. When scribed on the armature, these marks make a permanent record of the coil and commutator pitch for both lap or wave winding, progressive or retrogressive. Marking is done as follows (Fig. 15-11):

1. Mark the tooth on each side of the slot that contains the top half of a coil chosen for recording preliminary data, with two crosses or two center-punch marks ($\frac{x}{x}$ or :). Make the two crosses with a three-cornered file.
2. For a lap winding, center-punch one mark on the end of the commutator bar to which the lead from the slot previously marked is connected. For a wave winding, center-punch two marks (:).
3. Mark the teeth on each side of the slot that contains the bottom half of the same coil chosen for recording preliminary data, with one cross or one center-punch mark, (x or ⁎). Make the cross with a three-cornered file.
4. Center-punch one mark on the commutator bar, to which the lead from the slot previously marked is connected, with one cross or one center-punch mark. Do this for either a lap winding or a wave winding.

Removal of the Commutator

If the winding is deteriorated and brittle, it is advantageous to remove the commutator to acquire data while stripping. On large armatures the commutator is built upon a strong flange-like support or shell, bolted to the armature spider (Fig. 15-12), or

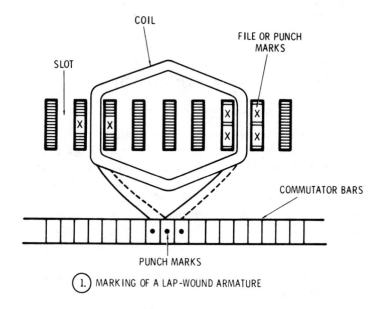

1. MARKING OF A LAP-WOUND ARMATURE

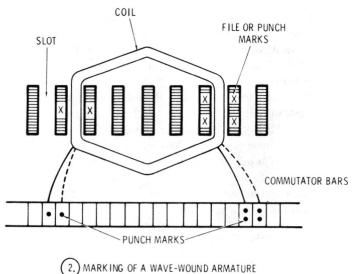

2. MARKING OF A WAVE-WOUND ARMATURE

Fig. 15-11. Marking of pitch and lead data on armature slot and commutator bars.

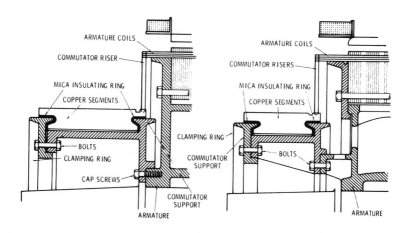

Fig. 15-12. Method of bolting a commutator spider to an armature spider.

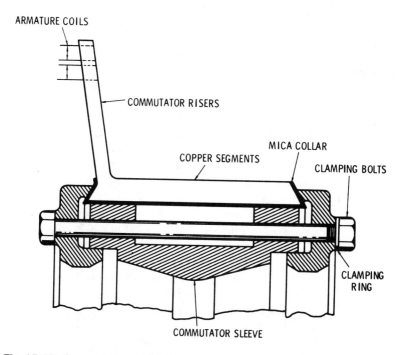

Fig. 15-13. Commutator spider for mounting on shaft.

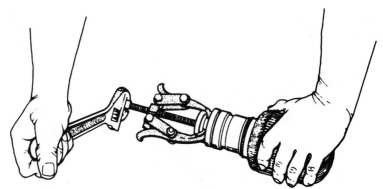

Fig. 15-14. Removing a ball bearing from a shaft using a hook-type puller.

mounted on a separate spider secured to the shaft (Fig. 15-13). On small machines, the commutator hub is usually of cast iron (Fig. 15-14) keyed and pressed onto the shaft. The following procedure will aid in removing a commutator and identifying the leads of a coil:

1. Remove all bands.
2. Locate a top coil; that is, a coil that is exposed across the rear end of the armature from slot to slot.
3. Remove the wedges from the slots that house this coil.
4. Cut or disconnect the coil lead or leads from the commutator bar to which it is connected for one coil side. It may be necessary to remove more than one lead from the commutator bar to locate the top coil lead.
5. Cut or disconnect the coil lead or leads from the commutator bar to which the other side of the located coil is connected.
6. Count and record the number of bars on the commutator spanned by the coil leads. This is the *commutator pitch.*
7. Check and cross-check the disconnected coils with an ohmmeter for a resistance check between the coil leads. The coil that registers the lowest resistance is the single coil.
8. Identify the single coil leads with tape or colored sleeving.
9. Count and record the number of slots spanned by the single coil. This is the *coil pitch.*
10. Record the wire size and type of wire insulation.
11. Remove all remaining leads from the commutator.

12. Remove all space fillers and tapes from behind the commutator risers until the shaft is exposed.
13. Record the distance of the shaft from the core to the commutator.
14. Remove the commutator with a hook type puller and arbor plates (Fig. 15-14). The procedures below will aid in placing and using the hook-type puller for this application.
    a. Place a pair of proper-fitting arbor plates behind the commutator.
    b. Adjust the hooks for the hook-type puller to the arbor plates and adjust the jackscrew to the armature shaft.
    c. Slowly tighten the jackscrew until the commutator starts to move from the shaft.
    d. Continue tightening the jackscrew until the commutator is finally removed from the shaft.
    e. Place the commutator in a safe place for a later check.

Wedge Removal

The wedges can be removed by using a piece of power-hacksaw blade with the sharp edges on the broken ends and the set of the teeth ground off. They can also be removed with a commercially manufactured wedge-removing tool.

1. Place the hacksaw blade or armature-wedge remover on top of the wedge in the slot so that the teeth will bite into the wedge.
2. Tap the hacksaw blade with a hammer in the direction the teeth point (Fig. 15-15), so that the teeth are embedded in the wedge.
3. Continue tapping the hacksaw blade with the hammer until the wedge is driven out.

After the Commutator is Removed

The remaining parts of the armature may now be subjected to either of two methods for loosening the impregnated varnish. One method is the application of heat with an acetylene torch flame, the other is the immersion of the entire unit in a caustic soda solution.

1. *Application of heat.* Use a large stem on the handle assembly

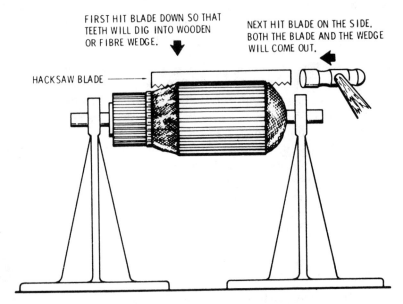

Fig. 15-15. Method of removing slot wedges from an armature.

of an acetylene brazing and soldering set for applying heat to the armature core. Follow the procedure set forth below:

    a. Remove all slot wedges.

    b. Place the armature (without the commutator) in a metal stand that is clear of all flammable materials.

    c. Light the torch and apply the flame to all parts of the armature core by slowly moving the torch flame. Concentrate application on the core slots and the windings. Do not permit the flame to stay too long on the shaft. If the shaft gets too hot, its temper may be affected.

    d. Continue applying the torch to the windings until the insulation has become charred and shrunken sufficiently for the coil conductors to become loosened.

    e. Allow the armature to cool.

  2. *Application of caustic soda.* Use of caustic-soda solution for loosening the windings:

    a. Remove all the wedges.

b. Place the armature (without the commutator) in a large pan or vat that is deep enough to hold sufficient caustic-soda solution to cover the armature.

c. Place the pan or vat on an electric hot plate or gas burner.

d. Slowly pour in the caustic-soda solution.

WARNING: Wear rubber gloves and handle caustic-soda solution carefully. Prevent it from splashing. Caustic-soda solution can cause serious burns or injuries.

e. Heat the solution until it attains boiling temperature.

f. Remove from the heat and allow the armature to soak until the varnish softens.

g. Remove the armature from the pan or vat. Soak and flush the armature with clean water until the caustic-soda solution has been weakened to the point where it is non-injurious.

h. Allow the armature to drain.

Testing the Commutator for Grounds and Short Circuits Between Bars

While waiting for the armature to cool or drain, it will be advantageous to check the commutator, which has been previously removed. Follow the procedure set forth below:

1. Clean out the commutator connection slots.

2. Blow out metal chips and dirt with compressed air.

3. Check for solder that may have lodged between bars and remove any that is found.

4. Use a test light or ohmmeter to check for grounds (Fig. 15-16).

5. Connect one lead of the tester to the commutator sleeve and touch the other lead to a commutator bar. If the lamp lights or the ohmmeter registers, the bar is grounded and the cause must be removed. Continue this check for every bar (Fig. 15-16).

6. Touch the one lead of a test light or ohmmeter to a bar and the other lead to an adjacent bar. If the lamp lights or the ohmmeter registers, one bar is short-circuited to the other. Remove the cause. Continue this check for every bar (Fig. 15-17).

7. After checking and servicing, store the commutator in a safe place until it is needed for reassembly.

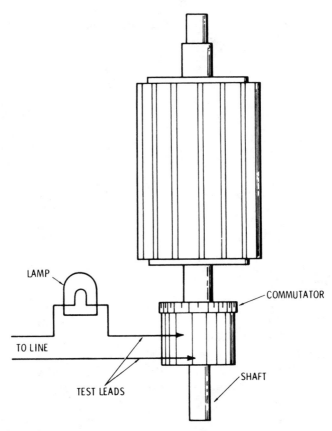

LAMP

COMMUTATOR

TO LINE

TEST LEADS

SHAFT

Fig. 15-16. Testing a commutator for grounds.

Stripping Procedure

After all bands and wedges have been removed and the top coil has been located, checked, recorded, and identified, proceed as follows:

1. Cut or disconnect the lead or leads of the next exposed coil from the commutator bar to which it is connected for one coil side.
2. Remove the wedges from the slots that house this coil.

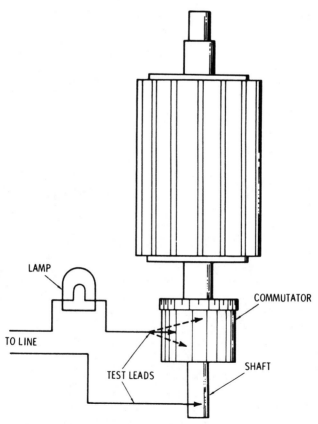

Fig. 15-17. Testing a commutator for shorts.

3. Check for the lead from the top coil by lifting the disconnect leads, one at a time, until one is found that can be lifted comparatively easily along the entire length of the slot.
4. Count and record the number of turns removed by continually unwinding the coil until it is entirely removed.
5. Check the turns count against the turns count of the previously removed coil. The count should not vary by more than two turns.
6. Check and record the type of winding, wave or lap, progressive or retrogressive.
7. Remove all slot wedges.

8. Cut or disconnect all remaining commutator connections.
9. Record the wire size and type of wire insulation.
10. Remove all space fillers and tapes from behind the commutator risers until the shaft is exposed.
11. Cut the coils at both ends.
12. Remove the coil conductors from the slots.
13. Remove all the old insulations.
14. Clean the slots with a file.
15. File all burrs smooth.
16. Clean the back of the commutator.
17. Clean excess solder and broken particles of copper out of the bar slots using a hacksaw blade that will make a cut in the bar no larger than the diameter of the new wire.
18. Test check the commutator.

Stripping Procedure with the Commutator Removed

If, after treating the armature windings as outlined in the stripping procedure with the commutator intact, it is found that the windings are still hard to remove, follow the following procedure:

1. With a hacksaw, cut the ends of the windings flush with the core, or place the armature in a lathe and use a V-shaped pointed tool to remove the end windings at each end of the core.
2. Remove the armature from the lathe or stand.
3. Pile up two rows of wooden blocks to a height that will accommodate the full length of the armature shaft. Have the core rest vertically upon the wooden blocks and the end of the shaft just clear of the workbench.
4. Drill a hole, larger than the width of the slot, into one set of the wooden blocks.
5. Align a slot with the hole in the wooden block. The purpose of aligning a slot with the drilled hole in the wooden blocks is to prevent the outside laminations of the core from becoming distorted or warped when the coil conductors are driven out.
6. Use a hammer and drift punch to knock out the coil conductors from the slots.
7. Follow steps 5 and 6 above for all the slots.

8. Check the drilled hole occasionally and clear it of the coil-conductor accumulations to provide clearance for successive coil conductors that are driven to it.
9. Remove all the old insulations.
10. Clean the slots with a file.
11. File all burrs smooth.

Summary

Armature windings are classified into two main groups: *lap* windings and *wave* windings. Either group can be arranged progressively or retrogressively and connected in simplex, duplex, or triplex.

Commutator pitch is the term used to refer to the number of bars between the coil leads. *Coil pitch* is the number of slots between coil sides. *Lead swing* is the direction a lead takes from a coil side to its connection on a commutator bar.

One of the most important tasks of any rewinding operation is the accurate recording of data concerning the armature being rewound. Slot and bar marking is an important operation, since incorrect lead throw causes sparking and poor operation. If the winding is deteriorated and brittle, it is advantageous to remove the commutator to acquire data while stripping.

The wedges can be removed by using a piece of hacksaw blade with the sharp edges on the broken ends and the set of the teeth ground off. After the commutator is removed, the remaining parts of the armature may be subjected to either of two methods for loosening the impregnated varnish. While waiting for the armature to cool or drain, it is advantageous to check the commutator, which has been previously removed. The commutator can be removed before stripping the armature or it can be stripped before it is removed.

Review Questions

1. What are the two main types of armature windings?
2. What is a simplex winding?
3. What is a simplex progressive lap winding?
4. What is commutator pitch?

5. What is coil pitch?
6. What is lead swing?
7. Why is data recording important before a rewinding operation?
8. What data should you take while stripping the armature?
9. Why should bars and slots be marked?
10. Why should a commutator be removed?
11. How are wedges removed?
12. How do you test a commutator for grounds and short circuits between bars?
13. How do you strip an armature with the commutator intact?
14. How do you strip an armature with the commutator removed?

CHAPTER 16

DC Armature Rewinding Procedure

The process of insulating an armature core, preparatory to rewinding, is one of the most important steps in the rewinding procedure. The selection of proper insulating materials, the method of application of this material, and the proper alignment of core laminations ensure the armature against premature insulation troubles. The procedures outlined below aid in covering all the important points of preparation, selection, and application of the insulations.

Cleaning Slots and Truing Up Laminations

After the old winding has been removed, the core slots, core ends, and the shaft must be thoroughly cleaned of all insulation and varnish.

1. Inspect the core for rough or irregular spots, sharp edges and burrs, and bent, broken, or burred lamination teeth.
2. Remove clinging particles of old insulation and varnish by scraping. A solution of 25 percent alcohol and 75 percent benzol will aid in removing stubborn particles. Do not apply any solution to the commutator. Use a cloth moistened, but not wet, with solution for cleaning the commutator.
3. Use a hammer and drift for straightening bent slot teeth and laminations.
4. Use an arbor press to force laminations together that may have flared or separated.
5. Use a file to smooth sharp edges and burrs.
6. Use a file to clear and true-up wedge grooves and band grooves. Use only enough pressure on the file to clear and true the grooves to their original dimensions. Removal of more metal than is necessary will cause wedges and bands to come loose after installation.
7. Use compressed air to blow out loose particles of metal, insulation, and varnish.

Insulating the Core Ends and Shaft

The insulation for the core ends are fiber washers about $1/16$ inch (1.6 mm) thick, punched to conform to the core-lamination slots, and having a tight-fitting hole for the shaft. This serves as protection for the windings that cross the ends of the core. When reinsulating a core and shaft, proceed as directed below:

1. Cut out two discs of $1/16$-inch (1.6 mm) fiber to the diameter of the core to be reinsulated.
2. Punch out a tight-fitting hole for the shaft on both sides.
3. Punch out slots to conform to the width and shape of the core slots on both discs.
3. Punch out slots to conform to the width and shape of the core slots on both discs.
4. Shellac on both ends of the core.
5. Shellac one side of each fiber disc.
6. Assemble and align both end washers to the core.
7. Wrap, half-lap, and shellac about four layers of $3/4$-inch-wide (20 mm) treated cloth tape (black or yellow cambric) on the

armature shaft. Start close to the rear-end washer at a point where the end windings will extend. Use masking tape to hold the wrapping down and to facilitate rewinding (Fig. 16-1).

8. Repeat step 7. Start up close to the front-end washer and wrap up close to the back of the commutator (Fig. 16-1).

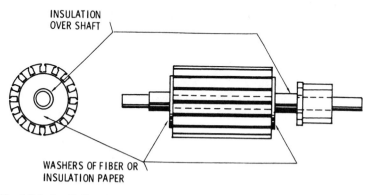

Fig. 16-1. Insulating the shaft between the commutator and the armature.

Replacing the Commutator

The commutator of a large armature merely requires bolting the commutator properly into place. However, the commutator of a small armature necessitates the use of an arbor press and proper-fitting arbor sleeve for the shaft and commutator. In the absence of an arbor press, a correctly sized arbor sleeve and hammer may be used. The procedure outlined below gives steps to follow when replacing a commutator:

1. Prepare a piece of pipe of sufficient length and diameter for use as an arbor sleeve. Do not use a pipe of an outside diameter that will bear on the commutator bars. The pipe should bear only on the cast iron hub of the commutator; otherwise irreparable damage may result to the commutator.

2. Determine the length for the pipe by measuring the distance along the front end of the shaft, from the front end of the armature core to the front end of the shaft. Then subtract from this measurement the commutator length and the re-

corded distance between the back of the commutator and the front end of the armature core. The result will be the proper length for the pipe.

3. Align the commutator with the shaft, key, and keyway of the commutator hub.
4. Slip the arbor sleeve (pipe) over the shaft, and align it with the hub of the commutator.
5. Tap the end of the pipe with a hammer to get the commutator started on the shaft shoulder.
6. Set the back end of the armature shaft on a block of soft iron, bronze, or brass of $1/4$- to $1/2$-inch (6 mm to 12 mm) thickness in an arbor press.
7. Center the ram of the arbor press on the pipe that is centered on the hub of the commutator or located on the front end of the armature shaft.
8. Apply gradual pressure until the commutator is driven into proper position. Be certain that the commutator is driven evenly on the shaft by constantly watching as pressure is applied.

After the commutator is replaced on the shaft, fill the hollow area between the core and the back of the commutator. Whether it is preferable to fill this space before or after winding, or fill part of it before and part of it after winding, will depend on the armature size. The procedure given below will aid in either case.

1. When the bottoms of the core slots are lower than the bottoms of the commutator riser slots:
   a. Use wide tape of cambric or cotton, or if wide tape is not available, cut wide strips of cambric.
   b. Wrap and fill the space up to the bottom level of the core slots.
   c. Secure the wrapping with a layer of masking tape and insulate and rewind the armature.
   d. After rewinding the armature, fill the space between the end windings and the bottom level of the commutator-riser slots with cambric or cotton tape. Secure the wrapping with a layer of masking tape.
2. When the bottom of the core slots are on a level with, or fairly close to, the bottom level of the commutator-riser slots:

a. Use cambric or cotton tape.
b. Wrap and fill the space up to the bottom level of the commutator-riser slots.
c. Secure the wrapping with a layer of masking tape, and insulate and rewind the armature.

Insulating the Slots

A good commercial grade of slot-insulation paper should be used. This paper is supplied in sheets, strips, or cuffed strips in various thicknesses. Plastic slot liners are common today. For 120-volt armatures with slots not over ³/₈ inch (10 mm) in width and wire for the coils not larger than No. 28 AWG gauge, 0.008 to 0.010 inch is the thickness of paper to use. For 240-volt armatures, use insulation no less than 0.010 thick with coil wire no greater than No. 28 AWG gauge. The insulation thickness will depend upon the available room in the slot. For slots larger than ³/₈ inch (10 mm) in width and coil wire that is as large as or larger than No. 18 AWG gauge, use slot insulation no less than 0.916 inch (about 23.25 mm) thick. If room is available in the slot, use thicker insulation and, if it is possible, use insulation of 0.023 inch thickness. Do not use insulation paper of two thicknesses to make up a required thickness. Thin insulations are weaker and tear more easily, especially in large slots.

The latest technology concerning paper and slot liners indicates a great deal of research has taken place since the early motors were wound with cloth that was later impregnated with varnish.

As far as materials are concerned, some service centers where motors are rewound use 10- to 14-mil oriented polyester film for slot and phase insulation and chlorosulfonated polyethylene for the leads on the Class B AC machines. They use aramid paper and silicone leads for Classes F and H. TI-180+ magnet wire enamel is used in all cases except when special requirements dictate otherwise. Many shops are changing to aramid (TI-180+) slot liners and phase insulation as standard.

1. Continuous-strip method of insulating slots follows (Fig. 16-2).
    a. Cut a strip of insulation ¹/₈-inch thick (3 mm) wider than the length of the slot, including the thickness of the fiber

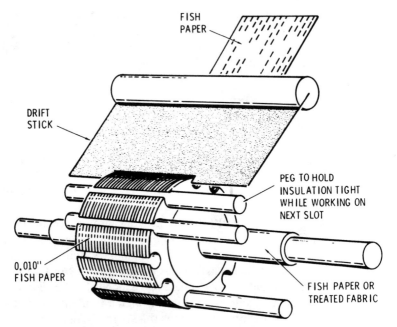

FISH PAPER

DRIFT STICK

PEG TO HOLD INSULATION TIGHT WHILE WORKING ON NEXT SLOT

0.010" FISH PAPER

FISH PAPER OR TREATED FABRIC

Fig. 16-2. Continuous-strip method of insulating slots.

end washers if they are encompassing the slots. Cut it long enough to pass entirely around the core and loop into every slot.

b. Push one end of the insulation into the first slot, with a stiff fiber plate, leaving about ½ inch (12 mm) of the end of insulation protruding from the slot. Continue to shape and fit the insulation into the slot with the fiber drift plate.

c. Insert a dowel into the first slot if the slot is circular, or insert a piece of fiber shaped to the slot if the slot is angular.

d. Insert the strip insulation into the next slot. Shape and fit the insulation to the slot with the fiber drift plate.

e. Insert a second dowel or preshaped fiber into this slot.

f. Insert the strip insulation into the next slot. Shape and fit the insulation to the slot with the fiber drift plate.

g. Remove the dowel or shaped fiber from the first slot and insert it into the third slot.

    h. Continue this operation for all the slots.

    i. Bend the insulation at the end slots (Fig. 16-3).

    j. Rewind according to winding procedure.

2. Cut-insulation method for small slots.

    a. Cut a strip of insulation to fit inside the periphery of the core slot, plus an allowance of ½ inch (12 mm).

    b. Center and crease the insulation strip length on the insulation former according to the inside dimensions of the bottom of the core slot.

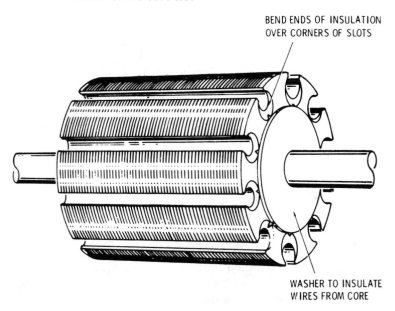

BEND ENDS OF INSULATION
OVER CORNERS OF SLOTS

WASHER TO INSULATE
WIRES FROM CORE

Fig. 16-3. Bending strip insulation at slot ends to protect conductor insulation.

    c. Cut the insulation strip length into sections that are dimensioned to the exact length of the slot.

    d. Insert the insulation sections into the slots with ⅛ inch (3 mm) protruding at each end (Fig. 16-4).

    e. Fold back the ¼-inch (6 mm) protruding ends of insulation along the length of the slots to reveal the opening and to facilitate threading the winding into the slot.

    f. Rewind according to winding procedure.

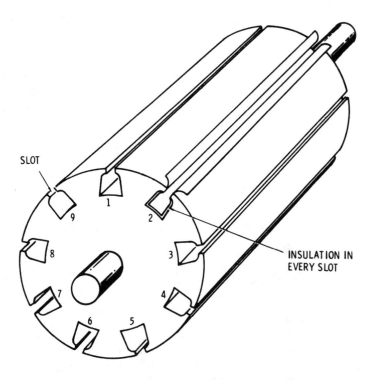

Fig. 16-4. Armature with cut insulation in place.

3. Cut-insulation method for large slots.
   a. Cut a strip of insulation that is dimensioned to fit the inside periphery of the core slot.
   b. Center and crease the insulation strip length on the insulation former according to the inside dimensions of the bottom of the core slot.
   c. Cut the insulation strip length into sections that are ½ inch (12 mm) longer than the length of the slot.
   d. Insert the insulation sections into the slots with ⅛ inch (3 mm) protruding at each end. In some instances (open slots) it may be necessary to shellac the insulation sections into place. This will depend on whether the insulation sections will remain in place in the slot while winding the armature. Do not use glue.

4. Preparation and placement of slot feeders.
    a. Cut a strip of thin insulation paper 0.007 to 0.010 inch thick, to fit the inside periphery of the core slot plus an allowance of ½ inch (12 mm).
    b. Center and crease the insulation strip length on the insulation former according to the inside dimensions of the bottom of the core slot.
    c. Cut the insulation strip length into sections that are dimensioned to the exact length of the slot.
    d. Insert the insulation sections into the preinsulated slots as the winding process progresses.
    e. Fold back the ¼-inch (6 mm) protruding ends to reveal the slot opening and to facilitate rewinding.
    f. Rewind according to winding procedure.
    g. Place a fiber stick or insulation paper between coils.
    h. Bend feeder into slot and wedge.

5. Use of slot feeders as insulation between windings. On armatures that operate on voltages higher than 110-120 volts, it is necessary to insulate between windings in the slits and between the overlapping end turns of the windings. The procedure outlined below gives the steps to be taken in using the insulation feeders for insulating between windings (Fig. 16-5). After a coil side is placed into a slot:
    a. Trim down the height of the coil-feeder insulation on both sides so that, when it is turned into the slot, it will be wide enough to cover the winding and fit into the slot (Fig. 16-5, step 1).
    b. Bend one side of the trimmed feeder insulation into the slot with a stiff fiber drift and press firmly into place (Fig. 16-5, step 2). Careful! Do not damage insulation on the wire.
    c. Bend the other side of the trimmed feeder insulation into the slot over the top of the first side and press firmly into place with a stiffer fiber drift (Fig. 16-5, step 2).
    d. Place a section of feeder insulation, cut to the proper dimensions, over the top of the first (Fig. 16-5, step 3.)
    e. Proceed with the second coil according to the type of winding required.
    f. Bend feeder insulation into the slot and wedge (Fig. 16-5, step 4).

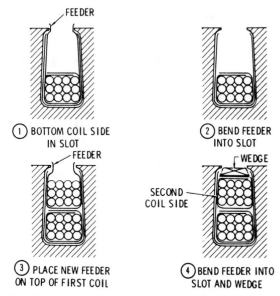

Fig. 16-5. Folding insulation into slot and wedging.

Hand Winding, Coil Forming, and Slot Wedging

It is impractical to use preformed winding on small armatures. End room is limited and the windings must be drawn up tightly to the armature core. In large-production shops, small armatures are machine-wound on special equipment designed for the purpose. It is expedient to use preformed coils on large armatures, and this method should be used whenever practicable. Any rewinding job requires the utmost care to prevent the windings from short-circuiting to the metal core.

General Methods of Hand Winding

A hand winding is defined according to the method by which it is wound. There are five general methods of hand winding: loop winding, chorded split-loop winding, split-V-loop winding, layer winding, and diametrically split winding. The term "loop" is used to indicate that the wire is not cut at the start and finish of each coil. The term "chorded" means that the coil pitch is not full pitch.

The advantages and disadvantages of these methods will be discussed under the appropriate headings, as will a general method of hand winding applicable to all of these windings.

Winding Procedures — First Coil

1. Set up the reel of wire, a proper wire size for the armature, on a reel rack with a suitable reel-tension device (Fig. 16-6).
2. Set the reel tension so that little mechanical resistance is offered to a pull on the wire. This resistance will cause the reel to stop and prevent the wire from becoming uncoiled when no longer pulled.
3. For armatures requiring sleeving for lead identification, push lengths of sleeving (dependent upon lead length) of alternate colors on the wire. The number of sleeve lengths will depend

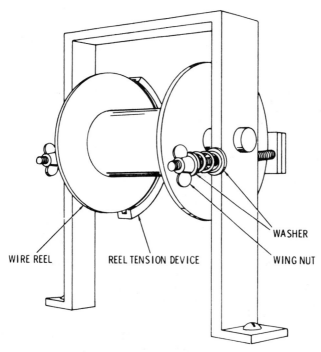

WASHER

WIRE REEL   REEL TENSION DEVICE   WING NUT

Fig. 16-6. Bench setup of a reel rack.

on the number of coils for the armature or the number of coils to be wound first. Use dissimilar colors for the beginning and ends of each coil. For example, red could be used for the beginning of a coil and white for the ending, and so on. Improvise a means for holding the sleeve back while drawing wire for the winding.

4. Mount a large armature in an armature stand.
5. Hold a small armature in one hand with the commutator end held toward the operator (Fig. 16-7).

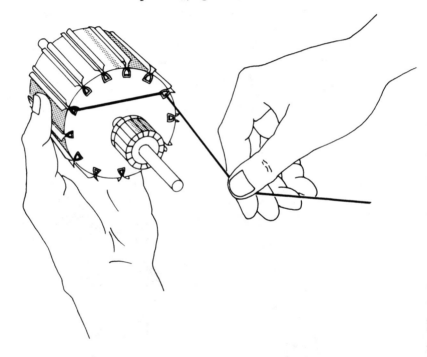

Fig. 16-7. Holding a small armature in one hand during winding.

6. Insert the wire into a slot (choose any slot and call it slot 1) and hold the end of the wire down on the commutator with the thumb.
7. Wind a few turns into the proper slots (coil pitch), pulling up each turn snug and firm but not to a point that the wire may snap or cut through the insulation.

8. Release the end of the wire under the thumb and continue winding the coil, pulling each turn snug and firm, until the required number of turns is inserted for any one of the winding methods outlined in *Loop Method of Winding with One Coil per Slot* through *Layer Winding* as appears later in this chapter. Pay strict attention to coil pitch.

9. Press down the windings with a stiff fiber drift plate to facilitate getting the required number of turns into a slot.

10. Use a coil tamper, when necessary, to tamp and pack the winding in the slot when it appears that no more turns can be accommodated. Exercise care when using the tamper so that the winding insulation will not be damaged. After the use of the coil tamper, a surprising number of turns may be inserted.

11. Secure the winding in the slot (see *Securing Windings in the Slots*). Exercise care when using the coil tamper so that the winding insulation will not be damaged.

12. Move a length of sleeving to the end of the winding and loop the wire. This is the start of the next coil.

13. Follow appropriate procedure according to method of winding the rest of the coils.

Securing Windings in the Slots (Older Method of Winding)

Slot wedges are used for securing windings in the slots. These wedges are of three different types. The use of a particular type is dependent upon the available remaining space in the slot. They may be made of $1/32$- or $1/16$-inch thick fiber or wood, and where space is limited, they can be made of 0.015- to 0.023-inch paper. The selection and methods for application of wedges are as follows:

1. *Paper wedges for well-filled slots.* When the required number of turns are in the slot and a wedge of an appreciable thickness cannot be driven in from the ends, a 0.023-inch paper wedge, cut just a trifle wider than the slot opening, is used and pressed into the slot from the top.

   a. Put one edge of the wedge under one overhanging edge of the slot.

b. Drive the other side of the wedge down under the opposite edge of the slot with a narrow-width steel drift and hammer. An old hacksaw blade with the teeth ground off can be used for a steel drift. This method also applies to open-slot armatures. The wedge makes a driving fit, and after the armature has been dipped in varnish and baked, this type of wedging will hold as well as wedges driven in from the end.

2. *Driving in paper, fiber, or wooden wedges.* The selection of any one of these wedges depends on the available room in the slot and on the availability of the materials. Many times it may be necessary to drive two wedges of either the same thickness or odd thicknesses into a slot to hold the winding firmly in place. Cut enough wedges of proper length and width for the total number of slots. (If long lengths of prepared wedges are available, select the proper width and cut to proper length). Start a wedge in the slot and and, if possible, push it all the way in by hand. The wedge must fit tightly. If it just slips in, a wedge of greater thickness must be used. If the wedge cannot be pushed in by hand, follow the procedure given below.

a. Use a coil tamper in the slot and force down the windings in the slot end.

b. Start the wedge in the slot. Place the coil tamper slightly in front of the wedge, to open the slot. Keep advancing the tamper and working the wedge along behind it until the wedge completely fills the slot.

c. If a wooden or hard fiber wedge is used and it is difficult to drive it in by hand, tap the wedge in with a rawhide or rubber mallet.

d. Use a wedge driver if the wedge bends when being driven into the slot. The wedge driver, however, is used as a last resort. If employed, the wedge is inserted into the outer shell of the wedge driver, leaving about $1/16$ inch (1.6 mm) of the wedge exposed. The exposed end of the wedge is started in the slot with the wedge driver's outer shell aligned with the slot. Then the driving pin is tapped with a rawhide or rubber mallet until the entire length of the wedge is in the slot.

Loop Method of Winding with One Coil per Slot

The most used winding method for small armatures is the loop method (Fig. 16-8). The advantage of this method is that it can be quickly applied and connected when correctly wound. The disadvantages are that the coils are not all alike, in that the first coil is smaller than the last. Also, where there is more than one coil per slot, the resistance of the last coil is greater than the first coil. As a result, the resistance of each coil is different. When the turns per

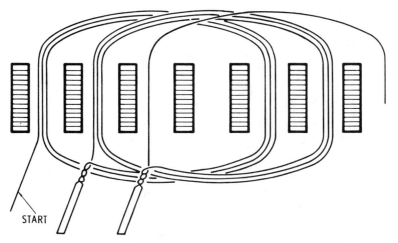

Fig. 16-8. A loop winding showing loops at the end of each coil.

coil are many or the wire size large, the ends of the coils pile up so that it is difficult and sometimes impossible to wind in the coils properly. Due to the unequal size of the coils, a loop winding is not balanced and requires extreme care in balancing after winding. The procedure given below will aid in winding a loop winding.

1. Follow 1 to 11 in *Winding Procedure — First Coil.*
2. For armatures that will operate from 110-120 volts or higher, place strips of insulation between the overlapping end turns of the windings at the back and front of the armature.
3. Wind the second coil into the slot adjacent to the first slot according to the coil pitch, and identify the end of the coil winding with color sleeving.
4. Make a loop of sufficient length to reach the commutator for

connection, and identify the beginning of the third coil with colored sleeving.

5. Wind in the third coil and follow the same procedure for all the coils of the armature.

6. Cut the wire from the reel when the last turn of the last coil is inserted into the last slot, then connect the end to the single lead of the first coil.

7. Lap windings (Fig. 16-9) can be wound left-hand or right-hand on either side of the shaft without affecting the direction of rotation of the motor or changing the brush polarity of a generator.

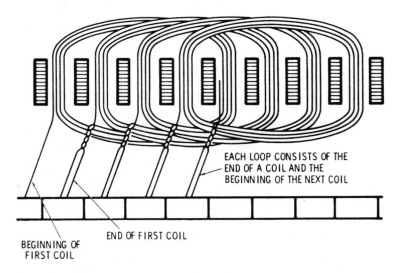

EACH LOOP CONSISTS OF THE END OF A COIL AND THE BEGINNING OF THE NEXT COIL

END OF FIRST COIL

BEGINNING OF FIRST COIL

Fig. 16-9. A lap winding with one coil per slot.

Loop Method of Winding with More Than One Coil per Slot

The winding procedure for this type of loop winding is the same as in *Winding Procedures — First Coil*, except that there will be more than one coil in every slot (Fig. 16-10). For example, a 12-slot armature with 36 commutator bars will require three coils per slot. Every loop of these three coils must be marked with different colored sleeving. Red, white, and blue will be satisfactory, or any

SLOT NO. 1   2   3   4   5   6   7   8

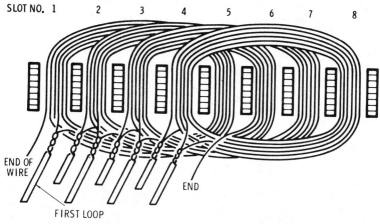

Fig. 16-10. A two-coil-per-slot winding with short and long loops for identification.

other combination of dissimilar colors. Sleeving can be used for one-fourth of the total number of coils of the armature and one-fourth the total number of coils loop-wound and identified. Then the wire is cut and identified. Winding and identification are continued for the next one-fourth of the total number of windings. Continue this process until the entire armature is wound. By this method each slot has three identified loops projecting, except at the end of each one-fourth of the total number of slots where there would be two identified loops and two identified ends.

Loop Method of Winding with More Than One Wire in Hand

With this method, the coils are wound in place using two or more wires in hand. This is dependent upon the ratio of the number of commutator bars to the slots. For example, if there are 12 slots and 24 bars, two wires are fed into the slots from two wire reels. If there are 36 bars, three wires are fed into the slots from three wire reels. With this process, the wire is cut at the end of each coil and identified with sleeving that indicates the end of each coil.

The wire is then brought around so the end turns are parallel to the starting leads. Following the procedure of *Winding Procedure — First Coil*, for armatures operating from 110-120 volts or higher, place strips of insulation between the overlapping end turns of the windings at the back and front of the armature.

Chorded, Split-Pitch Loop Winding

A feature of this winding (Fig. 16-11) is that two coils are wound in three slots on the same side of the shaft; two coil sides

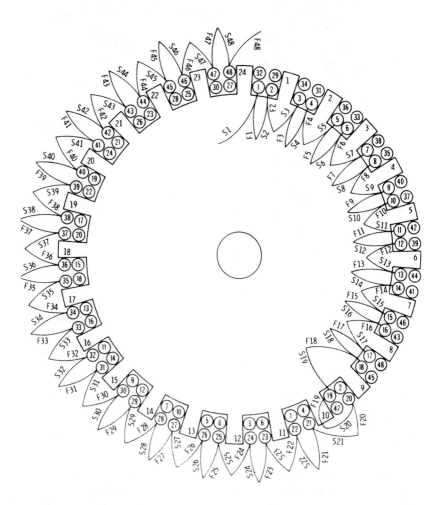

Fig. 16-11. A chorded, split-pitch loop winding.

together in one slot and the other two coils each having a different pitch. The advantages of the chorded, split-pitch winding are illustrated by a 5 hp, 1,100 rpm, two-pole motor having an armature 10 inches in diameter with 24 slots and 48 commutator bars. If this armature is wound full-pitch, 1 and 13, the winding would pile up on the ends of the armature and take up too much space. By the use of the chorded, split-pitch, loop-winding method, the coil pitch is reduced to 80 percent of the full pitch, eliminating some of the coil crossings and preventing the end turns from bending around the shaft. For example, if the full pitch equals 12, 80 percent of the full pitch equals 9.6 (0.80 $\times$ 12). This is equivalent to using slots 1 and 10.6. Since 0.6 of a slot is not possible, wind one coil in slots 1 and 11, with a pitch of 10, and one coil in slots 1 and 10, with a pitch of 9. The same procedure is carried out for all coil windings. This gives an average pitch of 9.5, which approximates the desired pitch of 9.6. With this particular type of winding, the long pitch, 1 and 11, has one more turn than the short pitch, 1 and 10, and can only be wound with one wire in hand. The loops are made where the two coil sides are in the same slot and are identified according to the method shown in *Identifying the Coil Leads When Winding*. For armatures operating from 110-120 volts or higher, place strips of insulation between the overlapping end turns of the winding at the back and front of the armature. Check later in this chapter for connecting to the commutator.

Chorded, Split-Loop Winding

The last coils wound on the armature are visible on each side of the shaft and are parallel to each other (Fig. 16-12). The advantages of a chorded, split-loop winding are better mechanical balance, more uniform coil resistance, and better coil distribution, resulting in more end room on the armature. The only disadvantage is that the wire is cut at the end of each coil and that connections to the commutator are slightly more complex in nature. The first coil is started as described in *Winding Procedure — First Coil*, steps 1 to 7. Cut the wire and identify the ends, then turn the armature 180° and repeat the process for the second coil in the slot adjacent and to the right of the slot filled previously. Turn the armature 180°, skip the slot on the right adjacent to the previously filled slot, and

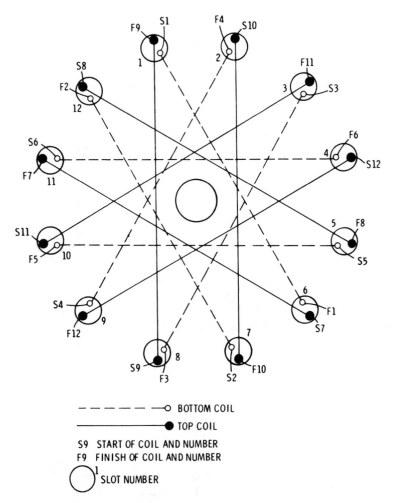

BOTTOM COIL
TOP COIL
S9 START OF COIL AND NUMBER
F9 FINISH OF COIL AND NUMBER
SLOT NUMBER

Fig. 16-12. A chorded, split-loop winding.

repeat the procedure for the third coil. Turn the armature 180° and repeat the procedure for the fourth coil in the slot and adjacent to the right of the slot filled previously. After winding in each coil, turn the armature 180°, repeating the aforementioned procedures until the entire armature is rewound. Place a strip of insulation between the group of end turns of one coil and the overlapping end turns of another coil at the back and front of the armature. Check later in this chapter for connections to the commutator.

Method of Putting on a Chorded, Split-Loop Winding When the Number of Slots is Not Divisible by Four

The method of starting the top layer is slightly different from that already described. After six coils have been put in place, there will be one empty slot on each side of the armature. To start the seventh coil, skip two slots, counting clockwise from the starting side of coil 5, then start winding in slot 7 to the empty slot adjacent to the finish side of the coil 2 on the right. The coil parallel to coil 7, which is coil 8, will start in the slot with the finish side of coil 2 and in the remaining empty slot, which is slot 6. The armature winding can now continue with the same procedure as was used on the bottom layer. A strip of insulation must be placed between the top layers and the bottom layers as described previously. The method of connecting this winding to the commutator will be described later in this chapter.

Split V-Loop Winding

A distinctive feature of this winding (Fig. 16-13) is that the slots are filled as the winding progresses. The variation in coil lengths is about the same as in the chorded, split-loop winding. Its advantage over the chorded, split-loop winding is that it can be wound on a core with any number of slots. The first coil is started as described earlier in *Winding Procedure — First Coil.* For armatures operating from 110-120 volts or higher, place strips of insulation between the overlapping end turns of the windings at the back and front of the armature. After the first coil is wound into the slots of proper pitch on the right-hand side of the shaft and identified, the armature is rotated until the finished side of the coil is on top. Wind the second coil into the slot containing the finish side of coil 1, to the same pitch. Rotate the armature until the finish side of the second coil is on top. Wind the third coil into the slot containing the finish side of coil 2, to the same pitch. Follow this procedure for every coil until the armature is completely wound. See later part of the chapter for attaching to the commutator segments.

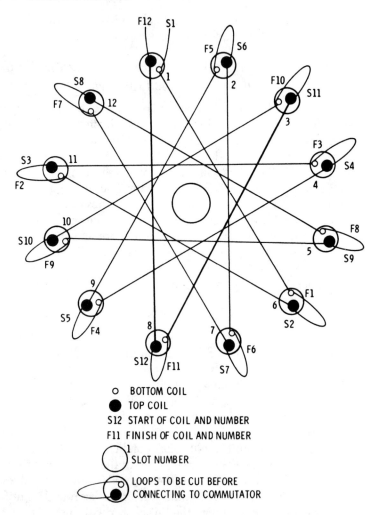

Fig. 16-13. A split V-loop winding.

Diametrically Split Winding

This type of winding is most suitable for an armature with coils that require a few turns of heavy wire (Fig. 16-14). It is recommended for machines that have an even number of slots for this type of winding. The hollow space between the shaft and windings

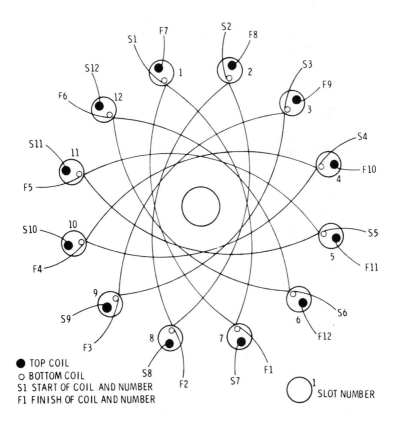

Fig. 16-14. A diametrically split winding.

is formed by using a wooden mandrel of the proper dimensions while winding the armature. Each coil is wound half on one side of the shaft and half on the other side of the shaft. To get an even distribution of the coil winding in the slots, wind two turns first on one side of the shaft, then two turns on the other side of the shaft; continue in this manner until the full number of turns for the coil are completed. Identify the start and finish of the coil and proceed to slot number 2. Follow the same procedure for every coil until the entire bottom layer is completed. There will be one identified lead at every slot for the bottom layer. The top layer is started in the slot that contains the finishing lead of the last coil in the bottom layer. Continue winding in the same manner, as outlined above, until every slot contains a beginning and an ending lead. A heavy cord has been placed so that the windings are on top of it. The ends of the cord are left out so they will serve as a means of tying the coils together. Tie the last coil down by means of the heavy cord threaded underneath the other coils.

Layer Winding

Layer winding is used where the coils have few turns of large wire and end room is limited, making it practically impossible to use for rewound coils (Fig. 16-15). It will be almost impossible to use this method if the following equation is not satisfied:

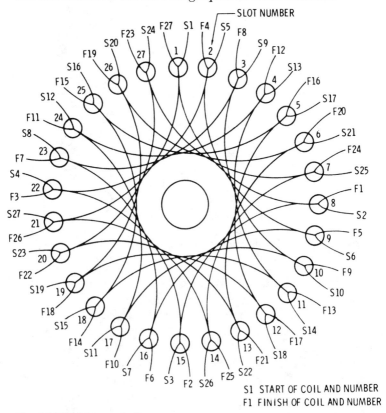

Fig. 16-15. A layer winding.

S1 START OF COIL AND NUMBER
F1 FINISH OF COIL AND NUMBER

Number of poles × pitch ±1 = number of slots

For example, to determine whether a four-pole armature with 27 or 29 slots is suitable for this type of winding, substitute the proper values in the equation (4 × 7) ± 1 = 27 or 29. With a coil pitch of seven, it is evident that layer winding can be used in this case. The 27-slot armature is progressively connected and the 29-slot armature is retrogressively connected. Use the procedure that folllows:

1. Set up the reel of wire, of proper size for the armature, on a reel rack with a suitable reel-tension device (Fig. 16-6).
2. Set the reel tension so that very little mechanical resistance is offered to pull on the wire. This resistance will cause the reel to stop and prevent the wire from becoming uncoiled when no longer pulled.
3. Shape the end turn of the first turn over the end of the armature, toward the shaft, to about the center of the space that all the end turns of the coil will occupy.
4. The following end turns are laid on opposite sides alternately; the second turn is laid to the right of the first end turn, the third turn is laid to the left of the first end turn, the fourth is laid to the right, and the fifth is laid to the left. Proceed alternately in this manner until the entire coil is wound. Allow the wires to bunch up in the slots, but start to fan them out as close to the slot ends as possible.
5. Proceed in the same manner as outlined in 1, 2, 3, and 4 for each coil until the entire armature is wound.
6. Refer to the later part of this chapter for connections to the commutator.

Identifying the Coil Leads When Winding

Leads must be properly identified; otherwise making connections will be highly complex and, in some cases, practically impossible. Below are listed the methods of identification for the various conditions encountered.

1. For armatures with the same number of bars on the commutator as there are slots, no identification is necessary.

2. For armatures with twice as many bars on the commutator as there are slots, there will be two loops per slot. To distinguish between the first and second loops of each slot, the second loop is made longer than the first, or sleeving of two different colors is put on the loops (Fig. 16-10). The sequence of color applied at the first slot must be carried out for every slot.

3. For armatures with three times as many bars on the commutator as there are slots, there will be three loops per slot. Use three dissimilar colors, one color for each loop of the slot. The sequence of color applied to the first slot must be carried out for every slot.

4. When coils are wound with more than one wire in hand, as in a case where the number of commutator bars exceed the number of slots, use one sleeve of one color over all of the start leads and one sleeve of another color over all of the finish leads.

Machine Winding

An armature-winding machine is used in production shops for winding small armatures. Machine winding is impractical for a shop that receives only a few small armatures to rewind. The machine consists of a drive motor mounted on a pedestal, a counting device, a handwheel on the drive shaft for controlling the last few turns of a coil, holding clamps to fit various sizes of armatures, a foot lever in the pedestal for controlling the motor, wide-tension devices, and wire-spool holders.

Form Winding

Very often a rewinding job is made easier by preforming the coils and then placing them in slots. This may be done in various ways.

1. Drive finishing nails into a board, leaving the nails protected with sleeving (Fig. 16-16). Space the nails at intervals identical with the size and shape of the coil. The size of the coil is usually obtained from an old one at the time the machine is disassembled. If it cannot be ascertained in this manner, wind a single piece of wire in the slots, allowing sufficient slack for later installation without crowding. Twist the two

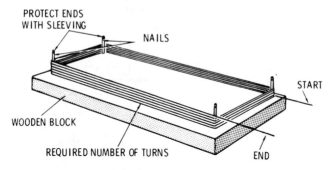

Fig. 16-16. Winding a skein on a form consisting of protected nails protruding from a board.

ends of the wire together, remove from the slots, and use for measuring the forming board (Fig. 16-16). After the required number of turns are wound on the form, tie the coil in several places and remove from the forming board.

2. Use a block or blocks of fiber insulation or wood shaped to the proper size of the coil. Clamp two side pieces of fiber insulation to the sides of the block form to retain the winding on the form, while the winding takes place. The form is rotated on a spindle, driven by a speed-controlled drive (Fig. 16-17). Wind the coil with the same number of turns and the same size of wire as the original coil. Obtain measurements as described in 1 above. Tie the coil in several places before removing it from the form. Coils may be shaped after winding by means of pull blocks, Fig. 16-18.

Coil-Winding Machine

A coil-winding machine (Fig. 16-19) generally consists of a stand, an electrical motor drive, a gear reduction unit, a handwheel on the output shaft for controlling the last few turns of a coil on the winder head, a turns-counting device, an adjustable coil-forming winder head, and a foot control. The general procedure for winding a coil or coils on this type of machine is:

1. Adjust the winder head to the size and shape of the coil

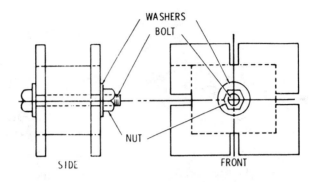

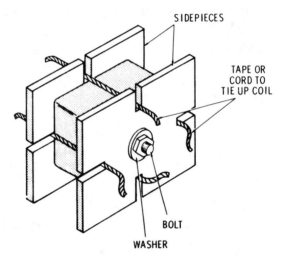

Fig. 16-17. A form for winding coils.

required; this is obtained from a coil removed from the armature in its entirety or from a wire loop adjusted to the coil
pitch and size of the core.

2. Set up a reel of the proper wire size for the coil or coils on a
reel rack with a suitable reel tension device on the floor.

3. Set the reel-rack tension so that very little mechanical resistance is offered to a pull on the wire and so that when pulling
is stopped, the reel will stop and thus prevent the wire from
becoming uncoiled.

4. Attach the end of the winding wire to one of the winder head

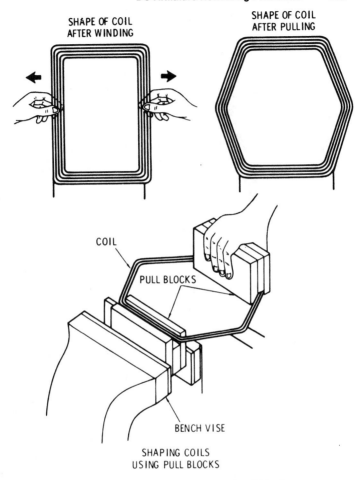

SHAPE OF COIL
AFTER WINDING

SHAPE OF COIL
AFTER PULLING

COIL

PULL BLOCKS

BENCH VISE

SHAPING COILS
USING PULL BLOCKS

Fig. 16-18. A method of shaping coils by means of pull blocks.

rod spools by looping the end once and giving the wire a couple of twists.

5. Start the motor by closing the motor-circuit switch.
6. Start the machine by depressing the foot pedal and control the starting speed with the left hand on the handwheel.
7. Guide the wire onto the winder head with the right hand.

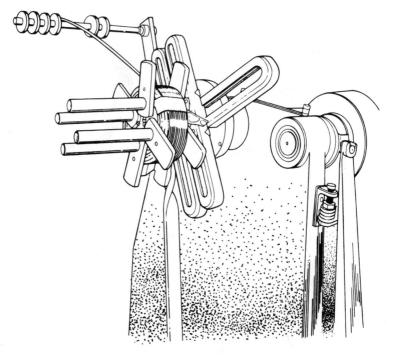

Fig. 16-19. Machine winder with head for winding multiple coils.

CAUTION: Protect the hand from the running wire by wearing a work glove or by using a folded piece of insulation held in the palm of the hand. Do not try to hold back the machine by increasing the hand tension on the wire. Use the handwheel on the output shaft for this purpose. If the folded insulation is squeezed to cause tension, the running wire may heat up and ruin the insulation. Use the folded insulation in the hand to guide the wire only.

8. Gradually let the coil winder attain the desired speed by relieving the pressure of the left hand on the handwheel.

9. Observe the counting device occasionally to determine the number of turns of wire being put on the coil form.

10. When the required number of turns have been wound onto the coil, release the foot pedal (by taking the foot pressure off) to stop the machine.

11. Tie the coils with twine at the top and bottom (to prevent unraveling).
12. Follow the procedure outline *Insulating Coils* for taping and insulating coils.
13. Identify the start and finish of the coils with sleeving.

Insulating Coils

Slots of armatures and stators are generally open and the coils are usually completely taped. Cotton tape is often used for this purpose, with varnished cambric and mica preferred for large motors and special applications. In medium-sized motors, the slots are generally semi-closed. The coils on such motors cannot be completely taped because the turns of the coil often must be fed into the slot one at a time. The part of the coil that is wrapped is that part that extends on either side of the slot. Methods are given below for taping and insulating coils when required.

1. Complete taping of a coil.
    a. Start taping near the finish lead, making sure to engage the sleeving.
    b. Continue around the coil, half-lapping each turn of tape over the preceding one until the start lead is reached.
    c. Engage the sleeving of the start lead and continue to the start of the taping.
    d. Fasten down the end of the wrapping with masking tape or twine.
2. Partial taping of a coil.
    a. Tape the back end of the coil first and only that portion that will be outside the slots (this is determined by measurement of the slot length).
    b. Half-lap each turn of tape over the preceding turn.
    c. Secure the tape end with masking tape or twine.
    d. Tape the front end by starting ahead of the start lead, at the point where the coil enters the slot (this is determined by measurement of the slot length).
    e. Half-lap each turn of tape over the preceding turn.
    f. Engage the sleeving of the start lead and continue taping to the finish lead.

    g. Engage the sleeving of the finish lead and continue to where the coil enters the slot.

    h. Secure the tape end with masking tape or twine.

3. Varnish-insulating a coil.

Refer to the procedures for reinsulating by applying insulating varnish. Normally, the insulated windings of a motor or generator require reinsulation when they appear dry, brittle, and faded. Reinsulation in this case takes the form of revarnishing. For small motors and generators, use a synthetic varnish of the baking type. The windings of large units take black baking varnish of the asphalt type. While satisfactory for insulating windings in small motors and generators, hard varnish should not be used for large-unit windings because of the risk of damaging the winding beyond use when it is removed from the frame.

The newer 100-percent solid impregnates have decided advantages over the varnish used previously. These include epoxy resins, which are somewhat more expensive than the standard polyesters but are physically superior. Epoxies give much better bond strengths. Impregnates are used to unitize the conductor and insulation and to fill voids.

Placing Armature Coils in Slots

The placement of preformed coils into the slots is dependent on the type of slot. Care must be exercised so that conductors are not scraped, wrappings are not disturbed, and insulations are kept intact. Follow the outlined procedures for inserting the coils after the slots have been prepared with insulation.

Semiclosed Slots

1. Spread or fan out the turns on one side of an untaped or partially taped coil (Fig. 16-20).

2. Hold the coil at an angle and introduce the turns into the first slot, one or a few at a time (depending on the space limitations of the slot opening) until all the turns are in the slot.

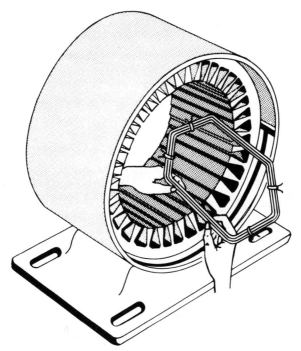

Fig. 16-20. Placing a preformed coil in a semi-closed slot.

3. Place one coil side in all the slots as outlined in step 2 above. Do not insert the second coil side until every slot has one coil side placed.

4. Place a strip of insulation in every slot on top of each inserted coil side. The strip should be 0.020 inch thick, and $\frac{1}{4}$ inch (6 mm) wider and 1 inch (25 mm) longer than the slot. Center it so that the width is evenly tucked around the coil side and extends $\frac{1}{2}$ inch (12 mm) on both ends. Another way is to bend the insulation feeder of every slot into the slot, one side overlapping the other (Fig. 16-5).

5. In the case where the slot insulation was bent and folded inward, place another slot feeder as shown in Fig. 16-5.

6. Insert the second side of a coil. Fit it on top of the first coil's side. This is done through the opening of the second feeder. Insert insulation. This may be done several slots away, depending on the coil pitch.

7. Make certain that each coil side extends beyond the slot at both ends and that it does not press against the iron core at the corners (Fig. 16-21).
8. Fold in the slot feeder as described in Fig. 16-5.
9. Wedge in the winding as described in *Securing Windings in the Slots*.

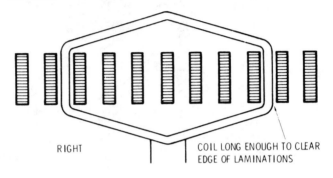

RIGHT   COIL LONG ENOUGH TO CLEAR EDGE OF LAMINATIONS

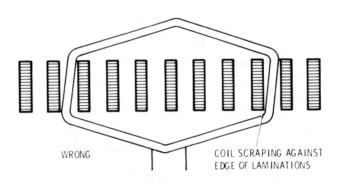

WRONG   COIL SCRAPING AGAINST EDGE OF LAMINATIONS

Fig. 16-21. The sides of each coil extending beyond the edge of the slot.

Open Slots

Completely taped coils are most often used in open slots, although in some instances partially taped coils are used; which ones are used depends on the available space within the slot. After the armature has been properly insulated, place coils in the slots according to the following procedure.

1. Place one coil side in all the slots. Do not insert the second coil side until every slot has one coil side placed.
2. Place the second side of a coil, fitting it on top of the first side of a coil several slots away, depending on the coil pitch.
3. Make certain that each coil side extends beyond the slot at both ends and that it does not press against the iron core at the corners.
4. Wedge the windings as described earlier in the chapter.

Connections

If the coil leads are properly identified, the process of connection is simplified. With the coil sides in their proper slots (coil pitch), which places them in the proper commutating field after connection, the process involves bringing the leads (lead swing) of the coils to the proper commutator bars (commutator pitch), and connecting each individual coil in the proper sequence. The procedures outlined below cover the various methods of connecting armature leads to commutators (Fig. 16-22).

Connecting the Loops of Windings of an Armature Which Has the Same Number of Commutator Bars as Slots

1. Fill the hollow between the coil ends and the back of the commutator if necessary (Fig. 16-9).
2. Determine the proper lead swing or follow the recorded data.
3. Scrape the insulation from the end of the coil-lead loop, coming out of the core slot, used to determine the proper lead swing.
4. Force the scraped coil-lead loop into the proper commutator slot according to the determined or recorded lead swing. If the scraped conductor fits loosely, use more care in the scraping operation so that only the insulation is removed and not the metal.
5. Proceed to the next coil-lead loop and remove the insulation.
6. Force this coil-lead loop into the bar slot adjacent to the bar in

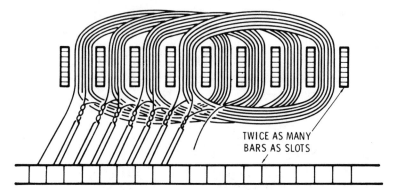

TWICE AS MANY
BARS AS SLOTS

Fig. 16-22. Method of connecting the loops of windings of an armature which has twice the number of commutator bars as slots.

which the first loop was placed, and to the right or left, depending on the lead swing.

7. Follow 5 and 6 above for all the coil-lead loops and commutator bars until all the loops are inserted in the commutator-bar slots progressing to the right, regardless of left or right lead swing.

8. Solder the connections.

9. Band the armature. Banding is shown later in this chapter.

Connecting Leads of a First Coil When There Is Only One Start and One Finish Lead for Every Slot (Simplex Lap Winding with One Coil per Slot) (Fig. 16-23)

1. Fill the hollow and determine proper lead swing.

2. Remove the insulation from the end of each coil lead.

3. Determine or follow the recorded lead swing.

4. Bend back all finish leads over the core.

5. Force a start coil lead into the proper commutator-bar slot according to lead swing.

6. Proceed to the next core slot on the right and force the start coil lead of this slot into the commutator-bar slot adjacent to the right of the first coil start lead.

7. Follow 6 above for every slot until commutator-bar slot has a start lead forced into it.

8. Wrap a layer of cotton or cambric tape over the top of all the

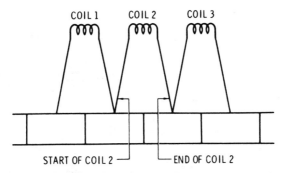

Fig. 16-23. Method of connecting the armature leads of a simplex lap winding with one coil per slot.

start leads between the end windings and the back of the commutator.

9. Secure the cotton or cambric tape with masking tape.
10. Bring a finish lead forward and down and force it into a bar slot adjacent to the right of the bar in which the start lead of this coil is placed.
11. Bring the next finish lead on the right forward and down and force it into the bar slot adjacent and to the right of the bar slot in which the previous finish lead was forced.
12. Follow 11 above until the commutator is filled.
13. Solder the connections.
14. Band the armature.

Connecting Start Leads of a First Coil When There Are Several Leads Identified with One Sleeve, as in the Case of Winding with More Than One Wire in Hand (Simplex Lap Winding with Two Coils per Slot)

In performing the connecting of start leads of a first coil follow the instructions below (Fig. 16-24):

1. Fill the hollow and determine proper lead swing.
2. Lift all the finish leads (as determined by the color used for identification of the finish leads) and bend them back over the armature core.
3. Select the start leads of the last coil on hand-wound armature or any group of start leads on an armature with preformed coils.

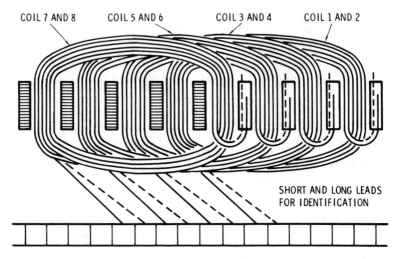

COIL 7 AND 8   COIL 5 AND 6   COIL 3 AND 4   COIL 1 AND 2

SHORT AND LONG LEADS
FOR IDENTIFICATION

**Fig. 16-24. Method of connecting start leads of a simplex lap winding
with two coils per slot.**

4. Scrape the insulation from the ends of all leads.
5. Determine the proper lead swing or follow the recorded data.
6. Force one start lead of a coil into the proper commutator slot according to the determined or recorded lead swing.
7. Force the remaining coil start lead of the same group of start leads into the adjacent commutator slot to the right of the first-placed coil start lead.
8. Proceed to the next core slot and force one of the coil start leads of the coil side into the commutator slot adjacent to the commutator slot that has the last-placed coil start lead of the coil start lead group of the core slot to the left.
9. Force the remaining coil start lead of the core slot into the adjacent commutator slot, progressing to the right.
10. Perform 8 above for every core slot until all the start leads are placed.
11. Wrap a layer of cotton or cambric tape over the top of all start leads between the end windings and the back of the commutator.
12. Secure this layer with masking tape.

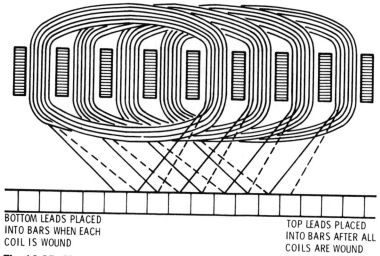

BOTTOM LEADS PLACED
INTO BARS WHEN EACH
COIL IS WOUND

TOP LEADS PLACED
INTO BARS AFTER ALL
COILS ARE WOUND

Fig. 16-25. Method of connecting the finish leads of a simplex lap winding with two coils per slot.

Connecting the Finish Leads of a Simplex Lap Winding with Two Coils per Slot

In performing the following steps, extreme care must be exercised to be sure that improper connection is avoided (Fig. 16-25).

1. Perform a continuity test for locating the coil finish leads of the coil start leads already placed. Continuity testing is done with a bench-test receptacle and lamp or an ohmmeter.
2. Touch one lead of the continuity tester to the commutator bar that has the first coil start lead connected to it. Touch the other continuity-tester lead to each of the finish leads of the coil (several slots away from the core slot where the start leads emerge, according to the coil pitch) until one is found that gives a closed-circuit indication.
3. Bring the located finish lead forward and down, and force it into the commutator-bar slot, two bars away from the commutator bar on the left, which has the first coil start lead connected.
4. Bring the remaining coil finish lead for the core slot forward and down, and force it into the bar slot on the left, adjacent to the left of the bar that has the first coil finish lead in it.

5. Keep the lead of the continuity tester on the same bar started with, advance to the next slot, and with the other continuity-tester lead, touch each of the coil finish leads until one is found that gives a closed-circuit indication.

6. Bring this lead forward and down, and force it into the bar slot on the left, two bars to the right of the last-placed coil finish lead.

7. Bring the remaining lead of the slot forward and down, and force it into the bar slot on the right, adjacent and to the right of the last-place finish lead.

8. Continue by following 6 and 7 above until all leads are down and placed, moving the continuity-tester lead on the commutator to the right only when necessary for convenience, but never past the commutator bars with connected coil finish leads.

9. Solder the connections.

10. Band the armature.

Connecting the Leads of a First Coil When There Are Several Leads Identified with One Sleeve, as in the Case of Winding with More Than One Wire in Hand (Simplex Lap Winding with Three Leads)

In this type of connection, follow the steps indicated below (Fig. 16-26):

1. Follow steps 1 to 12 of *Connecting Start Leads of a First Coil — With More Than One Wire in Hand.* Also follow steps 1 and 2 in the section immediately above. For 7 above, insert the two remaining coil start leads in adjacent bar slots, to the right of the first coil start leads.

2. Bring the located finish lead forward and down, and force it in the commutator bar slot three bars away from the commutator bar on the left that has the first coil start lead connected.

3. Keep the lead of the continuity tester on the commutator bar that has the first coil start lead connected to it, and touch the other continuity-tester lead to the remainder of the finish leads of the coil (several slots away from the core slot where

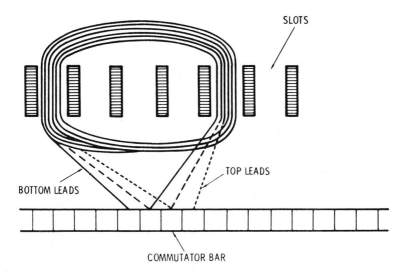

Fig. 16-26. Method of connecting the armature leads of a simplex lap winding with three coils per slot.

the start leads emerge, according to coil pitch) until one is found that gives a closed-circuit indication.

4. Bring the located finish lead forward and down, and force it into the commutator bar slot on the left, adjacent and to the left of the bar that has the first-located finish coil lead in it.

5. Bring the remaining finish lead of the slot forward and down, and force it into the bar slot on the left, adjacent and to the left of the bar that has the second-located finish coil lead in it.

6. Keep the lead of the continuity tester on the same bar started with, advance to the next slot, and, with the other continuity-tester lead, touch each of the coil finish leads until one is found that gives a closed-circuit indication.

7. Bring this lead forward and down, and force it into the bar slot on the left, three bars to the right of the last-placed coil finish lead.

8. Keep the lead of the continuity tester on the same bar started with, and, with the other test lead, touch each of the remaining coil finish leads until one is found that gives a closed circuit indication.

9. Bring this lead forward and down, and force it into the bar slot on the left, adjacent and to the left of the last-placed coil finish lead.

10. Bring the remaining lead of the slot forward and down, and force it into the bar slot on the left, adjacent and to the left of the last-placed finish lead.

11. Continue by following 1 to 10 above until all leads are down and placed, moving the continuity-tester lead on the commutator to the right only when necessary for convenience, but never past the commutator bars with connected coil finish leads.

12. Solder the connections.

13. Band the armature.

Connecting Leads of a First Coil When There Are Several Leads Identified with One Sleeve, as in the Case of Winding with More Than One Wire in Hand (Duplex Lap Winding with Two Coils per Slot)

In performing the following, extreme care must be exercised to be sure that improper connection is avoided (Fig. 16-27).

1. Perform a continuity test for locating the coil finish leads of the coil start leads already placed.

2. Touch one lead of the continuity tester to the commutator bar that has the first coil start lead connected to it. Touch the other continuity-tester lead to each of the finish leads

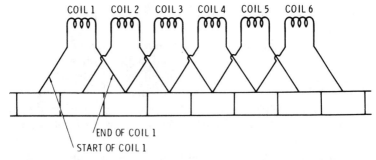

Fig. 16-27. Method of connecting the armature leads of a duplex lap winding with two coils per slot.

of the coil (several slots away from the core slot where the start leads emerge, according to the coil pitch), until one is found that gives a closed-circuit indication.

3. Bring the located finish lead forward and down, and force it into the commutator-bar-slot, two bars away from the commutator bar on the left that has the first coil finish lead in it.

4. Bring the remaining coil finish lead for the core slot forward and down, and force it into the bar slot on the left, adjacent and to the right of the bar that has the first coil finish lead in it.

5. Keep the lead of the continuity tester on the same bar started with, advance to the next slot, and, with the other continuity-tester lead, touch each of the coil finish leads until one is found that gives a closed-circuit indication.

6. Bring this lead forward and down, and force it into the bar slot on the left, adjacent and to the right of the last-placed coil finish lead.

7. Bring the remaining lead of the core slot forward and down, and force it into the bar slot on the left, adjacent and to the right of the last-placed coil finish lead.

8. Continue by following 6 and 7 above, until all leads are down and placed. Move the continuity-tester lead on the commutator to the right only when necessary for convenience, but never past the commutator bars with connected coil finish leads.

9. Solder the connections.

10. Band the armature.

Connecting Leads of a First Coil When There Are Several Leads Identified with One Sleeve, as in the Case of Winding with More Than One Wire in Hand (Triplex Lap Winding with Three Coils per Slot)

1. Follow steps 1 to 12 of *Duplex Lap Winding with Two Coils per Slot*, and steps 1 and 2 above. For step 7 above insert the two remaining coil start leads in adjacent bar slots to the right of the first coil start leads (Fig. 16-28).

2. Bring the located finish lead forward and down, and force it into the commutator-bar slot three bars away from the

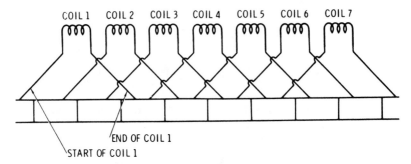

COIL 1   COIL 2   COIL 3   COIL 4   COIL 5   COIL 6   COIL 7

END OF COIL 1

START OF COIL 1

Fig. 16-28. Method of connecting the armature leads of a triplex lap winding with three coils per slot.

commutator bar on the left that has the first coil start lead connected.

3. Keep one lead of the continuity tester on the commutator bar that has the first coil start lead connected to it. Touch the other continuity-tester lead to the remainder of the coil finish leads (several slots away from the core slot where the start leads emerge, according to coil pitch) until one is found that gives a closed-circuit indication.

4. Bring the located finish lead forward and down, adjacent and to the right of the bar that has the first-located coil finish lead in it.

5. Bring the remaining finish lead of the slot forward and down, and force it into the bar slot on the left, adjacent and to the right of the last-placed coil finish lead.

6. Keep the first lead of the continuity tester on the same bar started with, advance to the next slot, and with the other continuity-tester lead, touch each of the coil finish leads until one is found that gives a closed-circuit indication.

7. Bring this lead forward and down, and force it into the bar slot on the left, adjacent and to the right of the last-placed coil finish lead.

8. Keep the lead of the continuity tester on the same bar started with, and, with the other test lead, touch each of the remaining coil finish leads until one is found that gives a closed-circuit indication.

9. Bring this lead forward and down, and force it into the bar

slot on the left, adjacent and to the right of the last-placed coil finish lead.

10. Bring the remaining lead of the slot forward and down, and force it into the bar slot on the left, adjacent and to the right of the last-placed coil finish lead.

11. Continue by following 6 to 10 above until all leads are down and placed, moving the continuity tester lead on the commutator to the right only when necessary for convenience, but never past the commutator bars with connected coil finish leads.

Connecting Wave-Wound Armatures

Wave-winding connections are made quite far apart on the commutator. For example, as already stated, for a four-pole motor, the winding end leads are connected on opposite sides of the commutator. Any four-pole, wave-wound armature must have an odd number of commutator bars. If the commutator has an even number of bars, two of them must be shorted. In all two-coil-per-slot, four-pole, wave-wound armatures, it is necessary to add a coil in the form of jumper lead. Where the number of bars is one more than the number of coils, and where the number of bars is even, there is a wound coil on the armature that is not connected. The dead coil must remain in the armature for mechanical balancing. The procedures outlined below are for the connection of typical wave-winding leads.

1. Connecting a simplex wave winding of a four-pole armature with 23 slots and 23 commutator bars. For a simplex progressive winding, see 1 of Fig. 16-29. For a simplex retrogressive winding of an armature, see Fig. 16-29.

   a. Fill the hollow between the coil ends and the back of the commutator if necessary.

   b. Remove the insulation from the end of each coil lead.

   c. Determine the proper lead swing or follow the recorded data.

   d. Bend back all finish leads over the core.

   e. Force a coil start lead into the proper commutator slot according to the determined or recorded lead swing.

   f. Proceed to the next core slot and place the coil start lead

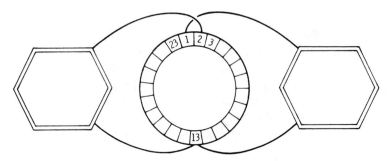

4-POLE, SIMPLEX, PROGRESSIVE WAVE WINDING WITH A COMMUTATOR PITCH OF 1 & 13

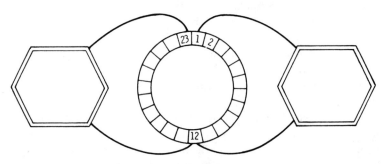

4-POLE, SIMPLEX, RETROGRESSIVE WAVE WINDING WITH A COMMUTATOR PITCH OF 1 & 12

Fig. 16-29. Method of connecting the leads of a simplex wave winding of a four-pole armature with 23 slots and 23 commutator bars.

in the bar, adjacent and to the right of the first-placed coil start lead.

g. Follow a through f above for every one of the coil start leads between the end windings and the back of the commutator.

h. Wrap a layer of cotton or cambric tape over the top of all the start leads between the end windings and the back of the commutator.

i. Secure the cotton or cambric tape with masking tape.

j. Bring a finish lead forward and down, and force it into a bar slot far to the right, according to the proper commutator pitch recorded or calculated.

k. Bring the next coil finish lead forward and down, and force it into the bar slot far to the right and adjacent to the right of the coil finish lead previously placed.

l. Follow k above until the commutator is filled.

m. Solder the connections.

n. Band the armature.

2. Connecting a simplex wave winding of a four-pole armature with 23 slots and 45 commutator bars or with 22 slots and 45 commutator bars. For a simplex progressive winding, see 2 of Fig. 16-30.

    a. Follow the procedure outlined in a to l of the previous connection steps.

    b. On a 23-slot armature there are one coil start lead and one coil finish lead that cannot be connected. Tape these leads and follow d and e below.

    c. On a 22-slot armature there are two commutator bars without any connection. Connect a jumper between the empty commutator bars and follow d and e below.

    d. Solder the connections.

    e. Band the armature.

Banding the Armature

Some open-slot types of armatures require bands to prevent the coils from flying out of the slots while the armature is rotating. Manmade fiber bands are placed on the front and back ends of the core, the number of bands depending on the size of the armature. This procedure is illustrated in 1 of Fig. 16-31.

1. Place and center the armature in a lathe.

2. Place insulating paper or mica in the band slot around the entire armature.

3. Tie the insulation in place with one turn of the cord.

4. Prepare small strips of tin or copper, of the appropriate width for the size of the armature and banding slot, and of sufficient length to overlap the banding wire.

5. Insert the strips under the cord equidistant around the armature-band slot.
6. Use the proper banding.
7. Place the reel of banding in a wire-reel stand with a suitable tensioning device to keep the reel from unraveling when the banding operation is at rest. Do not release the end of the banding until the wire spool is on the reel stand. Set the

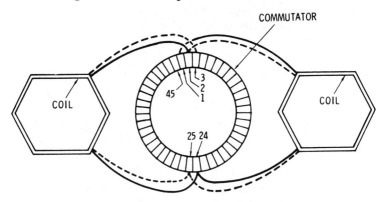

1 A 4-POLE, SIMPLEX, PROGRESSIVE WAVE WINDING

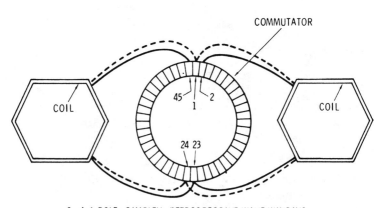

2 A 4-POLE, SIMPLEX, RETROGRESSIVE WAVE WINDING

Fig. 16-30. Method of connecting the leads of a simplex wave winding of an armature with 23 slots and 45 commutator bars or with 22 slots and 45 commutator bars.

tensioning device to prevent the spring of the banding from setting the spool in motion and causing the banding to snarl. Release the end of the banding, but do not let go of it. While working with the banding, constantly keep hold of it until it is firmly anchored on the armature.

8. Pass the end of the banding between the tension blocks.
9. Make one turn around a banding slot, placing the end of the banding over the top of the armature. Anchor the end of the banding, passing it over top of the wedge in the slot on the left and then bringing it down to the shaft. This is done by making two turns around the shaft and twisting the end back on the banding.
10. Set the tension blocks sufficiently to give firm and tight winding on the armature.
11. Remove the tied cord.
12. Start the lathe on slow speed in the direction away from you.
13. Guide the banding onto the armature with a folded piece of insulation paper held in the hand. A glove worn on the hand in contact with the banding would also suffice for protection.
14. Be sure that each turn comes up close to the preceding turn without overlapping.
15. Continue until the entire width of the banding slot is filled; then stop the lathe.
16. Bend over the right tab on all the strips while the lathe is at rest. Do not cut the steel wire until steps 17, 18, and 19 below have been completed.
17. Tin the top right tab with a soldering iron.
18. Bend the left tab over the top of the top right tab, bringing it down as close as possible.
19. Apply the soldering iron to the top of the tab, causing the tabs to sweat together.
20. Cut the banding close to the soldered clamp.
21. Let the end of the banding rest on the lathe bed.
22. Turn the lathe by hand and bring the tabs of another wire clamp on top.
23. Follow 18, 19, and 22 above until all the wire clamps are soldered.

24. Remove the anchoring end of the banding from around the shaft and trim it close to the clamp.
25. Repeat 2 to 5 and 9 to 24 above for each of the bands required for the armature.

For cord banding an armature, the following steps are to be taken. Use enough pressure in winding so that the band will be tight. Use the proper size of banding cord, determined by the size of the armature: light for small armatures, heavy for larger armatures. Refer to steps 1 through 7 of B, Fig. 16-31.

1. Start just behind the commutator risers, allowing about 6 inches (152 mm) of the free end of the cord to lie at right angles to the circumference of the armature, pointing toward the core slots.
2. Wind about eight turns onto the armature, each turn against the other firmly and tightly, progressing toward the core slots and over the free end at right angles to the winding.
3. Loop the free end under the winding and back on itself.
4. Wind more turns (at least as many as put on at first) over the loop ends, each turn against the other firmly and tightly, progressing toward the core slots.
5. Bring the winding end of the core through the loop and pull it up tightly.
6. Pull on the free end of the loop to bring the winding end under the turns of the cord and out between the turns.
7. Cut the two free ends, now emerging from between the turns, close to the turns; smooth out the turns so that the point where the free ends emerged will be even.

Testing

It is very important to test an armature after winding for shorts, opens, or grounds. If all of the precautions are observed and carried out in the rewinding procedure, the possibility of trouble is minimized.

Summary

The process of insulating an armature core, preparatory to rewinding, is one of the most important steps in the rewinding procedure. After the old winding has been removed, the core slots,

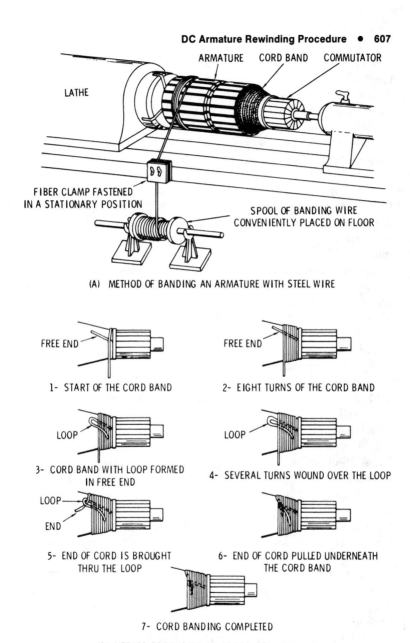

(A) METHOD OF BANDING AN ARMATURE WITH STEEL WIRE

1- START OF THE CORD BAND

2- EIGHT TURNS OF THE CORD BAND

3- CORD BAND WITH LOOP FORMED IN FREE END

4- SEVERAL TURNS WOUND OVER THE LOOP

5- END OF CORD IS BROUGHT THRU THE LOOP

6- END OF CORD PULLED UNDERNEATH THE CORD BAND

7- CORD BANDING COMPLETED

(B) METHOD OF BANDING AN ARMATURE WITH CORD

Fig. 16-31. Banding the armature.

core ends, and the shaft must be thoroughly cleaned of all insulation and varnish.

Core ends and shaft must be insulated. The commutator for a large armature merely requires bolting the commutator into place when replacing it. After the commutator is placed on the shaft, the hollow between the core and the back of the commutator must be filled. Insulating paper comes in sheets, strips, or cuffed strips in various thicknesses.

It is impractical to use preformed windings on small armatures. End room is limited and the windings must be drawn up tightly to the armature core. In large production shops the small armatures are machine wound on special equipment designed for the purpose.

A hand winding is defined according to the method by which it is wound. There are five methods of hand winding. Slot wedges are used for securing windings in slots. Leads must be properly identified when winding coils; otherwise making connections will be highly complex, and, in some cases, practically impossible.

Very often a rewinding job is made easier by preforming the coils and then placing them in slots. This may be done in a number of ways.

Slots of armatures and stators are generally open and the coils are usually completely taped. Tape is often used for this purpose, with varnished cambric and mica preferred for large motors and special applications. In medium-sized motors the slots are generally semi-closed.

The placement of preformed coils into the slot is dependent upon the type of slot. Care must be taken so that conductors are not scraped. Wrappings must stay intact and insulations undisturbed. There are semi-closed and open slots in motors.

If the coil leads are properly identified, the process of connection is simplified. With the coil sides in their proper slots, which places them in the proper commutating field after connection, the process involves bringing the leads (lead swing) of the coils to the proper commutator gears (commutator pitch), and connecting each individual coil in proper sequence.

Some open-slot types of armatures require banding to prevent the coils from flying out of the slots while the armature is rotating.

Banding is placed on the front and back ends of the core, the amount of banding depending on the size of the armature.

It is very important to test an armature after winding for shorts, opens, and for grounds. If all the precautions are observed and carried out in the rewinding procedure, the possibility of trouble is minimized.

Review Questions

1. What is the difference between the slots in a large and a medium-sized armature?
2. What determines the placement of preformed coils in slots?
3. Why do some armatures require banding?
4. Why should you test an armature before assembling a motor?
5. What is the most important step in rewinding a motor?
6. What kinds of insulating paper can be purchased for motor rewinding?
7. Why is it impractical to use preformed windings on small armatures?
8. Name the five methods of hand winding an armature.
9. What does the motor rewinder use to secure windings in slots?
10. How are leads identified during the rewinding process?

CHAPTER 17

Stator and Coil Winding

Recording data is one of the most important steps in rewinding a stator or coil. It consists of noticing certain specific information concerning the old winding so that no difficulty will be encountered when the stator or coil is rewound. As much information as possible is recorded before stripping, and the remainder is recorded during the stripping operation.

Another important item is the distance coils protrude from the stator. This is called end room. It is important that the new coils do not extend beyond the slots any further than this distance. If they do, the end bells may press against the coils causing a short circuit, or the end bells may not be able to be replaced at all. A form for recording data for a split-phase motor is shown in Fig. 17-1, and methods of recording the coil pitch data on the form are shown in Fig. 17-2. A form for recording data for a repulsion motor is shown in Fig. 17-3, and typical layout diagrams for recording the data on this form are shown in Fig. 17-4. A form for recording data for a polyphase motor is shown in Fig. 17-5. The information listed above must be recorded in such a manner as to enable any electric motor repairperson to rewind the motor or generator.

SPLIT PHASE MOTOR DATA

MAKE							
HP		RPM		VOLTS		AMPS	
CYCLE		TYPE		FRAME		STYLE	
TEMP		MODEL		SERIAL NO		PHASE	
NO OF POLES		END ROOM			NO OF SLOTS		
LEAD PITCH				COMMUTATOR PITCH			
WIRE INSULATION				WINDING (HAND, FORM AND SKEIN)			
SLOT INSULATION		TYPE		SIZE		THICKNESS	
TYPE CONNECTIONS			SWITCH		LINE		

WINDING	TYPE	SIZE AND KIND WIRE	NO OF CIRCUITS	COIL PITCH	TURNS
RUNNING					
STARTING					

SLOT NO	1 2 3 4 5 6 7 8 9 10 11 12 13 14 15 16 17 18 19 20 21 22 23 24 25 26 27 28 29 30 31 32 33 34 35 36 1
RUNNING	
STARTING	
ROTATION	CLOCKWISE COUNTER CLOCKWISE

Fig. 17-1. Form for recording split-phase motor data.

Data-Taking While Stripping

The importance of observing and recording data during the stripping procedure is not to be minimized. The following methods and procedures will aid in observing and recording the important details after preliminary data have been recorded.

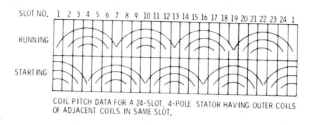

COIL PITCH DATA FOR A 24-SLOT, 4-POLE STATOR HAVING OUTER COILS
OF ADJACENT COILS IN SAME SLOT.

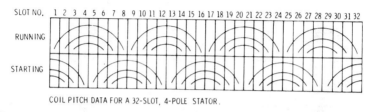

COIL PITCH DATA FOR A 32-SLOT, 4-POLE STATOR.

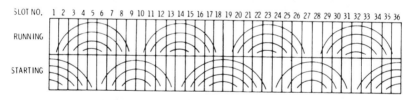

Fig. 17-2. Methods of recording coil pitch for data form of split-phase motor.

Split-Phase Stators

Sometimes it is only necessary to replace the starting winding. If care is exercised when removing the starting winding, the rest of the stator will not need a rewind job. Single-phase motors with more than four poles are a rarity, but if a motor with six or eight poles is encountered, be certain to record the number of poles for the starting winding (before stripping), since they will differ in many instances from that of the running winding. Follow the procedures outlined below for both starting and running windings.

1. Remove all wedges. Use a wedge remover or piece of hacksaw blade with the toothset ground off (Fig. 17-6).
2. Cut through the end turns of the windings of one pole on one side.

REPULSION MOTOR DATA

MAKE			
HP	RPM	VOLT	AMPS
CYCLE	TYPE	FRAME	STYLE
TEMP	MODEL	SERIAL NO	PHASE

ROTOR	BARS	SLOTS	COIL PITCH
LEAD PITCH	TURNS	COILS/SLOT	SIZE & KIND WIRE

WAVE   LAP

COMMUTATOR PITCH.	END ROOM	
WIRE INSULATION	WINDING (HAND, FORM, AND SKEIN)	
SLOT INSULATION	TYPE	SIZE	THICKNESS

TYPE CONNECTIONS	SWITCH	LINE
EQUALIZER	PITCH	

STATOR	POLES	SLOTS	SIZE & KIND WIRE	NO OF CIRCUITS

SLOT NO: 1 2 3 4 5 6 7 8 9 10 11 12 13 14 15 16 17 18 19 20 21 22 23 24 25 26 27 28 29 30 31 32 33 34 35 36 1

Fig. 17-3. Form for recording repulsion motor data.

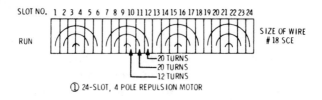

① 24-SLOT, 4 POLE REPULSION MOTOR

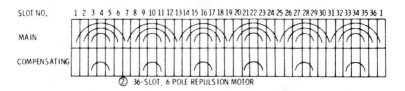

② 36-SLOT, 6 POLE REPULSION MOTOR

Fig. 17-4. Typical layout diagrams for data form of repulsion motor.

3. Count and record the number of wires entering the slots for every coil side of one side of a pole. The location of running-winding pole core slots, with respect to the frame, is to be recorded by punch-marking the center slot or slots of each pole. In many single-phase stator cores, the center of each main pole is identified by a change in slot size. This is sufficient for properly locating the poles when rewinding.

4. Strip the stator by using a pair of longnose pliers on the side of the stator on which the pole windings are not cut (Fig. 17-6). If it is difficult to remove the windings, the procedures for application of heat or caustic soda solution should be followed. Also, the windings may be connected across full-line current. This causes them to overheat and burn away varnishes and insulations, thereby becoming loosened.

MAKE

HP	RPM	VOLTS	AMPS
CYCLE	TYPE	FRAME	STYLE
TEMP	MODEL	SERIAL NO.	PHASE

NO. OF COILS	NO. OF SLOTS	END ROOM
SIZE OF COIL	TURNS PER COIL	NO. OF GROUPS
COILS/GROUP	NO. OF POLES	PITCH OF COIL

SIZE AND KIND OF WIRE	KIND OF WIRE INSULATION

TYPE OF CONNECTIONS: SWITCH   LINE

SLOT INSULATION: TYPE   SIZE   THICKNESS

TYPE OF WINDING (DIAMOND, GANG, AND SPECIAL)

RELATIVE POSITION OF EACH WINDING

TYPE OF WINDING CIRCUIT

Fig. 17-5. Form for recording polyphase motor data.

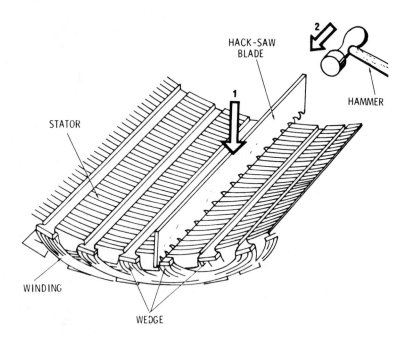

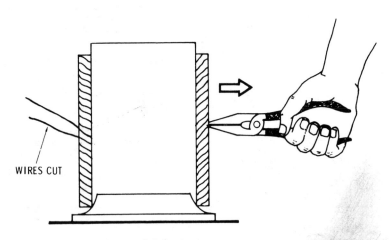

Fig. 17-6. A method of using a piece of hacksaw blade for removing wedges.

Stripping a Motor by Burnout

Burning out motors in ovens in connection with stripping them for rewind is a method long used. Modern ovens made for burn-offs operate on the controlled air principle. By restricting the air flow it is possible to restrict the combustion and temperature. The principle calls for smothering by limiting the oxygen available for burning. Pyrometers are usually attached so that the oven temperature can be monitored. This works with a circuit that will control the oxygen supply automatically when a preset temperature has been reached. When the temperature drops, the circuit calls for more oxygen and a valve is opened again.

All modern burnoff ovens include an integral afterburner. The afterburner is sized to add air to the outlet gas to bring the temperature to 1,400°F (760°C). It takes a temperature this high to completely burn off the insulation and cause it to oxidize. These high temperatures call for an oven and its stack both lined with refractory brick.

Laminations and Burnoff

Motor laminations are sometimes insulated with organic materials. These cannot stand prolonged exposure to elevated temperatures without damage. Unless you have positively identified the lamination insulation, be sure you use a controlled temperature starting at 300°F (149°C). Once a flame is observed, cut back on the air supply to control the slow burn until the insulation is substantially consumed. Temperature control is very important on T-frame stacks, which are more temperature-sensitive than the older ones.

Epoxy-Encapsulated Motors and Burnout

Epoxy-encapsulated motors have a resin that, when it burns, gives off an acrid black smoke and releases a variety of gases. Some of these gases are flammable and explosive. The byproducts include toluene, benzine, and acetone. The end pieces on larger motors are cut off so that the epoxy will not have to be burned off.

The Navy has experienced oven explosions during burnouts of large (over 50 hp) motors on board ships when epoxy motors are being processed.

Repulsion-Type Stators

Follow the same procedure as described above for split-phase stators.

Polyphase Stators

It is important, before stripping a polyphase stator, to examine and record the winding circuits for the particular unit that is being rewound. The use of the connection diagrams found in the appendix is an aid in this respect. After the stator winding circuits and preliminary data have been recorded, the electric motor repairperson can proceed to strip the stator according to the procedure outlined for *Split-Phase Motors*. A form for recording polyphase motor data is shown in Fig. 17-5.

Rewinding Procedure

The procedure for insulating the slots is similar to that used for armatures. It may be necessary to cut and fit $\frac{1}{16}$ inch (1.6-mm) fiber for the core ends and follow the procedure outlined in *Insulating the Core Ends and Shaft*, located in Chapter 16. Use steps number 1 and 3 through 6. The stator is supported on a bench block, its own stand, or an armature stand fitted with stator adaptors.

Coil Forming

Most single-phase stators are hand-wound. In very few instances a single-phase stator may be found that is composed of preformed coils and then, in the majority of cases, they are placed as discussed under *Polyphase Stators*. The various methods used for rewinding single-phase stators are listed below.

1. *Hand winding.* Hand winding may be used for both the starting and running windings. In this method, the wires are

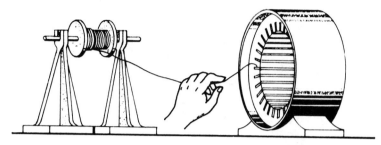

Fig. 17-7. Positioning of stator and wire spool during winding.

placed in the slots, one turn at a time, until the winding operation is completed (Fig. 17-7). With the stator still supported after completing the insulating, proceed as follows:

   a. Set up the reel of proper size wire for the stator on a reel rack with a suitable tensioning device.

   b. Insert the wire into a slot for the start of an inner coil (Fig. 17-8).

   c. Wind into the slot of proper coil pitch for the inner winding according to the recorded data.

   d. Continue to the next larger coil.

   e. Continue to the next larger coil until the entire pole is finished. Wooden dowels may be placed in empty slots to hold coils in position while winding (Fig. 17-9).

   f. Cut the wire and wedge the winding permanently in place if it is a top winding.

   g. If it is a bottom winding, push temporary, loose-fitting wooden or fiber wedges in place for each slot until the stator is removed.

   h. Remove the dowels.

   i. Continue b through h, above, for every pole until the bottom winding of the stator is wound or, if a top winding, the stator is wound completely.

   j. Set up a reel of proper size wire for the starting winding.

   k. Proceed to rewind the starting (tip) winding (after removing the temporary wedges) the same as the running (bottom) winding, making sure that each pole is wound 45° apart from the running winding for a four-pole stator, and 90° apart for a two-pole stator.

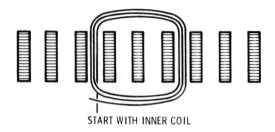

START WITH INNER COIL

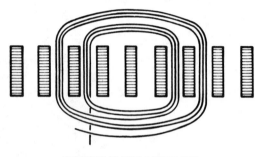

CONTINUE TO NEXT LARGER COIL

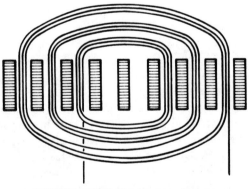

CONTINUE UNTIL ENTIRE POLE IS FINISHED

Fig. 17-8. The procedure for winding stator pole by hand.

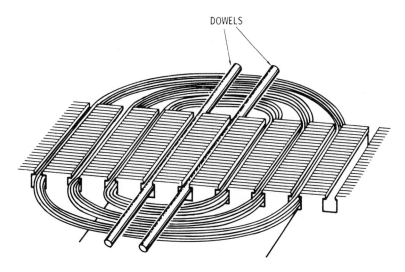

DOWELS

Fig. 17-9. Method of holding coils in position with dowels while winding.

 1. Wedge the windings in tightly.

 2. *Form Winding.* Check Figs. 17-10 and 17-11. Refer to *Form Winding* in Chapter 16.

 3. *Skein Winding.* Check Figs. 17-12 and 17-13. Also refer to Chapter 16 for machine and form winding of coils and their placement.

Polyphase Stators

Most polyphase fractional-horsepower motors are diamond-shaped, mush-wound coils arranged in lap winding. Usually there are as many coils as there are slots, or, in other words, two coil sides per slot, making what is known as a two-layer winding. All the coils are identical in number of turns and size of wire, and the phase coils usually are separated by a strip of insulation (a phase coil is a coil adjacent to a coil of a different phase). The coils are wound or connected in groups. A group is simply a given number of individual coils connected in series, with only two leads brought out. The stator winding is then connected from these groups. In a DC armature lap winding, all the coils are connected in series and a closed winding is formed. In polyphase stators and wound rotors, the coils are connected in a number of groups.

BLOCKS BOLTED TOGETHER

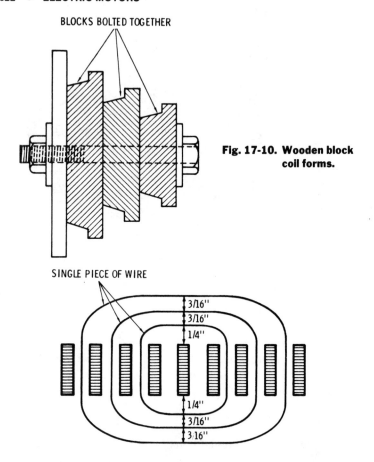

Fig. 17-10. Wooden block
coil forms.

SINGLE PIECE OF WIRE

3/16"
3/16"
1/4"

1/4"
3/16"
3/16"

Fig. 17-11. Use of single turns of wire to determine form and size.

1. *Coils for small stators.* In the wound coils used in fractional-horsepower motors, it is a fairly common practice to wind all the coils of a single group from a continuous strand of wire, if there are the same number of coils in all groups. This simplifies the final connections of a wound stator or rotor. If, however, the number of coils is not the same in all groups, it may be less confusing to bring out leads for each individual coil and to connect them in groups after the stator is wound.

2. *Coils for medium sized stators* (Fig. 17-14). Follow the same procedure in Step 1.

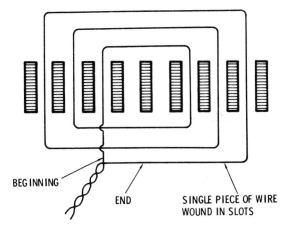

Fig. 17-12. Method of determining the skein size for a pole.

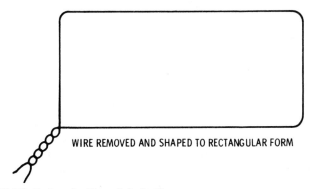

WIRE REMOVED AND SHAPED TO RECTANGULAR FORM

Fig. 17-13. Determination of skein size.

3. *Coils for large sized stators.* Coils for open-slot stators require a special form and must be wound to conform to the shape of the slot. Their sides must be completely rectangular. Such coils are completely taped.

Insulating Coils

The procedure for insulating coils for the various applications is found in Chapter 16 under *Insulating Coils.*

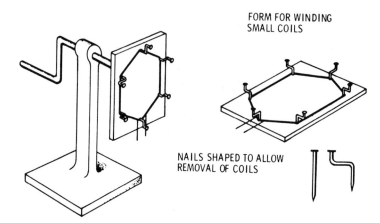

FORM FOR WINDING
SMALL COILS

NAILS SHAPED TO ALLOW
REMOVAL OF COILS

Fig. 17-14. Special form for winding coils of a medium or large-sized motor.

Placing Coils in Slots

Single-Phase Stators

1. *Distributed windings.* For distributed windings, follow the procedure outlined in the section on *Coil Forming* earlier in the chapter.

2. *Skein windings.* The coil is made large enough to be wound into all the slots necessary to complete the individual sections of a pole. The advantage of this method lies in the fact that many conductors may be placed in the slot at one time. After the skein coil is removed from the form, follow the outlined procedure:

   a. Place one end of the skein in the slots of the smallest pitch, as illustrated in 1, Fig. 17-15.
   b. Half-twist the coil in the opposite direction.
   c. Place the free sides of the skein into the slots of the next larger pitch, as illustrated in 2, Fig. 17-15.
   d. Half-twist the coil in the opposite direction.
   e. Place the free sides of the skein into the slots of the next larger pitch, as illustrated in 3, Fig. 17-15.
   f. Tightly wedge the winding.
   g. Proceed to the next pole and follow a through f above.

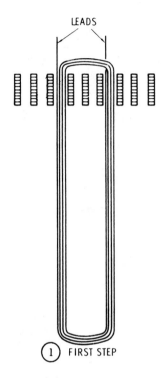

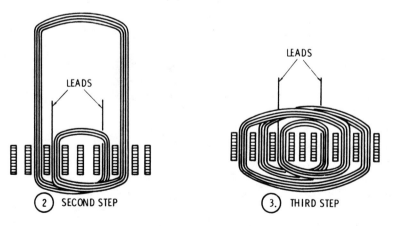

Fig. 17-15. Placing a skein-wound coil in the stator slots.

h. Follow a through f above for every pole until the stator is completed.

i. When a skein coil is placed in the same slot two or three times, follow the same procedure in a through h above, except place double or triple coils in slots as shown in Fig. 17-16.

Polyphase Stator

With one exception, the procedure in the section headed *Placing Armature Coils in Slots* in Chapter 16 can be used for placing coils in stator slots. Because of the spacing limitations inside the stator bore, all the first sides of a coil cannot be placed in the stator. Follow the added procedure outlined below.

1. Place first sides of coils into the slots in the amount of one or two more coils than the pitch of one coil.
2. Place a strip of insulation 0.020 inch back, ¼ inch (6 mm) wider, and 1 inch (25 mm) longer than the slot on top of the first coil side on which the second coil side of another coil, several away (according to coil pitch), is to be placed.
3. Place the second side of a coil (according to coil pitch) into the proper slot on top of the insulation on a first coil side.
4. Place the first side of another coil in the slot adjacent to the slot in which a second coil side has just been placed, and repeat 2 and 3 above.
5. Continue, following 2, 3, and 4 above, until the entire stator is completely wound.
6. Fold in all the slot feeders as described in Chapter 16.
7. Wedge the windings as described in Chapter 16.
8. For open slots using taped coils, it is not necessary to insulate between coil sides.

Connections

The connections must be made compact, mechanically secure, and well insulated. If stranded wires are connected to solid wires, care must be exercised to make certain that no strands are left unconnected or are severed or broken near the connection. Every strand of a stranded conductor must be incorporated in a connection for maximum current-carrying efficiency.

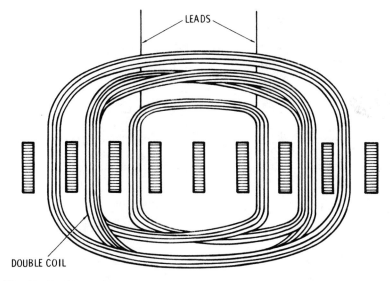

LEADS

DOUBLE COIL

Fig. 17-16. Skein winding with double coil in the center.

Connection of Coils

The procedure for interconnection of coils is given below:

1. *Polarity: Field Poles and Interpoles.* The field poles of a DC motor or generator or of an AC motor, generator, or synchronous motor must be alternately north and south around the pole faces of the machine. Where interpoles are used in DC generators, the interpole following the main pole, in the direction of rotation, is of opposite polarity to the main pole. However, in DC motors, the interpole following the main pole must have the same polarity as the main pole, this being opposite to the corresponding rotation in DC generators. The commutating flux thus obtained from the interpoles, together with the armature flux, produces the proper resultant flux by which sparkless commutation is obtained without shifting brushes (Fig. 17-17).

2. *Consequent Poles.* In some AC motors, the poles are connected so that the adjacent poles have the same polarity. The magnetic effect of this is to produce twice as many magnetic poles as there are wound poles. Each wound pole produces a

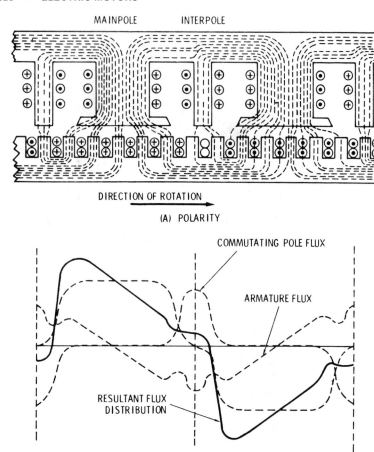

(A) POLARITY

(B) FLUX DISTRIBUTION

Fig. 17-17. Polarity of main poles and interpoles on DC motors.

pole of opposite polarity between the coils as a result of this magnetic flux, Fig. 17-18. When connected in this manner, they are called consequent poles. This principle is used to produce a two-speed motor by arranging the connections between the poles so that, when the switch is thrown in one direction, the wound poles will have alternate polarity, and when thrown in the opposite direction, they will have the same polarity. This principle is also used to double the number of poles in the starting winding of a split-phase motor.

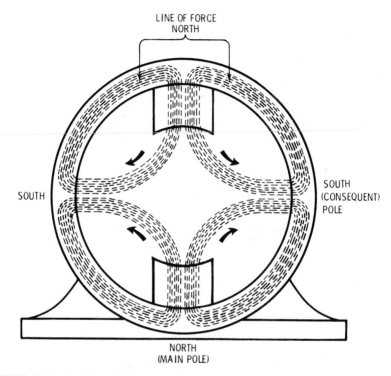

LINE OF FORCE
NORTH

SOUTH

SOUTH
(CONSEQUENT)
POLE

NORTH
(MAIN POLE)

Fig. 17-18. Polarity of a consequent-pole motor.

3. *Checking for Proper Connections.* Proper connections are directly related to proper polarity.

4. *Magnetic Compass Method of Checking Polarity of a Consequent-Pole Motor.* Fig. 17-19 shows a schematic diagram of a typical two-speed, split-phase motor with the running winding connected series-consequent.

   a. Connect line terminals A and C to a low-voltage DC source (about 20 percent of operating current). This connects the four coils of the running winding in series with the current flowing in the direction indicated by the arrows. The starting winding should be left open at B, D, and the centrifugal.

   b. Magnetic compass needle placed alternately at points 1 to 4, Fig. 17-19, should show similar polarity.

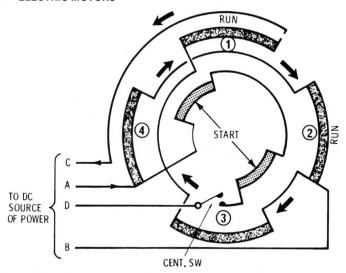

Fig. 17-19. Checking polarity of a consequent pole motor.

    c. If a similar polarity is not registered at the four poles, check to make sure windings are not reversed, and make proper connections.

Connection of Internal Switches

Some internal switches come equipped with binding screws. When making connection to these, be certain that the end of the conductor is bent around the screw in the direction in which the screw tightens, and that the screw is driven in as tight as possible without stripping the threads. Other internal switches are equipped with soldering lugs or leads. It is important at this time to use for reconnection the diagram developed during the disassembly of the unit.

Capacitor Types

Some of the many types of capacitor motors are listed below. Some of these types are designed to operate on one voltage, others on two voltages. Many of them are externally reversible; others can only be reversed internally. In any case, motor reversal on split-phase induction motors can be obtained only by reversing the relationship of the starting winding and the running winding, whether done externally or internally. The capacitor for starting is in series with the starting winding and, if and when changes are made for reversal of direction, it must be remembered to place the capacitor in series with the starting winding at some point of connection. The various types of capacitor-start motors are:

1. Single-voltage, externally reversible.
2. Single-voltage, nonreversible.
3. Single-voltage, reversible, with thermostat.
4. Single-voltage, nonreversible, with magnetic switch.
5. Two-voltage, reversible.
6. Two-voltage, nonreversible.
7. Two-voltage, reversible, with thermostat.
8. Single-voltage, three-lead, reversible.
9. Single-voltage, instantly reversible.
10. Single-voltage, nonreversible.
11. Two-speed.
12. Two-speed, two-capacitor.

Use of Electrolytic Capacitors

In all of the listed types, electrolytic capacitors are used. The two-speed, two-capacitor type is virtually two motors. In this type of motor there are two starting windings and two running windings. Each of the capacitors is connected in series with each of the starting windings and with a centrifugal switch common for both windings. At this time, the diagram developed during disassembly of a unit is used for reconnection at the time of motor re-assembly.

Testing

It is important to test a stator, after winding, for shorts, opens, or grounds. If all the precautions have been observed and carried out in the rewinding procedure, the possibility of trouble is minimized.

Summary

Recording data is one of the most important steps in rewinding a stator or coil. Forms are available for getting the correct information recorded. Single-phase motors with more than four poles are a rarity, but there are motors with six and eight poles. They deserve special attention before stripping for a rewind job.

The procedure for insulating the slots in a stator of a single-phase motor is similar to that used for armatures. Most single-phase motors are hand-wound. There are however, a number of ways to rewind single-phase motors. Form winding can be used, as can skein winding.

Most polyphase fractional-horsepower motors are diamond-shaped, mush-wound coils arranged in lap winding. Connections are made compact, mechanically secure, and well insulated. If stranded wires are connected to solid wires, care must be exercised to make certain that no strands are left unconnected, or are severed or broken near the connection.

There are eleven types of capacitor-start motors. Some are reversible and some are not. One type is virtually two motors in one frame since it has two capacitors and two starting windings.

Review Questions

1. What is the most important step in rewinding a stator or coil?
2. How many poles do single-phase motors usually have?
3. Why are most single-phase motors hand-wound?
4. What are the other types of winding which can be used to wind a single-phase motor?
5. What is a skein winding?
6. What three things should a connection be checked for?
7. What is a consequent pole?
8. List the eleven types of capacitor-start motors.

CHAPTER 18

Insulating and Reinsulating

Normally, the insulated windings of a motor or generator require reinsulation when they appear dry, brittle, and faded. Reinsulation in this case takes the form of revarnishing. For small motors and generators, use a synthetic varnish of the baking type. The windings of larger units take black baking varnish of the asphalt type. While quite satisfactory for insulating windings in small motors and generators, hard varnish should not be used for large-unit windings because of the risk of damaging the winding beyond use when it is removed from the frame. The following paragraphs give the procedure for preparing, applying, and baking both synthetic and asphalt varnishes. Newer materials increase the working temperature abuse a motor can handle. Laminates, sheets, tapes, and varnishes have all been improved.

Preparation

Prepare the varnish for application by checking its temperature, cleanliness, and viscosity.

1. Check the temperature in the tank. If the varnish is too hot, its thinning agents will dissolve and a skin will form on the surface, indicating that the varnish is too thick. A varnish that is too cold also becomes thick and does not spread well.
2. Make certain that the varnish is thoroughly mixed and free of dirt, skin, or other foreign matter. Strain the varnish through a fine wire mesh or several layers of cheesecloth if necessary.
3. The viscosity rate of insulating varnish must not exceed 70 seconds or be less than 20 seconds, as measured at 25°C (77°F) by viscosity cup. If the varnish is too thick for a given temperature, thin it with a diluting agent (thinner). Be careful not to add thinner to the varnish too rapidly lest the varnish become too thin for use.

Application

The most efficient method of applying insulating varnish to insulated windings is by dipping. With individual windings, this is a simple matter. However, special handling is required when dipping armatures and rotors.

1. *Insulating Windings.* Individual windings can be insulated by simple immersion in a varnish tank (Fig. 18-1). Splicing connections should be prepared as shown in Fig. 18-2, prior to immersion.
2. *Armatures and Rotors.* Cap, mask, or strip such parts of the armature or rotor that are to be kept free of varnish and allow sufficient time for stripping compound to dry before dipping. Suspend the armature or rotor, with the commutator or collector-ring end up, and lower into the tank until the varnish reaches the bottom of the commutator risers (Fig. 18-3). Keep it immersed for about 20 minutes, or until bubbling ceases. Subsequent dips should last about 10 minutes. If a rotor or armature is too large to be immersed, place the varnish in a pan and lower the winding, keeping

Fig. 18-1. A method of dipping individual coil windings into a tank of insulating varnish.

the shaft horizontal, until the varnish covers the bottom of the coil slots (Fig. 18-4). Then rotate the winding slowly until all parts are well covered and penetrated. After pan dipping, spray varnish heavily on front and back ends of the windings and on the back of the commutator risers not covered by dipping. Drain the winding, with commutator or collector-ring end up, for at least 30 minutes, or until dripping ceases. Room temperature should not be less than 20°C (68°F).

Baking

Windings of units of 200 horsepower or less are baked in an oven at from 130°C to 135°C (266° to 275°F). If only one dipping and one baking are required, bake the winding from 15 to 16 hours. More than one dipping requires baking from 6 to 8 hours

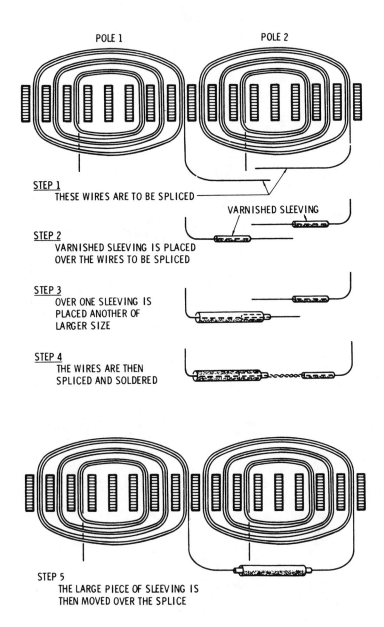

POLE 1

POLE 2

STEP 1
THESE WIRES ARE TO BE SPLICED

VARNISHED SLEEVING

STEP 2
VARNISHED SLEEVING IS PLACED
OVER THE WIRES TO BE SPLICED

STEP 3
OVER ONE SLEEVING IS
PLACED ANOTHER OF
LARGER SIZE

STEP 4
THE WIRES ARE THEN
SPLICED AND SOLDERED

STEP 5
THE LARGE PIECE OF SLEEVING IS
THEN MOVED OVER THE SPLICE

Fig. 18-2. A method of insulating a splicing connection.

Fig. 18-3. A method of dipping armature into tank of insulating varnish.

for each dip except the last, which needs at least 16 hours. Some windings necessitate a longer baking period in order to completely cure the varnish.

CAUTION: Make sure that the baking oven is provided with some means of circulating air and of removing the vapors.

Number of Dips and Bakes Required

For general-purpose units up to 200 horsepower (149.2 kW) use one dip and bake for armatures, rotors, or stators with formed coils, and two dips and bakes for mush (random-wound) coils. When operating conditions are severe and resistance to high humidity or acid is required, stators, rotors, or armatures should be

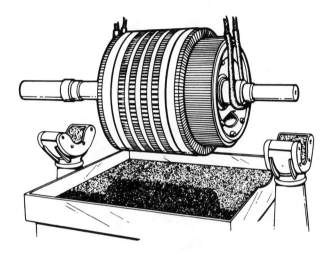

Fig. 18-4. A method of applying insulating varnish to armature windings.

given four dips and bakes. For general-purpose units, use two dips and bakes for an original winding or a rewound unit, one dip and bake for an old armature, not rebanded, and two dips and bakes for an old armature, rebanded.

Reinsulating by Pressure-Impregnation Method

Open-slot coils are, as a rule, thoroughly treated by vacuum- or pressure-impregnation, or by multiple dipping and baking, before being placed into the slots. All winding material used with these coils is also treated. Therefore, except in special cases, open-slot constructions do not require pressure impregnations. When operating conditions are such that vacuum- and pressure-impregnation is advisable and the necessary equipment is available, the instructions below should be followed. Use of the process is customarily limited to railway and relatively small, high-speed motors.

1. Place the properly supported stator, rotor, or armature into the tank. An armature should be supported in such a manner that it will stand vertically. Cap, mask, or strip armature or rotor parts.
2. Pump varnish into the tank until the varnish is approximately $1/2$ inch (12 mm) above the unit part or, in the case of an armature, until the varnish rises to about $1/4$ inch (6 mm) on the commutator risers.
3. Close the lid of the tank and tighten down the lid clamps.
4. Draw nitrogen directly from the cylinder through suitable pressure-reducing and regulating devices.
5. Apply the compressed nitrogen at a pressure of from 75 psi to 80 psi for about 1½ to 2 hours.
6. Force the varnish back into the varnish storage tank.
7. Relieve pressure to zero psi, loosen the lid clamps, raise the lid, and remove the unit part.
8. Drain the motor or generator part for at least 30 minutes.
9. Place in the baking oven. Follow previous baking instructions.

Precautions and Gas Substitution

1. In the absence of nitrogen gas, carbon dioxide can be used in gaseous form.
2. The use of oxygen or compressed air instead of gases specified is exceedingly dangerous and may result in fires and explosions.

   CAUTION: Threads of the valves of oxygen cylinders are the same as those of carbon dioxide cylinders, so the danger of connecting to an oxygen cylinder is present. Check before making connections.

3. When large volumes of carbon dioxide gas are required within a short time, position the cylinders of liquid carbon dioxide with the valves down, and vaporize in suitable equipment.

New Developments

Electric motors can now operate at higher temperatures because of the development of insulation systems capable of handling temperature extremes. The trend in motors is for reduced size and weight. This causes more heat to be generated by the operation of the motor. This higher heat must be handled by the insulation system without allowing motor life to be reduced significantly. The future may see a change inasmuch as aluminum may be used as the winding wire and the motor size will have to be increased to handle the larger (physical) size conductors. This will probably create a motor with more durability.

The insulation system, once the weakest link in the motor, no longer limits the designer of industrial motors. Newer materials may well be developed to allow for even longer life and the reduction of heat generated by motor operation.

Insulation Films and Sheets

The films and sheets available are of the same materials and kinds of plastics used for magnet wire enamels and varnishes. These sheets and films may run from a few mils in thickness up to about 1/16 inch. The thicker sheets are used for slot insulation on some random-wounds. The thicker sheets are used to wind around the lead wires.

Sheets are usually cast from unfilled polymers. Powdered mica is sometimes added to upgrade the electrical qualities of the sheets. When the mica is added, the physical qualities of the sheets sometimes suffer. Sheets made for ground insulation are made from Mylar®, in most cases with some additives. The film most commonly used on wire is Kapton®, which was introduced in the 1960s as a replacement for mica tapes on DC motors operating at up to 1,000 volts.

Laminates

Preformed laminates are popular. They increase strength by adding layers of insulation. The laminates are formed by cutting any number of insulations to length and placing them on top of one another. These are usually slot fillers, top sticks, and miscellaneous materials used for bracing. Laminates are available as rod, sheet, bar, and angle stock and are made to controlled dimensions. Coil insulation, such as laminates, must withstand winding abuse.

Tapes

Tapes are made for rewinds, for insulating end turns, and for tie cords. The latest tapes are pressure-sensitive adhesive (PSA). Occasionally they are used throughout the motor as winding aids to retain components.

PSA tapes are made of a layer of adhesive, a layer of primer, a layer of substrate, and a top or release coating. The substrate may be paper, cloth, plastic film, or a composite. The main advantage of the film portion of the tape is to increase tear resistance.

Mica tapes or papers are used on high-voltage machines. Glass has the most strength; mica is selected on the basis of considerations other than physical strength. Structural considerations are the primary concern for DC rotor coils. Unidirectional epoxy or polyester-impregnated glass fiber, or special tapes with cross-threads for added strength, are placed over the windings in some designs. The tape is also used around cores and to replace the metal banding of older days. The banding is applied under pressure to the required thickness and is then cured in an oven.

Varnishes

Electrical varnishes are used mainly to provide adhesion and increase buildup, and thus upgrade overall performance. They also serve to protect sensitive elements from the contaminants located in the environment in which the motor must operate.

Early electrical varnishes were oleoresinous and were cured by oxidation rather than heat. They were good for a reasonable time at temperatures that did not exceed 90°C (194°F).

Today's electric motor can be insulated with varnishes that have thermal stability for 100,000 hours, which is about 11½ years, at 220°C (428°F). This special insulation varnish, a polyamide, is used for small, tough-duty motors that must cope with serious environmental factors in the field. It can be found at service centers where motors are rewound.

Magnet Wire

For insulating magnet wire, the best choice for 180°C (356°F) is a coating of Armored Poly-Thermaleze 2000. This is an aramidimide over a thermosetting polyester. It will provide service for up to Class H motors.

The best and most expensive selection for insulation is the polyamide, provided the magnet wire coating is compatible. This particular process requires a number of dips and bakes. It should only be used for high-temperature polyester or aramidimide should be used for thermal stability. For best adhesion and chemical resistance, use an epoxy resin.

For most motors, solventless resins are preferred to varnishes for the aftertreatment.

Thermosetting plastics such as epoxies and polyesters are used in magnet wire coatings, varnishes, and 100 percent solid formulations for operations rated through Thermal Index (TI) 155. Crystalline thermoplastics are used in equipment through Class F. These crystalline thermoplastics include nylon-type polyamides and Mylar®-type saturated polyesters. Class C operation (motors operating in hot ambients and under high overloads) may be handled by a number of linear, aromatic thermoplastics based on aramids and polyamides. These are available as magnet wire coatings, varnishes, films, and fabrics.

Plastics for higher voltage machines are not used. They use mica tapes and papers instead. These will upgrade the operating class of the motor.

Summary

Most motors need reinsulation when their insulation windings appear dry, brittle, and faded. The windings of larger units take black baking varnish of the asphalt type. Hard varnish is satisfactory for use in small windings. The most efficient method of applying insulating varnish to insulated windings is by dipping. Armatures and rotors can also be dipped if properly handled.

Baking of dipped windings cures the varnish or insulating material. Some armatures, rotors, or coils require more than one dipping and more than one baking. The last baking is for about 16 hours and cures the varnish or insulating material.

The pressure-impregnation method of reinsulating is used to speed up the process of impregnation. It doesn't take as long as dipping. However, it should be remembered that the pressure can become dangerous and the wrong materials or pressure regulators can mean trouble.

Motors are decreasing in size and weight, but increasing in heat produced. Many insulation and wire-coating materials, developed to handle the heat buildup, have been put into use recently. They include the epoxies and polyamides. New tapes use mica and Mylar®. Newer methods of insulating the finished product have also been developed.

Review Questions

1. How can you tell when a motor or generator needs reinsulating?
2. What is used to reinsulate motors?
3. What methods are used to apply insulating materials to armatures, rotors, stators?
4. What is the most efficient method of applying insulating varnish to insulated windings?
5. What is the purpose of baking a winding after it has been dipped?
6. What was once the weakest link in the motor?
7. How thick are the sheets and films used for insulation?
8. What is a laminate?
9. What are preformed laminates?
10. What are tapes used for?
11. What are PSA tapes?
12. What is the purpose of an electrical varnish?
13. What is the best coating for a magnet wire to be used in a motor operating under high temperatures?
14. What is a thermosetting plastic?
15. Where are crystalline thermoplastics used in motors?

Appendix

Motor Connections

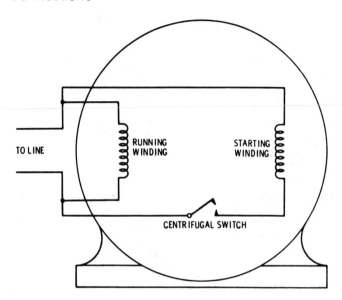

Fig. A-1. Schematic wiring diagram of a split-phase motor.

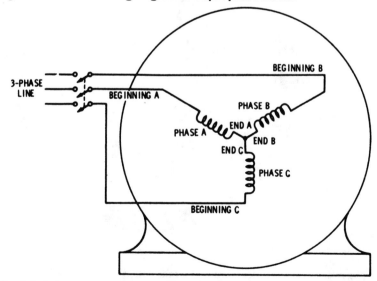

Fig. A-2. Schematic wiring diagram of a star-connected, polyphase motor.

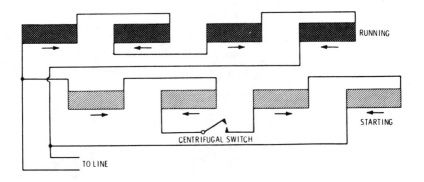

Fig. A-3. Block diagram (extended type) of a four-pole, split-phase motor.

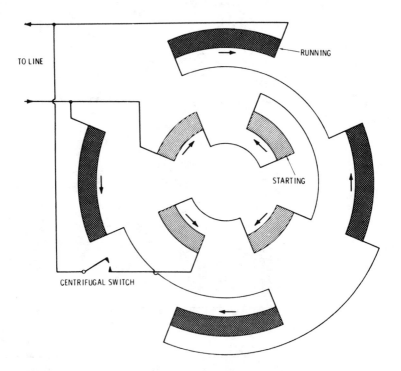

Fig. A-4. Circular block diagram of a four-pole, split-phase motor.

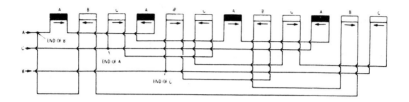

Fig. A-5. Block diagram (extended type) of three-phase, four-pole, series-delta motor.

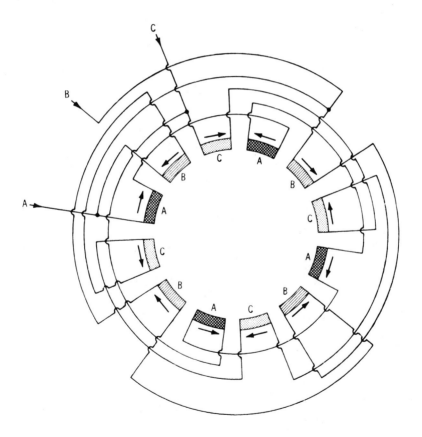

Fig. A-6. Circular block diagram of a four-pole, three-phase, series-delta motor.

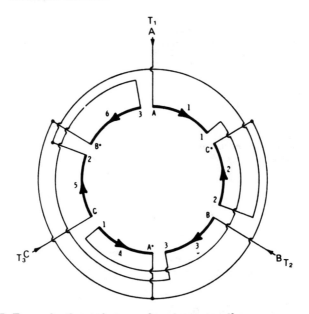

Fig. A-7. Two-pole, three-phase, series-star connection.

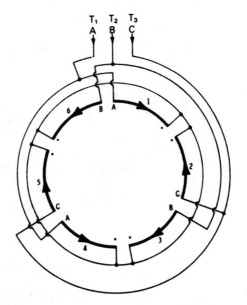

Fig. A-8. Two-pole, three-phase, two-parallel-star connection.

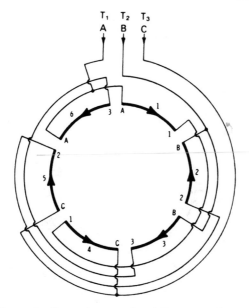

Fig. A-9. Two-pole, three-phase, series-delta connection.

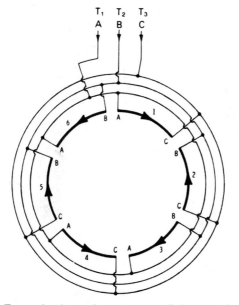

Fig. A-10. Two-pole, three-phase, two-parallel connection.

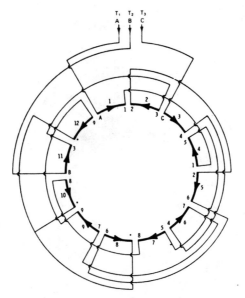

Fig. A-11. Four-pole, three-phase, series-star connection.

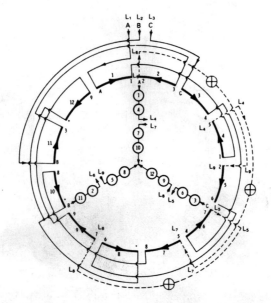

Fig. A-12. Four-pole, three-phase, with three or nine leads for series-star
or two-parallel-star connection.

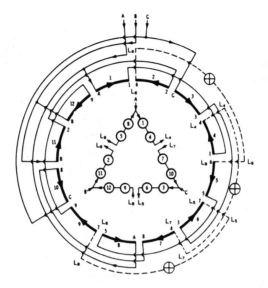

Fig. A-13. Four-pole, three-phase, with three or nine leads for series-delta or two-parallel-delta connection.

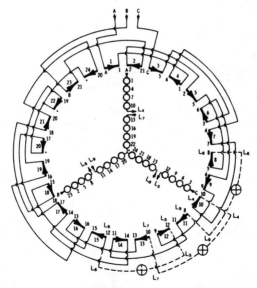

Fig. A-14. Eight-pole, three-phase, with three or nine leads for series-star or parallel-star connection.

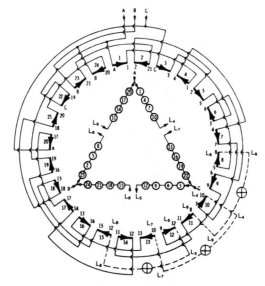

Fig. A-15. Eight-pole, three-phase, with three or nine leads for series-delta or parallel-delta connection.

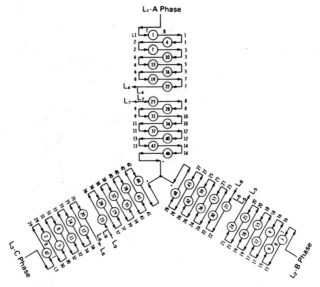

Fig. A-16. Sixteen-pole, three-phase, nine-lead, series-star or two-parallel-star connection.

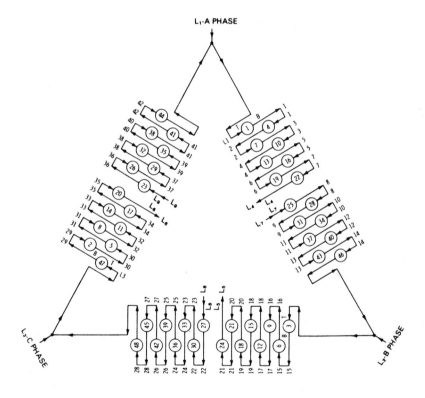

Fig. A-17. Sixteen-pole, three-phase, nine-lead, series-delta, or two-parallel-delta connection.

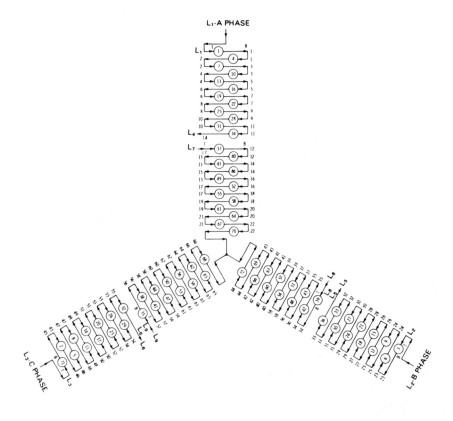

Fig. A-18. Twenty-four-pole, three-phase, nine-lead, series-star, or two-parallel-star connection.

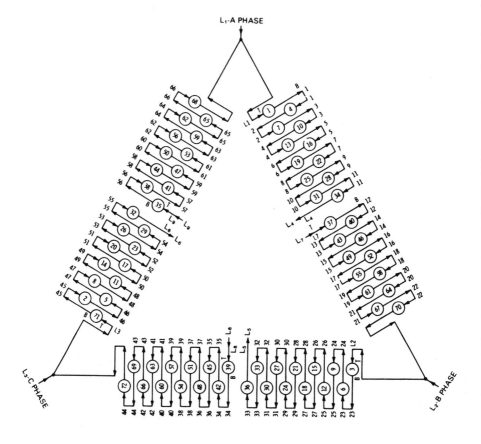

Fig. A-19. Twenty-four-pole, three-phase, nine-lead, series-delta, or two-parallel-delta connection.

Symbols Used in Motor Schematics

1.

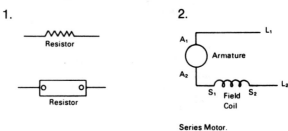

Series Motor.

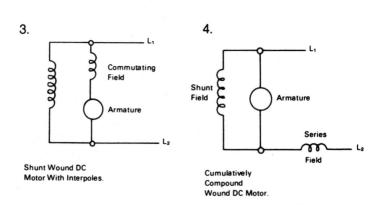

Shunt Wound DC
Motor With Interpoles.

Cumulatively
Compound
Wound DC Motor.

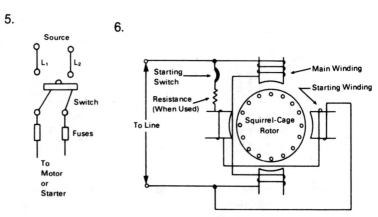

Location of
Switch in Line.

Split-Phase Induction Motor.

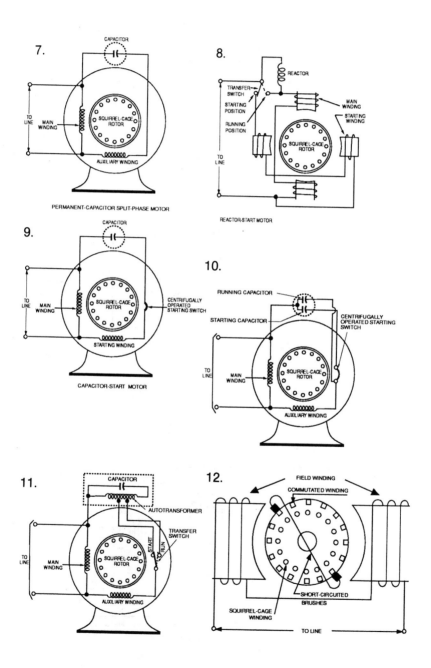

7. PERMANENT-CAPACITOR SPLIT-PHASE MOTOR

8. REACTOR-START MOTOR

9. CAPACITOR-START MOTOR

10.

11.

12.

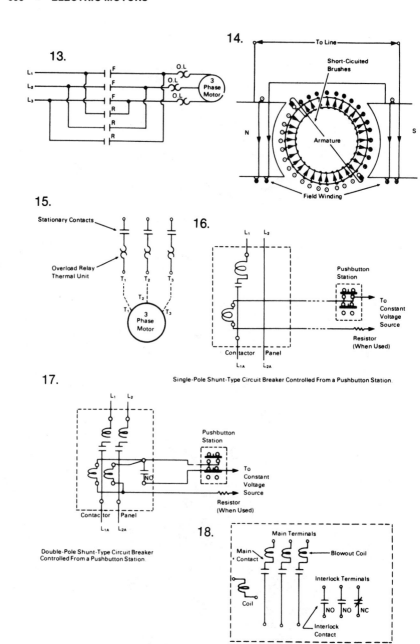

13.

14.

Single-Pole Shunt-Type Circuit Breaker Controlled From a Pushbutton Station.

15.

16.

17.

Double-Pole Shunt-Type Circuit Breaker Controlled From a Pushbutton Station.

18.

Three-Pole Magnetic Contactor With Two Normally Open (NO) and One Normally Closed (NC) Electrical Interlocks. Contactor Shown in its De-energized Position.

Metric Conversion Table

Fractions	Inches	mm	Fractions	Inches	mm
$1/64$	.0156	.3969	$33/64$	.5156	13.097
$1/32$	.0312	.7937	$17/32$	.5312	13.494
$3/64$	.0468	1.191	$35/64$	.5468	13.891
$1/16$	.0625	1.588	$9/16$	.5625	14.288
$5/64$	.0781	1.984	$37/64$	.5781	14.684
$3/32$	.0937	2.381	$19/32$	.5937	15.081
$7/64$	.1093	2.778	$39/64$	.6093	15.478
$1/8$	.125	3.175	$5/8$	.625	15.875
$9/64$	.1406	3.572	$41/64$	.6406	16.272
$5/32$	.1562	3.969	$21/32$	.6562	16.669
$11/64$	.1718	4.366	$43/64$	.6718	17.066
$3/16$	.1875	4.763	$11/16$	.6875	17.463
$13/64$	.2031	5.159	$45/64$	.7031	17.859
$7/32$	.2187	5.556	$23/32$	.7187	18.256
$15/64$	.2343	5.953	$47/64$	.7343	18.653
$1/4$	.25	6.350	$3/4$	.75	19.050
$17/64$	.2656	6.747	$49/64$	.7656	19.447
$9/32$	.2812	7.144	$25/32$	.7812	19.844
$19/64$	.2968	7.541	$51/64$	.7969	20.241
$5/16$	.3125	7.938	$13/16$	.8125	20.638
$21/64$	.3281	8.334	$53/64$	.8281	21.034
$11/32$	.3437	8.731	$27/32$	.8437	21.431
$23/64$	.3593	9.128	$55/64$	.8593	21.828
$3/8$	.375	9.525	$7/8$	.875	22.225
$25/64$	.3906	9.922	$57/64$	.8906	22.622
$13/32$	.4062	10.319	$29/32$	.9062	23.019
$27/64$	.4219	10.716	$59/64$	.9218	23.416
$7/16$	.4375	11.113	$15/16$	.9375	23.813
$29/64$	.4531	11.509	$61/64$	.9531	24.209
$15/32$	.4687	11.906	$31/32$	.9687	24.606
$31/64$	.4843	12.303	$63/64$	.9843	25.003
$1/2$	.5	12.700			

1 mm = .03937 inches
1 inch = 25.4 mm
1 meter = 3.2809 feet

SI (Metric) Conversion Table

	SI Unit	Imperial/Metric to SI	SI to Imperial/Metric
Length	meter (m)	1 inch = 2.54 × 10^{-2}m 1 foot = 0.305 m 1 yard = .914 m	1 m = 39.37 inches = 3.281 feet = 1.094 yards
Mass	kilogram (kg.)	1 ounce (mass) = 28.35 × 10^{-3} kg. 1 pound (mass) = 0.454 kg. 1 slug = 14.59 kg.	1 kg. = 35.27 ounces = 2.205 pounds = 68.521 × 10^{-3} slug
Area	square meter (m^2)	1 sq. in. = 6.45 × $10^{-4}m^2$ 1 sq. ft. = 0.93 × $10^{-1}m^2$ 1. sq. yd. = 0.836 m^2	1 m^2 = 1550 sq. in. = 10.76 sq. ft. = 1.196 sq. yd.
Volume	cubic meter (m^3)	1 cu. in. = 16.3 × $10^{-6}m^3$ 1 cu. ft. = 0.028 m^3	1 m^3 =6.102 × 10^4 cu. in. = 35.3 cu. ft.
Time	second (s)	same as Imperial/Metric	same as Imperial/Metric
Electric Current	ampere (A)	same as Imperial/Metric	same as Imperial/Metric
Plane Angle	radian (rad.)	1 angular deg. = 1.745 × 10^{-2}rad. 1 revolution = 6.283 rad.	1 r. = 57.296 rad.
Frequency	hertz (Hz.)	1 cycle/sec. = 1 Hz.	1 Hz. = 1 cps
Force	newton (N)	1 oz. (f) = 0.278 N 1 lb. (f) = 4.448 N 1 kilopond = 9.807 N 1 kgf = 9.807 N	1 N = 3.597 oz. (f) = 0.225 lb. (f) = 0.102 kp = 0.102 kgf
Energy (Work)	joule (J)	1 Btu = 1055.06 J 1 kWh = 3.6 × 10^6 J 1 Ws = 1 J 1 kcal = 4186.8 J	1 J = 9.478 × 10^{-4}Btu = 2.778 × 10^{-7}kWh = 1 Ws = 2.389 × 10^{-4}kcal
Power	watt (W)	1 hp (electric) = 746 W	1 W = 1.341 × 10^{-3}hp (electric)
Quantity of Electricity	coulomb (C)	same as Imperial/Metric	same as Imperial/Metric
EMF	volt (V)	same as Imperial/Metric	same as Imperial/Metric
Resistance	ohm (Ω)	same as Imperial/Metric	same as Imperial/Metric
Electric Capacitance	farad (F)	same as Imperial/Metric	same as Imperial/Metric
Electric Induction	henry (H)	same as Imperial/Metric	same as Imperial/Metric
Magnetic Flux	weber (Wb)	1 line = 10^{-8} Wb 1 Mx = 10^{-8} Wb 1 Vs = 1 Wb	1 Wb = 10^8 lines =10^8 Mx = 1 Vs
Magnetic Flux Density	tesla (T)	1 line/$in.^2$ = 1.55 × 10^{-5}T 1 gauss = 10^{-4}T	1 T = 6.452 × 10^4 lines/$in.^2$ = 10^4 gauss
Linear Velocity	meter/sec. (m/s)	1 inch/sec. = 2.54 × 10^{-2}m/s 1 mph = 1.609 km/s	1 m/s = 39.37 in./sec. = 3.281 ft./sec.
Linear Accel.	meter/$sec.^2$(m/s^2)	1 in./$sec.^2$ = 2.54 × $10^{-4}m/s^2$	1 m/s^2 = 39.37 in./$sec.^2$
Torque	newtonmeter (N·m)	1 lb. ft. = 1.356 N·m 1 oz. in. = 7.062 × 10^{-3}N·m 1 kilopondmeter = 9.807 N·m 1 lb. in. = 0.113 N·m	1 N·m = 0.738 lb. ft. = 8.851 lb. in. = 0.102 kpm = 141.61 oz. in.
Temperature	degree Celsius (°C)	F = (C × $^9/_5$) + 32	C = (F − 32) × $^5/_9$

NOTE: There is at press time no international equivalent for Revolutions Per Minute (RPM). Commonly used expressions are: RPM = r/min = t/min = U/min = Rev/min = min^{-1}
(f) = force

Horsepower to kW Equivalents	
Horsepower	**Kilowatt (kW)** [*]
$\frac{1}{20}$	0.025
	0.035
	0.05
	0.071
$\frac{1}{8}$	0.1
$\frac{1}{6}$	0.14
$\frac{1}{4}$	0.2
$\frac{1}{3}$	0.28
$\frac{1}{2}$	0.4
1	0.8
$1\frac{1}{2}$	1.1
2	1.6
3	2.5
5	4.0
7.5	5.6
10	8.0

[*]James W. Polk, *A Preview of Metric Motors,* Westinghouse Electric Corporation.

Horsepower/Watts vs. Torque Conversion Chart

hp	watts	@1125 rpm		@1200 rpm		@1425 rpm	
		Oz.-in.	mN·m	Oz.-in.	mN·m	Oz.-in.	mN·m
1/2000	0.373	0.4482	3.1649	0.4202	2.9670	0.3538	2.4986
1/1500	0.497	0.5976	4.2198	0.5602	3.9561	0.4718	3.3314
1/1000	0.746	0.8964	6.3297	0.8403	5.9341	0.7077	4.9971
1/750	0.994	1.1951	8.4396	1.1205	7.9121	0.9435	6.6628
1/500	1.49	1.7927	12.6594	1.6807	11.8682	1.4153	9.9943
1/200	3.73	4.4818	31.6485	4.2017	29.6705	3.5383	24.9857
1/150	4.97	5.9757	42.1980	5.6023	39.5608	4.7177	33.3142
1/100	7.46	8.9636	63.2970	8.4034	59.3409	7.0765	49.9713
1/75	9.94	11.9515	84.3960	11.2045	79.1212	9.4354	66.6284
1/70	10.70	12.8052	90.4243	12.0048	84.7727	10.1093	71.3876
1/60	12.40	14.9393	105.4950	14.0056	98.9015	11.7942	83.2855
1/50	14.90	17.9272	126.5940	16.8068	118.6818	14.1531	99.9426
1/40	18.60	22.4090	158.2425	21.0085	148.3523	17.6913	124.9283
1/30	24.90	29.8787	210.9899	28.0113	197.8031	23.5884	166.5710
1/25	29.80	35.8544	253.1879	33.6135	237.3637	28.3061	199.8852
1/20	37.30	44.8180	316.4849	42.0169	296.7046	35.3827	249.8565
1/15	49.70	59.7574	421.9799	56.0225	395.6061	47.1769	333.1420
1/12	62.10	74.6967	527.4748	70.0282	494.5077	58.9711	416.4275
1/10	74.6	89.6361	632.9698	84.0338	593.4092	70.7653	499.7130
1/8	93.2	112.0451	791.2123	105.0423	741.7615	88.4566	624.6413
1/6	124.0	149.3934	1054.9497	140.0563	989.0153	117.9422	832.8550
1/4	186.0	224.0902	1582.4245	210.0845	1483.5230	176.9133	1249.2825
1/3	249.0	298.7869	2109.8994	280.1127	1978.0307	235.8844	1665.7101
hp	watts	@1500 rpm		@1725 rpm		@1800 rpm	
		Oz.-in.	mN·m	Oz.-in.	mN·m	Oz.-in.	mN·m
1/2000	0.373	0.3361	2.3736	0.2923	2.0640	0.2801	1.9780
1/1500	0.497	0.4482	3.1648	0.3897	2.7520	0.3735	2.6374
1/1000	0.746	0.6723	4.7473	0.5846	4.1281	0.5602	3.9561
1/750	0.994	0.8964	6.3297	0.7794	5.5041	0.7470	5.2747
1/500	1.490	1.3445	9.4945	1.1692	8.2561	1.1205	7.9121
1/200	3.730	3.3614	23.7364	2.9229	20.6403	2.8011	19.7803
1/150	4.97	4.4818	31.6485	3.8972	27.5204	3.7348	26.3737
1/100	7.46	6.7227	47.4727	5.8458	41.2806	5.6023	39.5606
1/75	9.94	8.9636	63.2970	7.7944	55.0409	7.4697	52.7475
1/70	10.70	9.6039	67.8182	8.3512	58.9723	8.0032	56.5152
1/60	12.40	11.2045	79.1212	9.7431	68.8011	9.3371	65.9344
1/50	14.90	13.4454	94.9455	11.6917	82.5613	11.2045	79.1212
1/40	18.60	16.8068	118.6818	14.6146	103.2016	14.0056	98.9015
1/30	24.90	22.4090	158.2425	19.4861	137.6021	18.6742	131.8687
1/25	29.80	26.8908	185.9909	23.3833	165.1226	22.4090	158.2425
1/20	37.3	33.6135	237.3637	29.2292	206.4032	28.0113	197.8031
1/15	49.7	44.8180	316.4849	38.9722	275.2043	37.3484	263.7374
1/12	62.1	56.0225	395.6061	48.7153	344.0053	46.6854	329.6718
1/10	74.6	67.2270	474.7274	58.4583	412.8064	56.0225	395.6061
1/8	93.2	84.0338	593.4092	73.0729	516.0080	70.0282	494.5077
1/6	124.0	112.0451	791.2123	97.4305	688.0107	93.3709	659.3436
1/4	186.0	168.0676	1186.8184	146.1458	1032.0160	140.0563	989.0153
1/3	249.0	224.0902	1582.4245	194.8610	1376.0213	186.7418	1318.6871

Horsepower/Watts vs. Torque Conversion Chart (Cont'd)

hp	watts	@3000 rpm		@3450 rpm		@3600 rpm	
		Oz.-in.	mN·m	Oz.-in.	mN·m	Oz.-in.	mN·m
1/2000	0.373	0.1681	1.1868	0.1461	1.0320	0.1401	0.9890
1/1500	0.497	0.2241	1.5824	0.1949	1.3760	0.1867	1.3187
1/1000	0.746	0.3361	2.3736	0.2923	2.0640	0.2801	1.9780
1/750	0.994	0.4482	3.1648	0.3897	2.7520	0.3735	2.6374
1/500	1.490	0.6723	4.7473	0.5846	4.1281	0.5602	3.9561
1/200	3.730	1.6807	11.8682	1.4615	10.3202	1.4006	9.8902
1/150	4.97	2.2409	15.8242	1.9486	13.7602	1.8674	13.1869
1/100	7.46	3.3614	23.7364	2.9229	20.6403	2.8011	19.7803
1/75	9.94	4.4818	31.6485	3.8972	27.5204	3.7348	26.3737
1/70	10.70	4.8019	33.9091	4.1756	29.4862	4.0016	28.2576
1/60	12.40	5.6023	39.5606	4.8715	34.4005	4.6685	32.9672
1/50	14.90	6.7227	47.4727	5.8458	41.2806	5.6023	39.5606
1/40	18.6	8.4034	59.3409	7.3073	51.6008	7.0028	49.4508
1/30	24.9	11.2045	79.1212	9.7431	68.8011	9.3371	65.9344
1/25	29.8	13.4454	94.9455	11.6917	82.5613	11.2045	79.1212
1/20	37.3	16.8068	118.6818	14.6146	103.2016	14.0056	98.9015
1/15	49.7	22.4090	158.2425	19.4861	137.6021	18.6742	131.8687
1/12	62.1	28.0113	197.8031	24.3576	172.0027	23.3427	164.8359
1/10	74.6	33.6135	237.3637	29.2292	206.4032	28.0113	197.8031
1/8	93.2	42.0169	296.7046	36.5364	258.0040	35.0141	247.2538
1/6	124.0	56.0225	395.6061	48.7153	344.0053	46.6854	329.6718
1/4	186.0	84.0338	593.4092	73.0729	516.0080	70.0282	494.5077
1/3	249.0	112.0461	791.2123	97.4305	688.0107	93.3709	659.3436

hp	watts	@5000 rpm		@7500 rpm		@10,000 rpm	
		Oz.-in.	mN·m	Oz.-in.	mN·m	Oz.-in.	mN·m
1/2000	0.373	0.1008	0.7121	0.0672	0.4747	0.0504	0.3560
1/1500	0.497	0.1345	0.9495	0.0896	0.6330	0.0672	0.4747
1/1000	0.746	0.2017	1.4242	0.1345	0.9495	0.1008	0.7121
1/750	0.994	0.2689	1.8989	0.1793	1.2659	0.1345	0.9495
1/500	1.490	0.4034	2.8484	0.2689	1.8989	0.2017	1.4242
1/200	3.730	1.0084	7.1209	0.6723	4.7473	0.5042	3.5605
1/150	4.97	1.3445	9.4945	0.8964	6.3297	0.6723	4.7473
1/100	7.46	2.0168	14.2418	1.3445	9.4945	1.0084	7.1209
1/75	9.94	2.6891	18.9891	1.7927	12.6594	1.3445	9.4945
1/70	10.70	2.8812	20.3455	1.9208	13.5636	1.4406	10.1727
1/60	12.40	3.3614	23.7364	2.2409	15.8242	1.6807	11.8682
1/50	14.90	4.0336	28.4836	2.6891	18.9891	2.0168	14.2418
1/40	18.60	5.0420	35.6046	3.3614	23.7364	2.5210	17.8023
1/30	24.90	6.7227	47.4727	4.4818	31.6485	3.3614	23.7364
1/25	29.80	8.0672	56.9673	5.3782	37.9782	4.0336	28.4836
1/20	37.30	10.0841	71.2091	6.7227	47.4727	5.0420	35.6046
1/15	49.70	13.4454	94.9455	8.9636	63.2970	6.7227	47.4727
1/12	62.10	16.8068	118.6818	11.2045	79.1212	8.4034	59.3409
1/10	74.6	20.1681	142.4182	13.4454	94.9455	10.0841	71.2091
1/8	93.2	25.2101	178.0228	16.8068	118.6818	12.6051	89.0114
1/6	124.0	33.6135	237.3637	22.4090	158.2425	16.8068	118.6818
1/4	186.0	50.4203	356.0455	33.6135	237.3637	25.2101	178.0228
1/3	249.0	67.2270	474.7274	44.8180	316.4849	33.6135	237.3637

mNεm=Nεmω10⁻³

Glossary

Air Gap — The space between the rotating and stationary member in an electric motor.

Air Over (AO) — Motors intended for fan and blower service and cooled by the air stream from the fan or blower.

Alternating Current (AC) — The commonly available electric power supplied by an AC generator and distributed in one-, two-, or three-phase form. This is is the standard type of power supplied to homes, businesses, and industry as well as farms.

Ambient — For air-cooled rotating machinery, the ambient is considered the air surrounding the motor. The temperature of the space around the motor should not be over 40°C or 104°F.

Ampere — The constant current which, if maintained in two straight parallel conductors of infinite length and negligible cross section and spaced 1 meter apart in a vacuum, produces between them 2×10^{-7} Newton per meter of length; the unit of measurement for current. The ampere is abbreviated as A or, in some cases, amp.

Ampere Turn — The magnetomotive force produced by a current of one ampere in a coil of one turn.

Angular Velocity — Angular displacement per unit time, measured in degrees/time or radians/time.

Anti-Fungus Treatment — The anti-fungus requirement can be met by either supplying an insulation system which will not support fungus growth or a special anti-fungus varnish. Since either may be used, depending on the rating, do not specify anti-fungus varnish but request anti-fungus treatment.

Armature — The portion of the magnetic structure of a DC or universal motor which rotates.

Armature Reaction — The current that flows in the armature winding of a DC motor tends to produce magnetic flux in addition to that produced by the field current. This effect, which reduces the torque capacity, is called armature reaction and can affect the commutation and the magnitude of the motor's generated voltage.

— B —

Basic Speed — The speed which a motor develops at rated voltage with rated load applied.

Bearings (BRGS) — Sleeve-type bearings (SLV) are preferred where low noise level is important. The bearing resembles a short length of bronze tubing with grooves to direct oil flow. Ball bearings are used where higher load capacity is required or periodic lubrication is impractical.

Braking Torque — The torque required to bring a motor down from running speed to a standstill. The term is also used to describe the torque developed by a motor during dynamic braking conditions.

Breakdown Torque — The maximum torque a motor will develop at rated voltage without a relatively abrupt drop or loss in speed.

Brush — A piece of current-conducting material (usually carbon or graphite) which rides directly on the commutator of a commutated motor and conducts current from the power supply to the armature windings.

— C —

Canadian Standards Association (CSA) — Sets safety standards for motors and other electrical equipment used in Canada.

Cantilever Load — A load which tends to impose a radial force (perpendicular to the shaft axis) on a motor or gearmotor output shaft.

Capacitor — A device that, when connected in an alternating current circuit, causes the current to lead the voltage in time phase. The peak of the current wave is reached ahead of the voltage wave. This is the result of the successive storage and discharge of electric energy.

Center Ring — That part of a motor housing which supports the stator or field core.

Centrifugal Cut-Out Switch — A centrifugally operated automatic mechanism used in conjunction with split-phase and other types of induction motors. Centrifugal cut-out switches will open or disconnect the starting winding when the rotor has reached a predetermined speed, and reconnect it when the motor speed falls below it. Without such a device, the starting winding would be susceptible to rapid overheating and subsequent burnout.

Cogging — A term used to describe nonuniform angular velocity. It refers to rotation occurring in jerks or increments rather than smooth motion. When an armature coil enters the magnetic field produced by the field coils, it tends to speed up and slow down when leaving it. This effect becomes apparent at low speeds. The fewer the number of coils, the more noticeable it can be.

Commutator — A cylindrical device mounted on the armature shaft and consisting of a number of wedge-shaped copper segments arranged around the shaft (insulated from it and each other). The motor brushes ride on the periphery of the commutator and electrically connect and switch the armature coils to the power source.

Conductor — Any material which tends to make the flow of electrical current relatively easy (copper, aluminum, gold, silver, and others).

Counter Electromotive Force (CEMF) — The induced voltage in a motor armature caused by conductors moving through, or "cutting," field magnetic flux. This induced voltage opposes the armature current and tends to reduce it.

— D —

Direct Current (DC) — Type of power supply available from batteries or generators (not alternators) used for special-purpose applications. Current flows in one direction only; it does not alternate.

Duty Cycle — The relationship between the operating and rest time. A motor which can continue to operate within the temperature limits of its insulation system after it has reached normal operating (equilibrium) temperature is considered to have a continuous duty (CONT) rating. One which never reaches equilibrium temperature but is permitted to cool down between operations is operating under intermittent duty (INT) conditions.

Dynamic Unbalance — A noise-producing condition caused by the nonsymmetrical weight distribution of a rotating member. The lack of a uniform wire spacing in a wound armature or casting voids in a rotor or fan assembly can cause relatively high degrees of unbalance.

— E —

Eddy Current — Localized currents induced in an iron core by alternating magnetic flux. These currents translate into losses (heat), and their minimization is an important factor in lamination design.

Efficiency — The efficiency of a motor is the ratio of mechanical output to electrical input. It represents the effectiveness with which the motor converts electrical energy into mechanical energy.

Electrical Coupling — When two coils are so situated that some of the flux set up by either coil links some of the turns of the other, they are said to be electrically coupled.

Electromotive Force (emf) — A synonym for voltage, usually restricted to generated voltage.

Encapsulated Winding — A motor that has its winding structure completely coated with an insulating resin (such as epoxy). This construction type is more designed for exposure to severe atmospheric conditions than is the normal varnished winding.

Enclosure (ENCL) — The term used to describe the motor housing. Some of the more common types are:

Drip-Proof (DP) — Ventilation openings in end shells and shell placed so drops of liquid falling within an angle of 15° from vertical will not affect performance. Usually used indoors, in fairly clean, dry locations.

Open, Drip-Proof — Normal insulation treatment which consists of one or more dips and bakes of varnish. The insulation system is composed of materials which will not absorb or retain moisture.

Open, Drip-Proof with Extra Varnish Treatments — Same as standard open drip-proof except with extra dips and bakes to increase moisture resistance of the insulation system.

Explosion-Proof (EXP-PRF) — A special enclosed motor designed to withstand an internal explosion of specified gases or vapors and allow the internal flame or explosion to escape. Usually used in smaller ratings below $\frac{1}{3}$ hp if nonventilated (EPNV) and in fan-cooled (EPFC) in larger ratings.

Fan-Cooled (TEFC) — Includes an integral fan to blow cooling air over the motor.

Nonventilated (TENV) — Not equipped with a fan for external cooling. Depends on convection air for cooling.

Open (OP) — Ventilation openings in end shields and/or shell to permit passage of cooling air over and around the windings. Locations of openings not restricted. For use indoors, in fairly clean locations.

Totally Enclosed (TE) — No openings in the motor housing (but not airtight). Used in locations which are dirty, oily, and the like. The two types are:

Totally Enclosed, Severe Duty — Offers same enclosure as above, but with special features. Designed for use in chemical atmospheres or extremes of moisture and humidity. The table below is a guide to the proper selection of the enclosure and/or winding treatment, on applications where chemicals are encountered, and/or where mechanical protection from dust is required. Each construction is ranked as to its ability to withstand the particular condition. Note that the drip-proof motor with extra varnish treatment is well suited to high humidity.

Rating Enclosures

Enclosure and/or Winding Treatment	Humidity Resistance	Mechanical Protection	Resistance to Chemicals
Open, Drip-Proof	4	5	4
Open, Drip-Proof with Extra Varnish Treatment	2	4	3
Moisture-Sealed	1	3	3
Totally Enclosed	3	2	2
Severe Duty	2	1	1

Rank: 1 = High, 5 = Low

End Shield — That part of the motor housing which supports the bearing and acts as a protective guard to the electrical and rotating parts inside the motor. This part is frequently called the end bracket or end belt.

Excitation Current — A term applied to the current in the shunt field of a motor resulting from voltage applied across the field.

— F —

Farad — A unit of measurement for electrical capacitance. A capacitor has a capacitance of 1 farad when a potential difference of 1 volt will charge it with 1 coulomb of electricity.

Feedback — As it generally relates to motors and controls, feedback refers to the voltage information received by a feedback circuit. Depending on a predetermined potentiometer setting, a motor control can correct the voltage to deliver appropriate speed and/or torque.

Field — A term commonly used to describe the stationary (stator) member of a DC motor. The field provides the magnetic field with which the mechanically rotating (armature) member interacts.

Field Weakening — The introduction of resistance in series with the shunt-wound field of a motor to reduce the voltage and current that weakens the strength of the magnetic field and thereby increases the motor speed.

Flux — The magnetic field that is established around an energized conductor or permanent magnet.

Form Factor — A figure of merit that indicates how much rectified current departs from pure (nonpulsating) DC. A large departure from form factor (pure DC) increases the heating effect of the motor and reduces brush life.

Fractional-Horsepower Motor — A motor with continuous rating of less than 1 horsepower (hp), open construction at 1,700-1,800 rpm.

Frame (FR) — Usually refers to the NEMA system of standardization of motor-mounting dimensions.

Frequency — The rate at which alternating current reverses its direction of flow. Measured in hertz (Hz).

Full-Load Current — The current drawn from the line when the motor is operating at full-load torque and full-load speed at rated frequency and voltage.

Full-Load Torque — The torque necessary to produce rated horsepower at full-load speed.

— G —

Galvanometer — An extremely sensitive instrument used to measure small current and voltage in an electrical circuit.

Gearhead — The portion of a gearmotor which contains the actual gearing that converts the basic motor speed to the rated output speed.

— H —

Horsepower (hp) — Power rating of the motor. It takes 746 watts of electrical energy to produce 1 hp.

Hysteresis Loss — The resistance offered by materials to becoming magnetized. Reduced by using silicon steel laminations.

— I —

Impedance — The vectorial sum of both resistance and reactance in a motor; total opposition to current flow, measured in ohms. Z is the impedance symbol.

Inductance — The characteristic of a coil of wire to cause the current to lag the voltage in time phase. L is the symbol for inductance. Inductance is measured in henrys (H).

Inertial Load — A load (flywheel, fan, or the like) which tends to cause the motor shaft to continue to rotate after the power has been removed.

Insulation (INSUL) — In motors, usually classified by maximum allowable operating temperature:

Class A: 221°F (105°C)
Class B: 266°F (130°C)
Class F: 311°F (155°C)
Class H: 365°F (180°C)

Insulator — A material which tends to resist the flow of electric current.

Integral Horsepower Motor — In terms of horsepower, a motor built in a frame having continuous rating of 1 hp or more, open construction at 1,700-1,800 rpm. In terms of size, an integral horsepower motor is usually greater than 9 inches in diameter, although it can be as small as 6 inches.

— L —

Line Voltage — Voltage supplied by the power company or voltage supplied as input to the device.

Locked-Rotor Current — Steady-state current taken from the line with the rotor at standstill. Steady-state current means the current does not vary in intensity, but remains constant.

Locked-Rotor Torque — The minimum torque that a motor will develop at rest for all angular positions of the rotor.

— M —

Magnetomotive Force (mmf) — The magnetic energy supplied with the establishment of flux between the poles of a magnet.

Mechanical Degree — The popular physical understanding of degrees (360° = 1 rotation).

Mechanical Protection — Where clogging materials are present in severe proportions, the air gap of open motors may become clogged. Therefore, the recommendation is a totally enclosed motor.

Motor Types — Classified by operating characteristics and/or type of power required. *Induction Motors* include single-phase and three-phase motors. Direct-current motors are further classified as *shunt, series,* and *compound.*

— N —

National Electrical Code (NEC) — A code for the purpose of practical safeguarding of persons and property from the hazards arising from the use of electricity. It is sponsored by the National Fire Protection Institute. It is used to serve as a guide for governmental bodies whose duty is to regulate building codes.

NEMA — The National Electrical Manufacturers Association. This organization establishes certain voluntary industry standards relating to motors. These standards refer to the operating characteristics, terminology, basic dimensions, ratings, and testing.

— O —

Open Circuit — An open circuit in a motor is a defect that causes an interruption in the path through which the electrical current normally flows.

— P —

Phase — A term that indicates the space relationship of windings and changing values of the recurring cycles of AC.

Phase Displacement — Mechanical and electrical angle by which phases in a polyphase motor or main and capacitor (or starting) windings in an induction motor are displaced from one another.

Plug Reversal — Reconnecting a motor's windings to reverse its direction of rotation while running.

Polarities — Terms (positive, negative, north, and south) that indicate the direction of current and flux flow in electrical and magnetic circuits at any given instant.

Power Factor — A measurement of the time-phase difference between the voltage and current in an AC circuit. It is represented by the cosine of the angle of the phase difference. Zero degrees has a power factor of 100 percent. That means the watts and volt-amperes are equal and there is nothing more than resistance in the circuit. Ninety degrees of angle represents nothing in the way of resistance and only inductance in the circuit. PF is also found by the formula:

$$\frac{\text{True Power (TP)}}{\text{Apparent Power (AP)}}$$

Prony Brake — A simple mechanical device, normally made of wood with an adjustable leather strap, that is used to test for the torque output of a motor. The prony brake loads the motor and a spring scale attached to it gives a relatively accurate measurement or torque.

Pull-In Torque — The maximum constant torque that a synchronous motor will accelerate into synchronism at rated voltage and frequency.

Pull-Up Torque — The minimum torque delivered by an AC motor during the period of acceleration from zero to the speed at which breakdown occurs. For motors which do not have a definite breakdown torque, the pull-up torque is the minimum torque developed during the process of getting up to rated speed.

— R —

Rectifier — An electronic circuit which converts alternating current into direct current.

Reluctance — The characteristic of a magnetic material which resists the flow of magnetic lines of force through it.

Resilient Mounting — A suspension system or cushioned mounting designed to reduce the transmission of normal motor noise and vibration to the mounting surface.

Resistance — The degree of obstacle presented by a material to the flow of electrical current is known as resistance. Resistance is measured in ohms. R is the symbol for resistance.

Resistance to Chemicals — The first recommendation for any type of atmosphere containing chemicals, acids, bases, solvents, etc., should be severe-duty enclosed construction.

Rotor — The rotating member of an induction motor in a single-phase device. Current that is normally induced in the rotor reacts with the magnetic field produced by the stator. This produces torque and rotation.

— S —

Salient Pole — A motor has salient poles when its stator or field poles are concentrated into confined arcs and the winding is wrapped around them (as opposed to distributing them in a series of slots).

Secondary Winding — The secondary winding of a motor is a winding that is not connected to the power source but carries current induced in it through its magnetic linkage with the primary winding.

Semiconductor — A material, usually silicon or germanium, that permits limited current flow.

Service Factor (SF) — A measure of the overload capacity designed into a motor. A 1.15 SF means the motor can deliver 15 percent more than the rated horsepower without injurious overheating. A 1.0 SF motor should not be overloaded beyond its rated horsepower. Service factors will vary for various horsepower motors and motor speeds are shown in the table below for easy reference.

Service Factors

Horsepower	Service Factor			
	Synchronous Speed			
	3600	1800	1200	900
$1/20$, $1/12$, $1/8$	1.40	1.40	1.40	1.40
$1/6$, $1/4$, $1/3$	1.35	1.35	1.35	1.35
$1/2$	1.25	1.25	1.25	1.15
$3/4$	1.25	1.25	1.15	1.15
1	1.25	1.15	1.15	1.15
1½ and up	1.15	1.15	1.15	1.15

Short Circuit — A defect in a winding that causes part of the normal electrical circuit to be bypassed.

Skew — Arrangement of laminations on a rotor or armature to provide a slight diagonal pattern of their slots with respect to the shaft axis. This pattern helps to eliminate low-speed cogging effects in an armature and minimize induced vibration in a rotor.

Slip — The difference between the speed of the rotating magnetic field (which is always synchronous) and the rotor in a non-synchronous induction motor. Slip is expressed as a percentage of a synchronous speed and generally increases with an increase in load.

Slip Ring — A conductor band, mounted on an armature and insulated from it. A conductor strip slides on the band as the armature rotates. The function of the slip ring system is essentially the same as a commutator and brushes. Slip rings are also used to transmit current from the armature in a generator application.

Starting Torque — The torque or twisting force delivered by a motor when energized.

Stator — That part of an induction motor's magnetic structure that does not rotate. It usually contains the primary winding.

Synchronous Speed — The speed of the rotating magnetic field set up by an energized stator winding. In synchronous motors, the rotor locks into synchronism with the field, and is said to run at synchronous speed.

— T —

Tachometer — A small generator normally used as a velocity-sensing device. Tachometers are typically attached to the output shaft of DC servo motors requiring close speed regulation. The tachometer feeds its signal to a control which adjusts its output to the DC motor accordingly (called "closed loop feedback" control).

Temperature Rise — The amount by which a motor, operating under rated conditions, is hotter than its surrounding ambient temperature.

Thermal Protector — A protective device, built into the motor, that disconnects the motor from its power source if the temperature becomes excessive for any reason.

Thermocouple — A junction of two dissimilar materials which generates a minute voltage in proportion to its temperature. Such devices may be used as signal source in indicating instruments and control equipment.

Torque — Turning force delivered by a motor or gearmotor shaft, usually expressed in ounce-inches or Newton-meters. There are three types of torque associated with electric motors: **Starting Torque, Full-Load Torque,** and **Breakdown Torque.**

Tropical Protection — Specifications for motors to be used in a hot and humid location may call for tropical protection or tropical insulation. This will be assumed to mean that the windings must be specially protected against moisture and fungus and able to operate with normal life expectancy in higher-than-normal ambient temperatures up to 65°C.

Motors specified for tropical use will be supplied with extra dips and bakes of insulating varnish and a 65°C ambient insulation system. Special treatment for fungus proofing is not required, as the materials used in the insulation system are resistant to fungus growth.

— U —

Underwriters Laboratories, Inc. (UL) — An independent testing organization that sets safety standards for motors and other electrical equipment.

— V —

Variable Resistor — A resistor, connected in series with a motor, that can be adjusted to vary the amount of current available and thereby alter motor speed.

Voltage — The force that causes a current to flow in an electrical circuit. Analogous to pressure in hydraulics, voltage is often referred to as electrical pressure. Voltage is measured in volts (V).

Voltage Drop — Loss encountered across a circuit impedance. Voltage drop across a resistor takes the form of heat released into the air at the point of resistance.

— W —

Watt — The amount of power required to maintain a current of 1 ampere at a pressure of 1 volt. One horsepower is equal to 746 watts. The symbol for watt is W.

(Glossary courtesy of Bodine and General Electric.)

Index

**Over a Century of Excellence
for the Professional
and
Vocational Trades and the Crafts**

Order now from your local bookstore
or use the convenient order form
at the back of this book.

AUDEL

These fully illustrated, up-to-date guides and manuals mean a better job done for mechanics, engineers, electricians, plumbers, carpenters, and all skilled workers.

CONTENTS

ELECTRICAL

House Wiring (Seventh Edition)
ROLAND E. PALMQUIST;
revised by PAUL ROSENBERG
*5 1/2 x 8 1/4 Hardcover 248 pp. 150 Illus.
ISBN: 0-02-594692-7 $22.95*
Rules and regulations of the current 1990 National Electrical Code for residential wiring fully explained and illustrated.

Practical Electricity
(Fifth Edition)
ROBERT G. MIDDLETON;
revised by L. DONALD MEYERS
*5 1/2 x 8 1/4 Hardcover 512 pp. 335 Illus.
ISBN: 0-02-584561-6 $19.95*
The fundamentals of electricity for electrical workers, apprentices, and others requiring concise information about electric principles and their practical applications.

Guide to the 1990 National Electrical Code
ROLAND E. PALMQUIST;
revised by PAUL ROSENBERG
*5 1/2 x 8 1/4 Hardcover 664 pp. 230 Illus.
ISBN: 0-02-594565-3 $24.95*
The most authoritative guide available to interpreting the National Electrical Code for electricians, contractors, electrical inspectors, and homeowners. Examples and illustrations.

New Book for 1991!
Installation Requirements of the 1990 National Electrical Code
PAUL ROSENBERG
*5 1/2 x 8 1/4 Hardcover 240 pp.
ISBN: 0-02-604941-4 $24.95*
Field guide for installation requirements makes understanding the 1990 Electrical Code simple while on the job. Applications and easy-to-understand tables make this the perfect working companion.

Mathematics for Electricians and Electronics Technicians
REX MILLER
*5 1/2 x 8 1/4 Hardcover 312 pp. 115 Illus.
ISBN: 0-8161-1700-4 $14.95*
Mathematical concepts, formulas, and problem-solving techniques utilized on-the-job by electricians and those in electronics and related fields.

Fractional-Horsepower Electric Motors
REX MILLER and
MARK RICHARD MILLER
*5 1/2 x 8 1/4 Hardcover 436 pp. 285 Illus.
ISBN: 0-672-23410-6 $15.95*
The installation, operation, maintenance, repair, and replacement of the small-to-moderate-size electric motors that power home appliances and industrial equipment.

Electric Motors (Fifth Edition)

EDWIN P. ANDERSON
and REX MILLER

5 1/2 x 8 1/4   Hardcover   696 pp.
Photos/line art
ISBN: 0-02-501920-1 $35.00

Complete reference guide for electricians, industrial maintenance personnel, and installers. Contains both theoretical and practical
descriptions.

Home Appliance Servicing
(Fourth Edition)

EDWIN P. ANDERSON;
revised by REX MILLER

5 1/2 x 8 1/4   Hardcover   640 pp.   345 Illus.
ISBN: 0-672-23379-7 $22.50

The essentials of testing, maintaining, and
repairing all types of home appliances.

Television Service Manual
(Fifth Edition)

ROBERT G. MIDDLETON;
revised by JOSEPH G. BARRILE

5 1/2 x 8 1/4   Hardcover   512 pp.   395 Illus.
ISBN: 0-672-23395-9 $16.95

A guide to all aspects of television transmission and reception, including the operating
principles of black and white and color re-
ceivers. Step-by-step maintenance and repair procedures.

Electrical Course for Apprentices and Journeymen
(Third Edition)

ROLAND E. PALMQUIST

5 1/2 x 8 1/4   Hardcover   478 pp.   290 Illus.
ISBN: 0-02-594550-5 $19.95

This practical course in electricity for those
in formal training programs or learning on
their own provides a thorough understanding
of operational theory and its applications on
the job.

Questions and Answers for Electricians Examinations
(Tenth Edition)

Revised by PAUL ROSENBERG

5 1/2 x 8 1/4   Hardcover   316 pp.   110 Illus.
ISBN: 0-02-604955-4 $22.95

Based on the 1990 National Electrical Code,
this book reviews the subjects included in the
various electricians examinations—apprentice, journeyman, and master.

> # MACHINE SHOP AND MECHANICAL TRADES

Machinists Library
(Fourth Edition, 3 Vols.)

REX MILLER

5 1/2 x 8 1/4 Hardcover 1,352 pp. 1120 Illus.
ISBN: 0-672-23380-0 $52.95

An indispensable three-volume reference set
for machinists, tool and die makers, machine
operators, metal workers, and those with home
workshops. The principles and methods of
the entire field are covered in an up-to-date
text, photographs, diagrams, and tables.

Volume I: Basic Machine Shop

REX MILLER

5 1/2 x 8 1/4   Hardcover   392 pp.   375 Illus.
ISBN: 0-672-23381-9 $17.95

Volume II: Machine Shop

REX MILLER

5 1/2 x 8 1/4   Hardcover   528 pp.   445 Illus.
ISBN: 0-672-23382-7 $19.95

Volume III: Toolmakers Handy Book

REX MILLER

5 1/2 x 8 1/4   Hardcover   432 pp.   300 Illus.
ISBN: 0-672-23383-5 $14.95

Mathematics for Mechanical Technicians and Technologists

JOHN D. BIES

5 1/2 x 8 1/4   Hardcover   342 pp.   190 Illus.
ISBN: 0-02-510620-1 $17.95

The mathematical concepts, formulas, and
problem-solving techniques utilized on the
job by engineers, technicians, and other
workers in industrial and mechanical technology and related fields.

Millwrights and Mechanics Guide (Fourth Edition)

CARL A. NELSON

5 1/2 x 8 1/4   Hardcover   1,040 pp.   880 Illus.
ISBN: 0-02-588591-x $29.95

The most comprehensive and authoritative
guide available for millwrights, mechanics,
maintenance workers, riggers, shop workers, foremen, inspectors, and superinten-
dents on plant installation, operation, and
maintenance.

Welders Guide (Third Edition)
JAMES E. BRUMBAUGH

*5 1/2 x 8 1/4   Hardcover   960 pp.   615 Illus.
ISBN: 0-672-23374-6 $23.95*

The theory, operation, and maintenance of all welding machines. Covers gas welding equipment, supplies, and process; arc welding equipment, supplies, and process; TIG and MIG welding; and much more.

Welders/Fitters Guide
HARRY L. STEWART

*8 1/2 x 11   Paperback   160 pp.   195 Illus.
ISBN: 0-672-23325-8 $7.95*

Step-by-step instruction for those training to become welders/fitters who have some knowledge of welding and the ability to read blueprints.

Sheet Metal Work
JOHN D. BIES

*5 1/2 x 8 1/4   Hardcover   456 pp.   215 Illus.
ISBN: 0-8161-1706-3 $19.95*

An on-the-job guide for workers in the manufacturing and construction industries and for those with home workshops. All facets of sheet metal work detailed and illustrated by drawings, photographs, and tables.

Power Plant Engineers Guide
(Third Edition)
FRANK D. GRAHAM;
revised by CHARLIE BUFFINGTON

*5 1/2 x 8 1/4   Hardcover   960 pp.   530 Illus.
ISBN: 0-672-23329-0 $27.50*

This all-inclusive, one-volume guide is perfect for engineers, firemen, water tenders, oilers, operators of steam and diesel-power engines, and those applying for engineer's and firemen's licenses.

Mechanical Trades Pocket Manual (Third Edition)
CARL A. NELSON

*4 x 6   Paperback   364 pp.   255 Illus.
ISBN: 0-02-588665-7 $14.95*

A handbook for workers in the industrial and mechanical trades on methods, tools, equipment, and procedures. Pocket-sized for easy reference and fully illustrated.

PLUMBING

Plumbers and Pipe Fitters Library (Fourth Edition, 3 Vols.)
CHARLES N. McCONNELL

*5 1/2 x 8 1/4   Hardcover   952 pp.   560 Illus.
ISBN: 0-02-582914-9 $68.45*

This comprehensive three-volume set contains the most up-to-date information available for master plumbers, journeymen, apprentices, engineers, and those in the building trades. A detailed text and clear diagrams, photographs, and charts and tables treat all aspects of the plumbing, heating, and air conditioning trades.

Volume I: Materials, Tools, Roughing-In
CHARLES N. McCONNELL;
revised by TOM PHILBIN

*5 1/2 x 8 1/4   Hardcover   304 pp.   240 Illus.
ISBN: 0-02-582911-4 $20.95*

**Volume II: Welding, Heating,
Air Conditioning**
CHARLES N. McCONNELL;
revised by TOM PHILBIN

*5 1/2 x 8 1/4   Hardcover   384 pp.   220 Illus.
ISBN: 0-02-582912-2 $22.95*

**Volume III: Water Supply, Drainage,
Calculations**
CHARLES N. McCONNELL;
revised by TOM PHILBIN

*5 1/2 x 8 1/4   Hardcover   264 pp.   100 Illus.
ISBN: 0-02-582913-0 $20.95*

Home Plumbing Handbook
(Third Edition)
CHARLES N. McCONNELL

*8 1/2 x 11   Paperback   200 pp.   100 Illus.
ISBN: 0-672-23413-0 $14.95*

An up-to-date guide to home plumbing installation and repair.

The Plumbers Handbook
(Eighth Edition)
JOSEPH P. ALMOND, SR.;
revised by REX MILLER

*4 × 6   Paperback   368 pp.   170 Illus.
ISBN: 0-02-501570-2 $19.95*

Comprehensive and handy guide for plumbers and pipefitters—fits in the toolbox or pocket. For apprentices, journeymen, or experts.

Questions and Answers for Plumbers' Examinations
(Third Edition)
JULES ORAVETZ;
revised by REX MILLER
5 1/2 x 8 1/4 Paperback 288 pp. 145 Illus.
ISBN: 0-02-593510-0 $14.95
Complete guide to preparation for the plumbers' exams given by local licensing authorities. Includes requirements of the National Bureau of Standards.

HVAC

Air Conditioning: Home and Commercial (Fourth Edition)
EDWIN P. ANDERSON;
revised by REX MILLER
5 1/2 x 8 1/4 Hardcover 528 pp. 180 Illus.
ISBN: 0-02-584885-2 $29.95
A guide to the construction, installation, operation, maintenance, and repair of home, commercial, and industrial air conditioning systems.

Heating, Ventilating, and Air Conditioning Library
(Second Edition, 3 Vols.)
JAMES E. BRUMBAUGH
5 1/2 x 8 1/4 Hardcover 1,840 pp. 1,275 Illus.
ISBN: 0-672-23388-6 $53.85
An authoritative three-volume reference library for those who install, operate, maintain, and repair HVAC equipment commercially, industrially, or at home.

Volume I: Heating Fundamentals, Furnaces, Boilers, Boiler Conversions
JAMES E. BRUMBAUGH
5 1/2 x 8 1/4 Hardcover 656 pp. 405 Illus.
ISBN: 0-672-23389-4 $17.95

Volume II: Oil, Gas and Coal Burners, Controls, Ducts, Piping, Valves
JAMES E. BRUMBAUGH
5 1/2 x 8 1/4 Hardcover 592 pp. 455 Illus.
ISBN: 0-672-23390-8 $17.95

Volume III: Radiant Heating, Water Heaters, Ventilation, Air Conditioning, Heat Pumps, Air Cleaners
JAMES E. BRUMBAUGH
5 1/2 x 8 1/4 Hardcover 592 pp. 415 Illus.
ISBN: 0-672-23391-6 $17.95

Oil Burners (Fifth Edition)
EDWIN M. FIELD
5 1/2 x 8 1/4 Hardcover 360 pp. 170 Illus.
ISBN: 0-02-537745-0 $29.95
An up-to-date sourcebook on the construction, installation, operation, testing, servicing, and repair of all types of oil burners, both industrial and domestic.

Refrigeration: Home and Commercial (Fourth Edition)
EDWIN P. ANDERSON;
revised by REX MILLER
5 1/2 x 8 1/4 Hardcover 768 pp. 285 Illus.
ISBN: 0-02-584875-5 $34.95
A reference for technicians, plant engineers, and the homeowner on the installation, operation, servicing, and repair of everything from single refrigeration units to commercial and industrial systems.

PNEUMATICS AND HYDRAULICS

Hydraulics for Off-the-Road Equipment (Second Edition)
HARRY L. STEWART;
revised by TOM PHILBIN
5 1/2 x 8 1/4 Hardcover 256 pp. 175 Illus.
ISBN: 0-8161-1701-2 $13.95
This complete reference manual on heavy equipment covers hydraulic pumps, accumulators, and motors; force components; hydraulic control components; filters and filtration, lines and fittings, and fluids; hydrostatic transmissions; maintenance; and troubleshooting.

Pneumatics and Hydraulics
(Fourth Edition)
HARRY L. STEWART;
revised by TOM STEWART
5 1/2 x 8 1/4 Hardcover 512 pp. 315 Illus.
ISBN: 0-672-23412-2 $19.95
The principles and applications of fluid power. Covers pressure, work, and power; general features of machines; hydraulic and pneumatic symbols; pressure boosters; air compressors and accessories; and much more.

Pumps (Fifth Edition)
HARRY L. STEWART;
revised by REX MILLER

5 1/2 x 8 1/4   Hardcover   552 pp.   360 Illus.
ISBN: 0-02-614725-4 $35.00

The practical guide to operating principles of pumps, controls, and hydraulics. Covers installation and day-to-day service.

CARPENTRY AND CONSTRUCTION

Carpenters and Builders Library
(Fifth Edition, 4 Vols.)
JOHN E. BALL;
revised by TOM PHILBIN

5 1/2 x 8 1/4 Hardcover 1,224 pp. 1,010 Illus.
ISBN: 0-02-506450-9 $43.95

This comprehensive four-vol~~u~~~~ry has~~ set the professional stand~~~~s for carpenters, joiners, a~~~~

Volume I: Tool~~s~~ ~~~~ry
JOHN E. ~~B~~
revise~~~~

NEW EDITION FOR 1991–92

5 1/~~~~ver 384 pp. 345 Illus.
ISBN:~~~~~~5-7 $10.95~~

Volume~~~~Builders Math, Plans, Specifications
JOHN E. BALL;
revised by TOM PHILBIN

5 1/2 x 8 1/4   Hardcover 304 pp. 205 Illus.
ISBN: 0-672-23366-5 $10.95

Volume III: Layouts, Foundati~~o~~~~ming
JOHN E. BALL;
revised by TOM PH~~~~

NEW EDITION FOR 1991–92

5 1/2 x 8 1/4   H~~~~~~15 Illus.
ISBN: 0-672-~~~~

Volum~~e~~~~ Tools, Painting
JOH~~N~~
revise~~~~PHILBIN

5 1/2 x 8~~~~Hardcover 344 pp. 245 Illus.
ISBN: 0-672-23368-1 $10.95

Complete Building Construction
(Second Edition)
JOHN PHELPS;
revised by TOM PHILBIN

5 1/2 x 8 1/4   Hardcover 744 pp. 645 Illus.
ISBN: 0-672-23377-0 $22.50

Constructing a frame or brick building from the footings to the ridge. Whether the building project is a tool shed, garage, or a com-

---

plete home, this single fully illustrated volume provides all the necessary information.

Complete Roofing Handbook
JAMES E. BRUMBAUGH

5 1/2 x 8 1/4   Hardcover   536 pp.   510 Illus.
ISBN: 0-02-517850-4 $29.95

Covers types of roofs; roofing and reroofing; roof and attic insulation and ventilation; skylights and roof openings; dormer construction; roof flashing details; and much more.

Complete Siding Handbook
JAMES E. BRUMBAUGH

5 1/2 x 8 1/4   Hardcover   512 pp.   450 Illus.
ISBN: 0-02-517880-6 $24.95

This companion volume to the *Complete Roofing Handbook* includes comprehensive step-by-step instructions and accompanying line drawings on every aspect of siding a building.

Masons and Builders Library
(Second Edition, 2 Vols.)
LOUIS M. DEZETTEL;
revised by TOM PHILBIN

5 1/2 x 8 1/4   Hardcover   688 pp.   500 Illus.
ISBN: 0-672-23401-7 $27.95

This two-volume set provides practical instruction in bricklaying and masonry. Covers brick; mortar; tools; bonding; corners, openings, and arches; chimneys and fireplaces; structural clay tile and glass block; brick walls; and much more.

Volume 1: Concrete, Block, Tile, Terrazzo
LOUIS M. DEZETTEL;
revised by TOM PHILBIN

5 1/2 x 8 1/4   Hardcover   304 pp.   190 Illus.
ISBN: 0-672-23402-5 $14.95

Volume 2: Bricklaying, Plastering, Rock Masonry, Clay Tile
LOUIS M. DEZETTEL;
revised by TOM PHILBIN

5 1/2 x 8 1/4   Hardcover   384 pp.   310 Illus.
ISBN: 0-672-23403-3 $14.95

WOODWORKING

Wood Furniture: Finishing, Refinishing, Repairing
(Second Edition)
JAMES E. BRUMBAUGH

5 1/2 x 8 1/4   Hardcover   352 pp.   185 Illus.
ISBN: 0-672-23409-2 $12.95

A fully illustrated guide to repairing furniture and finishing and refinishing wood surfaces. Covers tools and supplies; types of wood; veneering; inlaying; repairing, restoring, and stripping; wood preparation; and much more.

Woodworking and Cabinetmaking

F. RICHARD BOLLER

5 1/2 x 8 1/4   Hardcover   360 pp.   455 Illus.
ISBN: 0-02-512800-0 $18.95

Essential information on all aspects of working with wood. Step-by-step procedures for woodworking projects are accompanied by detailed drawings and photographs.

MAINTENANCE AND REPAIR

Building Maintenance
(Second Edition)

JULES ORAVETZ

5 1/2 x 8 1/4   Paperback   384 pp.   210 Illus.
ISBN: 0-672-23278-2 $11.95

Professional maintenance procedures used in office, educational, and commercial buildings. Covers painting and decorating; plumbing and pipe fitting; concrete and masonry; and much more.

Gardening, Landscaping and Grounds Maintenance
(Third Edition)

JULES ORAVETZ

5 1/2 x 8 1/4   Hardcover   424 pp.   340 Illus.
ISBN: 0-672-23417-3 $15.95

Maintaining lawns and gardens as well as industrial, municipal, and estate grounds.

Home Maintenance and Repair: Walls, Ceilings and Floors

GARY D. BRANSON

8 1/2 x 11   Paperback   80 pp.   80 Illus.
ISBN: 0-672-23281-2 $6.95

The do-it-yourselfer's guide to interior remodeling with professional results.

Painting and Decorating

REX MILLER and GLEN E. BAKER

5 1/2 x 8 1/4   Hardcover   464 pp.   325 Illus.
ISBN: 0-672-23405-x $18.95

A practical guide for painters, decorators, and homeowners to the most up-to-date materials and techniques in the field.

Tree Care (Second Edition)

JOHN M. HALLER

8 1/2 x 11   Paperback   224 pp.   305 Illus.
ISBN: 0-02-062870-6 $16.95

The standard in the field. A comprehensive guide for growers, nursery owners, foresters, landscapers, and homeowners to planting, nurturing and protecting trees.

Upholstering (Updated)

JAMES E. BRUMBAUGH

5 1/2 x 8 1/4   Hardcover   400 pp.   380 Illus.
ISBN: 0-672-23372-x $15.95

The esentials of upholstering fully explained and illustrated for the professional, the apprentice, and the hobbyist.

AUTOMOTIVE AND ENGINES

Diesel Engine Manual
(Fourth Edition)

PERRY O. BLACK;
revised by WILLIAM E. SCAHILL

5 1/2 x 8 1/4   Hardcover   512 pp.   255 Illus.
ISBN: 0-672-23371-1 $15.95

The principles, design, operation, and maintenance of today's diesel engines. All aspects of typical two- and four-cycle engines are thoroughly explained and illustrated by photographs, line drawings, and charts and tables.

Gas Engine Manual
(Third Edition)

EDWIN P. ANDERSON;
revised by CHARLES G. FACKLAM

5 1/2 x 8 1/4   Hardcover   424 pp.   225 Illus.
ISBN: 0-8161-1707-1 $12.95

How to operate, maintain, and repair gas engines of all types and sizes. All engine parts and step-by-step procedures are illustrated by photographs, diagrams, and troubleshooting charts.

Small Gasoline Engines

REX MILLER and
MARK RICHARD MILLER

5 1/2 x 8 1/4   Hardcover   640 pp.   525 Illus.
ISBN: 0-672-23414-9 $16.95

Practical information for those who repair, maintain, and overhaul two- and four-cycle engines—including lawn mowers, edgers, grass sweepers, snowblowers, emergency electrical generators, outboard motors, and other equipment with engines of up to ten horsepower.

Truck Guide Library (3 Vols.)
JAMES E. BRUMBAUGH

5 1/2 x 8 1/4   2,144 pp.   1,715 Illus.
ISBN: 0-672-23392-4 $50.95

This three-volume set provides the most comprehensive, profusely illustrated collection of information available on truck operation and maintenance.

Volume 1: Engines
JAMES E. BRUMBAUGH

5 1/2 x 8 1/4   Hardcover   416 pp.   290 Illus.
ISBN: 0-672-23356-8 $16.95

Volume 2: Engine Auxiliary Systems
JAMES E. BRUMBAUGH

5 1/2 x 8 1/4   Hardcover   704 pp.   520 Illus.
ISBN: 0-672-23357-6 $16.95

Volume 3: Transmissions, Steering, and Brakes
JAMES E. BRUMBAUGH

5 1/2 x 8 1/4   Hardcover   1,024 pp.   905 Illus.
ISBN: 0-672-23406-8 $16.95

DRAFTING

Industrial Drafting
JOHN D. BIES

5 1/2 x 8 1/4   Hardcover   544 pp.   Illus.
ISBN: 0-02-510610-4 $24.95

Professional-level introductory guide for practicing drafters, engineers, managers, and technical workers in all industries who use or prepare working drawings.

Answers on Blueprint Reading
(Fourth Edition)
ROLAND PALMQUIST;
revised by THOMAS J. MORRISEY

5 1/2 x 8 1/4   Hardcover   320 pp.   275 Illus.
ISBN: 0-8161-1704-7 $12.95

Understanding blueprints of machines and tools, electrical systems, and architecture. Question and answer format.

HOBBIES

Complete Course in Stained Glass
PEPE MENDEZ

8 1/2 x 11   Paperback   80 pp.   50 Illus.
ISBN: 0-672-23287-1 $8.95

The tools, materials, and techniques of the art of working with stained glass.

Prices are subject to change without notice.